NEUROMETHODS

Series Editor
Wolfgang Walz
University of Saskatchewan
Saskatoon, SK, Canada

For further volumes:
http://www.springer.com/series/7657

Synaptosomes

Edited by

Kathryn M. Murphy

Department of Psychology, Neuroscience & Behavior, McMaster University, Hamilton, ON, Canada

Humana Press

Editor
Kathryn M. Murphy
Department of Psychology, Neuroscience & Behavior
McMaster University
Hamilton, ON, Canada

ISSN 0893-2336 ISSN 1940-6045 (electronic)
Neuromethods
ISBN 978-1-4939-9379-6 ISBN 978-1-4939-8739-9 (eBook)
https://doi.org/10.1007/978-1-4939-8739-9

This Humana Press imprint is published by the registered company Springer Science+Business Media, LLC part of Springer Nature.
The registered company address is: 233 Spring Street, New York, NY 10013, U.S.A.

Dedication

To my parents:
Mary and Gerald Murphy
And my family:
David and Coco

Preface to the Series

Experimental life sciences have two basic foundations: concepts and tools. The *Neuromethods* series focuses on the tools and techniques unique to the investigation of the nervous system and excitable cells. It will not, however, shortchange the concept side of things as care has been taken to integrate these tools within the context of the concepts and questions under investigation. In this way, the series is unique in that it not only collects protocols but also includes theoretical background information and critiques which led to the methods and their development. Thus it gives the reader a better understanding of the origin of the techniques and their potential future development. The *Neuromethods* publishing program strikes a balance between recent and exciting developments like those concerning new animal models of disease, imaging, in vivo methods, and more established techniques, including, for example, immunocytochemistry and electrophysiological technologies. New trainees in neurosciences still need a sound footing in these older methods in order to apply a critical approach to their results.

Under the guidance of its founders, Alan Boulton and Glen Baker, the *Neuromethods* series has been a success since its first volume published through Humana Press in 1985. The series continues to flourish through many changes over the years. It is now published under the umbrella of Springer Protocols. While methods involving brain research have changed a lot since the series started, the publishing environment and technology have changed even more radically. Neuromethods has the distinct layout and style of the Springer Protocols program, designed specifically for readability and ease of reference in a laboratory setting.

The careful application of methods is potentially the most important step in the process of scientific inquiry. In the past, new methodologies led the way in developing new disciplines in the biological and medical sciences. For example, Physiology emerged out of Anatomy in the nineteenth century by harnessing new methods based on the newly discovered phenomenon of electricity. Nowadays, the relationships between disciplines and methods are more complex. Methods are now widely shared between disciplines and research areas. New developments in electronic publishing make it possible for scientists that encounter new methods to quickly find sources of information electronically. The design of individual volumes and chapters in this series takes this new access technology into account. Springer Protocols makes it possible to download single protocols separately. In addition, Springer makes its print-on-demand technology available globally. A print copy can therefore be acquired quickly and for a competitive price anywhere in the world.

Saskatoon, SK, Canada *Wolfgang Walz*

Preface

I came to edit this book of synaptosome protocols because of my fascination with studying how the synaptic environment develops and supports plasticity. In my laboratory, we study the development of human cortex, and few methods exist to study the molecular environment of human synapses, but the synaptosome is the perfect tool. My students and I have used this tool to chart how the expression of presynaptic and postsynaptic proteins changes in human cortex across the lifespan. Using crude synaptic fractions, synaptoneurosomes, we have found new aspects of human cortical development. Here, I sought out authors doing cutting-edge research about synaptic functions and molecules using synaptosomes and asked them to share their protocols. In the book, top neuroscientists describe different ways that synaptosomes are used to answer diverse questions about synapses and provide detailed protocols to make synaptosomes.

Synaptosomes are made by subcellular fractionation of brain tissue that enriches for detached nerve terminals where the presynaptic and postsynaptic plasma membranes reseal and contain all of the molecular machinery at the synapse. At the EM level, those terminals look like a "snowman" where the presynaptic and postsynaptic snowballs are joined together by the synapse. That "snowman" retains synaptic functions including releasing and responding to neurotransmitters. As the field of neuroscience moves toward mapping all the connections in the brain, the synaptosome remains front and center as an important tool for answering questions about how synaptic connections work.

The chapters describe the history that led to making synaptosomes as well as modern protocols for using synaptosomes to answer current questions about the synapse. I thank each of the authors who contributed a chapter to the book. I am particularly thankful to my colleagues Dr. Elizabeth Quinlan who introduced my lab to the synaptosome and Dr. David Jones who has provided the computational insights for analyzing our synaptosome results. I have been very fortunate to have a talented group of past and current trainees using and modifying the synaptosome preparation to address our questions about experience-dependent development and plasticity in the visual cortex. These include Justin Balsor, Simon Beshara, Brett Beston, Katie Clayworth, Trang Dam, Kyle Hornby, David Irwin, Vashti Latchman, Steven Mancini, Robyn McKenzie, David Pinto, Joshua Pinto, Grayson Roumeliotis, Caitlin Siu, Lilia Tcharnaia, Madeline Verhovsek, and Kate Williams.

Hamilton, ON, Canada *Kathryn M. Murphy*

Contents

Contributors

MARIA FLORENCIA ANGELO • *Univ. Bordeaux, IINS, UMR 5297, Bordeaux, France; CNRS, IINS, UMR 5297, Bordeaux, France*
NICOLAI BACHE • *Children's Medical Research Institute, The University of Sydney, Wentworthville, NSW, Australia*
JUSTIN L. BALSOR • *McMaster Integrative Neuroscience Discovery and Study (MiNDS) Program, McMaster University, Hamilton, ON, Canada*
OXANA V. BARANOVA • *Neuroapoptosis and Translational Therapeutics Laboratory, Department of Neurological Surgery, University of Pittsburgh, Pittsburgh, PA, USA*
SERGEI V. BARANOV • *Neuroapoptosis and Translational Therapeutics Laboratory, Department of Neurological Surgery, University of Pittsburgh, Pittsburgh, PA, USA*
SIMON P. BESHARA • *McMaster Integrated Neuroscience Discovery and Study (MiNDS), McMaster University, Hamilton, ON, Canada*
AMRIT S. BOESE • *Zoonotic Diseases and Special Pathogens, Public Health Agency of Canada, National Microbiology Laboratory, Winnipeg, MB, Canada; Department of Medical Microbiology and Infectious Diseases, Rady Faculty of Health Sciences, University of Manitoba, Winnipeg, MB, Canada*
STEPHANIE A. BOOTH • *Zoonotic Diseases and Special Pathogens, Public Health Agency of Canada, National Microbiology Laboratory, Winnipeg, MB, Canada; Department of Medical Microbiology and Infectious Diseases, Rady Faculty of Health Sciences, University of Manitoba, Winnipeg, MB, Canada*
JUANITA BUSTAMANTE • *Centro de Altos Estudios en Ciencias de la Sauld (CAECIHS), Universidad Abierta Interamericana, Buenos Aires, Argentina*
DIANE L. CARLISLE • *Neuroapoptosis and Translational Therapeutics Laboratory, Department of Neurological Surgery, University of Pittsburgh, Pittsburgh, PA, USA*
MARIANNE CILLUFFO • *Brain Research Institute, UCLA, Los Angeles, CA, USA*
JEFFREY CLARK • *Department of Neurobiology, Weinberg College of Arts & Sciences, Northwestern University, Evanston, IL, USA*
LAETITIA DARRIET • *Univ. Bordeaux, IINS, UMR 5297, Bordeaux, France; CNRS, IINS, UMR 5297, Bordeaux, France; Lycée Fernand Daguin, Académie de Bordeaux, Mérignac, France*
MARTIN DARVAS • *Department of Pathology, University of Washington, Seattle, WA, USA*
BIREN DAVE • *Department of Biochemistry and Biomedical Sciences, Michael G. DeGroote School of Medicine, Faculty of Health Sciences, McMaster University, Hamilton, ON, Canada*
VÉRONIQUE DE-SMEDT-PEYRUSSE • *INRA, Nutrition et Neurobiologie Intégrée, UMR 1286, Bordeaux, France; Univ. Bordeaux, Nutrition et Neurobiologie Intégrée, UMR 1286, Bordeaux, France*
THOMAS DICHIARA • *Department of Neurobiology, Weinberg College of Arts & Sciences, Northwestern University, Evanston, IL, USA*

C. DIRK KEENE • *Department of Pathology, University of Washington, Seattle, WA, USA*
PETER R. DUNKLEY • *School of Biomedical Sciences and Pharmacy, Hunter Medical Research Institute, The University of Newcastle, Callaghan, NSW, Australia*
KASPER ENGHOLM-KELLER • *Children's Medical Research Institute, The University of Sydney, Wentworthville, NSW, Australia; Department of Biochemistry and Molecular Biology, University of Southern Denmark, Odense, Denmark*
ROBERT M. FRIEDLANDER • *Neuroapoptosis and Translational Therapeutics Laboratory, Department of Neurological Surgery, University of Pittsburgh, Pittsburgh, PA, USA; University of Pittsburgh Medical Center, Pittsburgh, PA, USA*
MARK E. GRAHAM • *Children's Medical Research Institute, The University of Sydney, Wentworthville, NSW, Australia*
ETIENNE HERZOG • *Univ. Bordeaux, IINS, UMR 5297, Bordeaux, France; CNRS, IINS, UMR 5297, Bordeaux, France*
BHANU P. JENA • *Department of Physiology, Wayne State University School of Medicine, Detroit, MI, USA*
ANALÍA G. KARADAYIAN • *Facultad de Farmacia y Bioquimica, Fisicoquimica, Universidad de Buenos Aires, Buenos Aires, Argentina; Instituto de Bioquimica y Medicina Molecular (IBIMOL), CONICET-Universidad de Buenos Aires, Buenos Aires, Argentina*
NICOLAS K. KHATTAR • *Neuroapoptosis and Translational Therapeutics Laboratory, Department of Neurological Surgery, University of Pittsburgh, Pittsburgh, PA, USA; Department of Neurological Surgery, University of Louisville School of Medicine, Louisville, KY, USA*
WILLIAM L. KLEIN • *Department of Neurobiology, Weinberg College of Arts & Sciences, Northwestern University, Evanston, IL, USA; Department of Neurology, Feinberg School of Medicine, Northwestern University, Evanston, IL, USA*
ERIC S. KRETZ • *Neuroapoptosis and Translational Therapeutics Laboratory, Department of Neurological Surgery, University of Pittsburgh, Pittsburgh, PA, USA*
VICKIE KWAN • *Department of Biochemistry and Biomedical Sciences, Michael G. DeGroote School of Medicine, Faculty of Health Sciences, McMaster University, Hamilton, ON, Canada*
TIMOTHY M. LARKIN • *Neuroapoptosis and Translational Therapeutics Laboratory, Department of Neurological Surgery, University of Pittsburgh, Pittsburgh, PA, USA*
MARTIN R. LARSEN • *Department of Biochemistry and Molecular Biology, University of Southern Denmark, Odense, Denmark*
CAITLIN S. LATIMER • *Department of Pathology, University of Washington, Seattle, WA, USA*
RIANA LO BU • *Department of Neurobiology, Weinberg College of Arts & Sciences, Northwestern University, Evanston, IL, USA*
SILVIA LORES-ARNAIZ • *Facultad de Farmacia y Bioquimica, Fisicoquimica, Universidad de Buenos Aires, Buenos Aires, Argentina; Instituto de Bioquimica y Medicina Molecular (IBIMOL), CONICET-Universidad de Buenos Aires, Buenos Aires, Argentina*
ANNA MAJER • *Viral Diseases Division, Public Health Agency of Canada, National Microbiology Laboratory, Winnipeg, MB, Canada*
STEVE J. MANCINI • *McMaster Integrated Neuroscience Discovery and Study (MiNDS), McMaster University, Hamilton, ON, Canada*
KATHLEEN S. MONTINE • *Department of Pathology, Stanford University, Stanford, CA, USA*

THOMAS J. MONTINE • *Department of Pathology, Stanford University, Stanford, CA, USA*
KATHRYN M. MURPHY • *McMaster Integrative Neuroscience Discovery and Study (MiNDS) Program, McMaster University, Hamilton, ON, Canada; Department of Psychology, Neuroscience & Behavior, McMaster University, Hamilton, ON, Canada*
NADEEM MURTAZA • *Department of Biochemistry and Biomedical Sciences, Michael G. DeGroote School of Medicine, Faculty of Health Sciences, McMaster University, Hamilton, ON, Canada*
DAVID G. NICHOLLS • *Buck Institute for Research on Aging, Novato, CA, USA*
NADIA O. POSTUPNA • *Department of Pathology, University of Washington, Seattle, WA, USA*
ANNIE QUAN • *Cell Signalling Unit, Children's Medical Research Institute, The University of Sydney, Wentworthville, NSW, Australia*
SUSHMA R. RAO • *Children's Medical Research Institute, The University of Sydney, Wentworthville, NSW, Australia*
PHILLIP J. ROBINSON • *Cell Signalling Unit, Children's Medical Research Institute, The University of Sydney, Wentworthville, NSW, Australia*
GEORGINA RODIGUEZ DE LORES ARNAIZ • *Instituto de Biología Celular y Neurociencias "Prof. E. De Robertis" (IBCN), CONICET-Universidad de Buenos Aires, Buenos Aires, Argentina; Facultad de Medicina, Universidad de Buenos Aires, Buenos Aires, Argentina*
THEODORE A. SARAFIAN • *Department of Psychiatry and Biobehavioral Sciences, David Geffen School of Medicine, UCLA, Los Angeles, CA, USA*
KARUN K. SINGH • *Department of Biochemistry and Biomedical Sciences, Michael G. DeGroote School of Medicine, Faculty of Health Sciences, McMaster University, Hamilton, ON, Canada*
CAITLIN R. SIU • *McMaster Integrated Neuroscience Discovery and Study (MiNDS), McMaster University, Hamilton, ON, Canada*
STEPHEN G. SLIGAR • *Department of Biochemistry, University of Illinois, Urbana-Champagne, Champaign, IL, USA*
PIERRE TRIFILIEFF • *INRA, Nutrition et Neurobiologie Intégrée, UMR 1286, Bordeaux, France; Univ. Bordeaux, Nutrition et Neurobiologie Intégrée, UMR 1286, Bordeaux, France*
JESSE R. WARK • *Children's Medical Research Institute, The University of Sydney, Wentworthville, NSW, Australia*
JOSEPH B. WATSON • *Brain Research Institute, UCLA, Los Angeles, CA, USA; Department of Psychiatry and Biobehavioral Sciences, David Geffen School of Medicine, UCLA, Los Angeles, CA, USA*
JING XUE • *Cell Signalling Unit, Children's Medical Research Institute, The University of Sydney, Wentworthville, NSW, Australia*
SVITLANA YABLONSKA • *Neuroapoptosis and Translational Therapeutics Laboratory, Department of Neurological Surgery, University of Pittsburgh, Pittsburgh, PA, USA*
HERBERT ZIMMERMANN • *Institute of Cell Biology and Neuroscience, Goethe University, Frankfurt am Main, Germany*

Part I

Introduction

Chapter 1

Introduction to Synaptosomes

Kathryn M. Murphy

Abstract

Long before the synapse was first visualized Cajal theorized that dendrites and axons communicate by contacts he called *"protoplasmic kisses."* Now, the synapse is a central topic of neuroscience research with thousands of publications each year focused on understanding how synapses function. Isolating synapses by making synaptosomes has become an important tool for studying synapses. The synaptosome is a *"halfway house"* between neurochemistry and neurophysiology that strips away everything except the synapse, and its molecular machinery to function. This book introduces the reader to a collection of protocols for making synaptosomes. The chapter covers the history of developing synaptosome protocols and 15 detailed protocols for making and using synaptosomes in modern neuroscience research. Here the protocols cover the spectrum from crude synaptic fractions to highly purified functional synaptosomes. Importantly, each chapter describes how the authors have used synaptosomes to address current research questions and gives insights about applying synaptosome protocols to answer new questions about the synapse.

Key words Synapse, Protocols, Synaptosome, Synaptoneurosome

The synapse has taken on a vaulted position in science and society; it is at the core of understanding who we are and of staving off diseases that can rob us of the ability to think, feel, and communicate. This recognition about the importance of the synapse has spurred on neuroscience research.

> "Every president needs a blockbuster science project to crown his time in the White House. While John Kennedy reached for the stars, Barack Obama decided to reach for the synapses." [1]

As neuroscientists pursue that presidential goal, studies using the synaptosome are proving to be an important part of the pathway to mapping synaptic function. There is a pressing need to sort through the thousands of proteins that reside at the synapse, unpack every aspect of synaptic function, and identify all the changes that occur when synaptic functions go awry. Synaptosomal preparations are ideal for tackling those questions because they are a kind of *"halfway house"* between neurochemistry and electrophysiology (David

Kathryn M. Murphy (ed.), *Synaptosomes*, Neuromethods, vol. 141,
https://doi.org/10.1007/978-1-4939-8739-9_1,

Nicholls Chap. 7) that help piece together the structures and functions of the synapse.

A quote from Santiago Ramon y Cajal beautifully captures his neuron theory that dendrites and axons communicate by contact, the *kisses* that we now know are synapses.

> "What mysterious forces precede the appearance of the processes [dendrites and axon], promoting their growth and ramification, provoking the coherent migration of the cells and fibres in predetermined directions, as if obeying a wise architectonic plan, and finally establishing those protoplasmic kisses, the intercellular articulations [synapses] that appear to constitute the final ecstasy of an epic love story?" (Recuerdos de mi Vida, Moya, Madrid, 1917, translated by Javier DeFelipe [2])

The ultrastructure of the synapse was first seen in the 1950s when the advent of electron microscopy revealed the details of the synapse. Over the next decade Victor Whittaker and colleagues [3–8] perfected a method to isolate synapses and called those particles *"synaptosomes"* [7].

The tools to visualize and isolate synapses changed the direction of neuroscience research. Whether studying development, plasticity, degeneration, learning, memory, or any other aspect of brain function the research questions include the central role of the synapse. The synapse has even entered the modern lexicon where business, politicians, and even music lyrics use the word synapse to express ideas about the brain, its complexity, how it changes with experience or disease, and the hopes of harnessing the remarkable plasticity of the synapse to improve human health.

Synaptosome preparations are simple and robust methods that pare down brain tissue to the bare essentials of the synapse. The complexity of neural circuitry is stripped away leaving *"snowman"*-shaped particles of fused pre- and postsynaptic membranes that are ready to be exploited to reveal the secrets of the synapse. These protocols can be used to uncover functional diversity among individual synapses as well as large-scale proteomics across a population of synapses.

For all these reasons I'm excited to introduce you to the chapters in this book that describe synaptosome protocols for a range of experiments from the crude synaptoneurosomes made by filtration steps to the high purity and functionality afforded by Percoll gradient methods. The chapters have clear protocols so that you can produce the right kind of synaptosomes for your study. As a note about nomenclature: Victor Whittaker called the nerve endings isolated using Percoll gradients methods "synaptosomes" [7] while others used the term "synaptoneurosome" for crude synaptic fractions made by filtration and centrifugation steps

[9]. The chapters in this book use both terms, and the authors have selected the term that is the best fit for their protocol.

In this book, we explore a collection of protocols for making synaptosomes that address a range of research questions about the synapse. The chapters share detailed protocols for making synaptosomes so that you can use these preparations in your studies. The book begins in Chap. 2 with some of the histories that led up to isolating synaptosomes. Chapters 3–6 focus on how to prepare synaptosomes with Chap. 2 including a blueprint for selecting the right protocol for an application. These chapters explain different approaches for making synaptosomes including a high-throughput filtered protocol to make synaptoneurosomes (Chap. 4), making synaptosomes from small samples (Chap. 5), and an optimized protocol for isolating functional synaptosomes (Chap. 6). Synaptic energy and metabolism is the topic for Chaps. 7–9 and those chapters describe protocols for studying the bioenergetics of neurotransmitter release (Chap. 7) and synaptic mitochondrial energetics and free radicals in aging (Chap. 8), and making highly purified synaptosomal mitochondria to characterize the mitochondrial proteome (Chap. 9). The protocols in Chaps. 10–12 are optimized for studying presynaptic functions including phosphoproteome signaling (Chap. 10), presynaptic protein regulation of neurotransmitter release (Chap. 11), and prosome-mediated discharge of synaptic vesicles (Chap. 12). Chapter 13 describes using synaptosomes to make nanodiscs that contain single synaptic proteins and a protocol for the simultaneous detection of synaptic proteins and miRNA in synaptosomes is explained in Chap. 14. Chapters 15–17 focus on protocols for making synaptosomes from human tissue. Synaptosome protocols are described for proteomic analysis of frozen human postmortem cortical tissue (Chap. 15) and for making synaptosomes from human induced pluripotent stem cell neurons (Chap. 16). Finally, Chap. 17 describes a protocol to isolate crude synaptosomes from human tissue, label particles inside the synaptosomes, and use flow cytometry for fast analysis of 1000s of synaptic particles.

This collection of protocols gives a broad look into how synaptosomes are made and used to answer questions about synaptic plasticity, neurotransmission, drug discovery, and synaptic function in health and disease. Some of the advantages of using synaptosomes are the interpretation of synaptic functions and the robust measurement of synaptic molecules in human tissue. Another important advantage is the ease with which these protocols can be adapted for use with new and emerging methods like biological barcoding and ultrahigh-resolution imaging. The synaptosome will continue to be an important tool for deepening our understanding of Cajal's *"protoplasmic kisses."*

References

1. Ahuja A (2014) Obama's moonshot probes the space inside our skulls. Financial times
2. DeFelipe J (2010) From the connectome to the synaptome: an epic love story. Science 330:1198–1201. https://doi.org/10.1126/science.1193378
3. Derobertis E, Pellegrinodeiraldi A, Rodriguezdelores G, Zieher LM (1965) Synaptic vesicles from the rat hypothalmus. Isolation and norepinephrine content. Life Sci (1962) 4:193–201
4. Whittaker VP, Dowe GH (1964) The identification of the acetylcholine-like substance in synatosomes derived from Guinea-pig brain as acetylcholine itself. Int J Neuropharmacol 3:593–597
5. Gray EG, Whittaker VP (1962) The isolation of nerve endings from brain: an electron-microscopic study of cell fragments derived by homogenization and centrifugation. J Anat 96:79–88
6. Whittaker VP, Michaelson IA, Kirkland RJ (1963) The separation of synaptic vesicles from disrupted nervending particles. Biochem Pharmacol 12:300–302
7. Whittaker VP, Michaelson IA, Kirkland RJ (1964) The separation of synaptic vesicles from nerve-ending particles ("synaptosomes"). Biochem J 90:293–303
8. Whittaker VP, Gray EG (1962) The synapse: biology and morphology. Br Med Bull 18:223–228
9. Hollingsworth EB, McNeal ET, Burton JL et al (1985) Biochemical characterization of a filtered synaptoneurosome preparation from Guinea pig cerebral cortex: cyclic adenosine 3':5'-monophosphate-generating systems, receptors, and enzymes. J Neurosci 5:2240–2253

Part II

History of Synaptosomes

Chapter 2

Victor P. Whittaker: The Discovery of the Synaptosome and Its Implications

Herbert Zimmermann

Abstract

The isolation of synaptosomes and synaptic vesicles from mammalian brain was a major breakthrough in the study of the cellular and molecular function of the nervous system. It laid the foundations for the present-day molecular biological analysis of all other functional compartments of the synapse. Synaptosomes are still much in use for the study of synaptic transmission. This work was performed at the beginning of the 1960s by Victor P. Whittaker and coworkers at the Agricultural Research Council Institute of Animal Physiology in Babraham, near Cambridge, UK. This chapter describes the history of the discovery of the synaptosome and of synaptic vesicles and its implications and highlights the very personal engagement of Victor Whittaker in this field.

Key words Discovery, History, Synaptosome, Synaptic vesicle, Victor Whittaker

In his 1993 essay "Thirty years of synaptosome research" [1] Victor Whittaker (1919–2016) quotes the following sentence from a review of David Nicholls [2] ".... the synaptosome is the simplest system in which the complete chain of events from plasma membrane depolarization to transmitter release can be studied." And he continues, "Even after 30 years of sustained research the preparation still retains its usefulness as a model system in which to study almost all aspects of synaptic function." This is still true today when PubMed lists 142 entries of the term synaptosome for 2017 and more than 12,000 altogether.

The discovery, description, and first functional characterization of synaptosomes resulted from a rather random observation when Whittaker actually aimed to isolate a subcellular compartment containing so-called bound acetylcholine which he thought might correspond to synaptic vesicles. This pool of acetylcholine was not liberated during the homogenization of brain tissue under iso-osmotic conditions and was apparently segregated in an occluded membrane-bound compartment. As it happened, what Whittaker first isolated turned out to be detached and sealed nerve endings (which he later named synaptosomes) containing synaptic

Kathryn M. Murphy (ed.), *Synaptosomes*, Neuromethods, vol. 141,
https://doi.org/10.1007/978-1-4939-8739-9_2,

Fig. 1 Victor P. Whittaker around 1970

vesicles and mitochondria. This was very soon followed by the isolation of synaptic vesicles from osmotically lysed synaptosomes. Both discoveries had a tremendous impact on the further development of the biochemistry and more recently cellular and molecular biology of the nervous system.

Some of the major stations in the life of Victor Whittaker (Fig. 1) have previously been summarized [3]. He obtained his degree in chemistry and biochemistry and his PhD at Oxford University where he continued as Demonstrator and Lecturer in Biochemistry until 1951. From 1951 to 1955 he was Assistant Professor of Physiology at the University of Cincinnati, College of Medicine. There he worked on the identification of naturally occurring and pharmacologically active homologues of acetylcholine [4]. In 1955 he returned to England with his family, his wife Margaret, an MD, and three children, to take the position of Principal Scientific Officer (1955–1959) and Senior Principal Scientific Officer (1959–1966) at the Agricultural Research Council Institute of Animal Physiology in Babraham (now Babraham Institute). Babraham is a little village close to Cambridge. This institute is the cradle of the synaptosome. There Whittaker performed his breakthrough experiments on the subcellular fractionation of brain tissue around 1960. In 1966 he was offered the Sir William Dunn Readership in Biochemistry at the University of Cambridge. Since laboratory space and financial support were limited he took up in parallel the position of Chief Research Scientist at the New York State Institute for Basic Research in Mental Retardation in Staten Island (1967–1971) and was Visiting Professor at the City University of New York (1968–1972). He managed two laboratories requiring regular visits between the continents. Laboratory space and financial support in Cambridge again became an issue. At that time the German Max Planck Society came up with the offer to set up his own Department of Neurochemistry at the Max-Planck-Institute for Biophysical Chemistry in Göttingen, including generous financial and logistic support. Together with his entire group he moved to Göttingen in 1973 where he worked until his retirement in 1987. He remained active as an emeritus first in Göttingen and then at the Medical School of the University of Mainz (1991–1992) before he returned to Cambridge.

The author of this chapter joined Whittaker's laboratory in the Department of Biochemistry in Cambridge in January 1972 and stayed with him in Göttingen until December 1979. Recollections of the isolation of synaptosomes and work on synaptic vesicles are derived from own experience, many discussions in the laboratory and with Victor Whittaker, and also volumes 3 [5] and 4 [6] of Whittaker's memoirs.

The account of the discovery of the synaptosome and its prerequisites requires some elucidation of the state of the art of the synaptic transmission process at the time. Major impacts had come from two scientific fields, neurophysiological and fine structural investigations of the synapse, and from novel technical developments in ultracentrifugation and subcellular fractionation. At the beginning of the 1950s considerable process had been achieved in understanding the physiology and pharmacology of cholinergic synaptic transmission. Since the demonstration that acetylcholine is the neurotransmitter at the neuromuscular junction by Dale, Feldberg and Vogt at the National Institute for Medical Research, London, in 1936 [7], this peripheral organ preparation had become a model system for the study of synaptic transmission. The brain was still out of reach. It had been known that nerve stimulation induces the release of the neurotransmitter acetylcholine at the neuromuscular junction and induces a depolarization of the muscle cell, called end-plate potential (EPP). Using a sophisticated intracellular recording protocol, Bernard Katz and coworkers at the University College London discovered at the beginning of the 1950s that the resting nerve muscle preparation of the frog spontaneously generates all or no small random postsynaptic depolarizations of standard size (miniature end-plate potentials, mEPPs). They allocated the mEPPs to the presynaptic release of quantal (standard size) multimolecular packages of acetylcholine. They also suggested that the bulk release evoked by nerve stimulation reflects simultaneously discharged packages of neurotransmitter with the dimensions of mEPPs, generating an EPP (hypothesis of quantal neurotransmitter release) [8–11].

At about the same time electron microscopic techniques and specimen preparation for electron microscopy had been developed to a stage which allowed the fine structural analysis of biological tissues. It could be shown that nerve terminals contained a large number of small electron-lucent vesicular structures of about 50 nm diameter [12–16]. As a result of this, Del Castillo and Katz suggested in 1956 that synaptic vesicles could be the structural basis for the quantal release of the neurotransmitter, whereby the vesicles lose their acetylcholine content when they reach the plasma membrane. They actually used rather careful wording. "One might suppose that when a "critical" collision occurs between an intracellular "ACh carrier" and the membranes of the nerve terminal, the two barriers are opened simultaneously and the ACh contents of the carrier particle are suddenly discharged" [11]. This anticipated today's SNARE hypothesis of vesicle exocytosis. After completing a fine structural analysis of synapses in the central nervous system, Sanford Palay from the Department of Anatomy at Yale University came up in the same year with a suggestion complementary to that of Del Castillo and Katz. He suggested "that the small synaptic vesicles provide the morphological representation of the prejunctional, subcellular units of neurohumoral discharge at the synapse demanded by physiological evidence" [17]. The vesicle hypothesis of the release of the neurotransmitter acetylcholine was born. But it was still to be demonstrated that the neurotransmitter was actually contained inside synaptic vesicles.

The other field of equal relevance for the discovery of the synaptosome was the development of the preparative ultracentrifuge and the resulting advances in the isolation of subcellular organelles. This was pioneered by Albert Claude at Rockefeller University (then Rockefeller Institute) in the mid-1940s who disrupted liver cells by the application of liquid shear force in cold iso-osmotic sucrose, leading to the separation amongst others of microsomes and mitochondria [18, 19]. In the first half of the 1950s future application of these techniques led to the isolation of chromaffin granules from the adrenal medulla by Blaschko and Hillarp and coworkers [20, 21] and of lysosomes by De Duve and coworkers [22].

When Whittaker moved to the Agricultural Research Council Institute of Animal Physiology in Babraham in 1955, outstanding scientists like Sir Rudolph Peters, Whittaker's former chief in Oxford days and now head of the Department of Biochemistry, Sir John Gaddum, Catherine Hebb, Marthe Vogt, and Krešimir Krnjević made this a unique place for studying neuroscience. Whittaker's interest in cellular pools of acetylcholine started at that time (see also his extensive review on the application of subcellular fractionation techniques to the study of brain function [23]). He was intrigued by a subcellular fractionation compartment referred to as bound acetylcholine. Extraction of brain tissue by isotonic media released only part of the acetylcholine present. When membranes were fully destroyed the yield of the recovered acetylcholine was two to three times higher. This suggested that an occluded cellular pool of acetylcholine existed, similar to the occlusion of hydrolytic enzymes in lysosomes or of adrenaline in chromaffin granules. A fine overview of the subcellular aspects of acetylcholine metabolism in the mammalian brain has been provided by Louis A. Barker [24], a former collaborator of Whittaker at the Institute for Basic Research in Mental Retardation, Staten Island, NY. The pressing question thus was whether bound acetylcholine represented acetylcholine sequestered in synaptic vesicles. And could these vesicles be isolated? Up to then the evidence for synaptic acetylcholine was derived from the neuromuscular junction. But the yield of acetylcholine in muscle was far too low and the tissue too tough for gentle homogenization. Brain would thus be the better tissue of choice.

Whittaker teamed up with Catherine Hebb (1904–2005), the head of the Department of Physiology, who studied the biosynthesis of acetylcholine in nervous tissue [25] and who actually lived in the Babraham cottage next to the Whittaker family. She was interested in the subcellular localization of choline acetylase (later named choline acetyltransferase). Using differential centrifugation of rabbit brain tissue submitted to homogenization in 0.25 M sucrose solution she had recently shown (1956) that 50–70% of this enzyme was contained in a bound and inactive state in the so-called mitochondrial fraction from where it could be released by membrane disruption [26]. The authors were wondering whether the enzyme was actually contained inside mitochondria. Hebb and Whittaker went one step further and submitted the crude mitochondrial fraction which contained the highest bound acetylcholine content to sucrose density gradient centrifugation with sucrose concentrations from 0.8 to 1.6 M (1958) [27]. Of help were the Model L preparative ultracentrifuge developed by Spinco (later Beckman) which allowed to sediment also very small subcellular fractionation particles and the introduction of the swing-bucket rotor which opened the doorway to density gradient centrifugation. They determined the distribution in the gradient of the mitochondrial enzyme succinate dehydrogenase and of choline acetyltransferase and acetylcholine. Interestingly succinate dehydrogenase was located in a denser part of the gradient than acetylcholine and choline acetyltransferase, providing first evidence that bound acetylcholine resides in a particle different from mitochondria.

As Catherine Hebb went on to an extended holiday in Canada, Whittaker decided to press on without her. This time he used forebrain tissue (rather than total brain) from rabbit and guinea pig as this was assumed to be rich in synapses of seemingly uniform size. Technical specifications included a 10% homogenate in ice-cold 0.32 M sucrose and a Perspex and glass homogenizer (Aldridge homogenizer). He simplified the sucrose gradient to two steps in density, 0.8 and 1.2 M sucrose. At the interface between the layers of 0.8 and 1.2 M sucrose he obtained a greyish particulate material rich in bound acetylcholine which later turned out to be the synaptosome fraction. It differed from the tan-colored pellet below the layer of 1.2 M sucrose rich in mitochondrial enzyme activity. A challenging goal now was the ultrastructural identification of the material obtained. Being a dilettante in this field Whittaker played with various fixation protocols that did not really provide much insight—even if the findings were published in 1959 in the Biochemical Journal [28]. After fixing and staining fractions in OsO_4 and embedding them in methyl methacrylate resin, mitochondria could be identified but his "acetylcholine particles" refused to reveal their identity. Whittaker actually thought that these would represent synaptic vesicles. However, osmotic lysis experiments already revealed that part of the acetylcholine contained in these particles could be readily released whereas a more stable fraction remained—foreshadowing the synaptic vesicle-bound pool of acetylcholine. The next and very important step is quoted directly from Whittaker's memoirs [5]: "I consulted Eileen Evans whom I had met at Woods Hole and who was working in the Anatomy Department in University College (London); she recommended that I approach George Gray who, following J.D. Robertson's (see ref. [16]) return to the USA, was now head of the electron microscopy unit, and although a morphologist was possibly young and flexible enough to be willing to examine fractions of "smashed-up tissue." This proved to be so; George visited me at Babraham in March 1960 and was horrified by the way I had maltreated my fractions. "You must treat your pellets as though they were pieces of tissue. I wouldn't expect brittle osmicated membranes to stand up to repeated suspensions in alcohol," he told me. He was not impressed by my images of mitochondria either, and I had to admit that they looked somewhat distressed. He took some samples with him to London and promised to contact me again in about 2 weeks' time when he would have completed his electron microscopy using less destructive araldite as the embedding medium. George was as good as his word and I drove down to Cambridge station to meet his train and bring him back to Babraham. I shall never forget this meeting! George's first words to me were "Victor, you have isolated synaptic vesicles, but inside nerve endings". When looking at the micrographs it became clear to them that the homogenization had sheared the club-like presynaptic nerve endings from their axonal connections. As he had to press on with his own work, Gray taught Whittaker to do electron microscopy by himself which is reflected by many structural analyses in Whittaker's further biochemical studies. His Siemens Elmiskop I ÜM 11 electron microscope, considered to be user friendly, required—according to Whittaker—the evolution of a new type of human being, *Homo sapiens siemensiensis*, who had very long arms to reach the adjustment knobs at the top of the instrument while still remaining seated in front of the fluorescence screen. At this point it has to be mentioned that Edward George Gray, FRS (1924–1999), has produced excellent structural research on the cerebral cortex. He showed that dendritic spines are major targets of synaptic input and described the spine apparatus. He also performed a detailed structural analysis of the *Octopus* brain. He is perhaps best remembered for his differentiation of "type 1" and "type 2" synapses related to excitatory and inhibitory synapses, respectively [29, 30].

Whittaker gave the first presentation of his new data in April 1960 at the Department of Pharmacology in the Philadelphia University School of Medicine, following an invitation of George Koelle, with whom he shared his interest in cholinergic biochemistry and physiology. The first printed publication of the novel data followed a Meeting of the Physiological Society in July 1960, depicting pictures of pinched-off nerve endings [31]. The publication of the full paper of Whittaker's and Gray's findings met several obstacles and was therefore delayed. First it was rejected by the Journal of Physiology as "unsuitable for publication" in this journal. The paper included eight pages of electron micrographs and thus the authors did not dare to submit to the Biochemical Journal where Whittaker had published his first attempts to subfractionate nervous tissue [28]. Following Gray's suggestion they finally submitted the paper to the Journal of Anatomy where it was accepted. However the paper did not appear in print until January 1962 [32]. This citation classic provides the full biochemical and morphological characterization of the fractions obtained by density gradient fractionation derived from guinea pig brain, including myelin fragments sedimenting on top, then a fraction which they initially referred to as "nerve ending particles" or also "pinched-off nerve endings," and the denser mitochondria. The electron micrographs depict beautiful "nerve ending particles", including those with attached postsynaptic membrane. For a more handy term, Whittaker introduced NPPs as an acronym for nerve ending particles [33]. But he was not happy with this either and thought of alternatives. In his memoirs [5] he writes: "I pondered this problem while relaxing in my bath one Saturday morning in my cottage in Babraham while a drone bee I had kept alive all winter by feeding it syrup buzzed lazily on the window still. The word 'synaptosome' suddenly occurred to me, from 'synapse' and 'soma', a body. Its predigree was impeccable. Like all respectable scientific words it was derived from a classical language (in this case Greek), the word 'soma' already occurred in other scientific words (e.g. 'microsome', 'lysosome', 'ribosome') and it was euphonious." The word synaptosome then first appeared in print in the title of the paper which describes the isolation of synaptic vesicles, published in 1964 [34]. The editors insisted, however, that the word synaptosome is placed within inverted commas. Whittaker took some pride in the fact that the term synaptosome was later included in the Oxford English Dictionary and that it entered all languages in which neurochemical results were reported. He later would have wished to have patented the synaptosome. But, according to Whittaker, the emergence of the Yuppy scientist still lay in the future.

In the early 1960s, the approach of subcellular fractionation to the study of synaptic function and other functional aspects of the nervous system was still in its infancy. It is therefore not surprising that physiologists and morphologists who used to study intact organs or animals were not convinced that one could learn anything meaningful from smashed-up tissues. Whittaker felt that the British biological establishment failed to accept the impact of his work for several years which limited access to financial resources he needed to further compete in the field.

The next step was obvious; the liberation of synaptic vesicles occluded inside synaptosomes and their purification. This goal was achieved in 1963/1964 together with Arthur Michaelson, who had already worked with Whittaker in Cincinnati and Jeanette Kirkland, an Australian who had just completed her PhD in Biochemistry at the University of Cambridge [34, 35]. It also involved two technicians, Gordon Dowe, his senior technician who had joined Whittaker's lab right from school, and Nick Fincham, the junior technician (Fig. 2). Gordon Dowe (1941–2016) had golden hands in the laboratory. He was very loyal to Whittaker and not only later moved with him to the University of Cambridge and spent some months with him in New York, but also finally followed him to Göttingen where he stayed in his laboratory until Whittaker's retirement. He subsequently worked in another laboratory of the Max-Planck-Institute of Biophysical Chemistry until his retirement and stayed in Germany with his family. Dowe is co-author of numerous papers with Whittaker between 1964 and 1989. The trick for liberating synaptic vesicles was to expose the synaptosomes isolated from guinea pig forebrain to mild hypo-osmotic shock by suspending them in a small volume of water. This would disrupt these osmotically labile miniature cell compartments and would release synaptic vesicles without doing damage to their vesicular content. Sucrose gradient centrifugation was again applied, this time using sucrose steps from 0.4 to 1.2 M sucrose. Besides the biochemical characterization, negative staining of the previously formaldehyde-fixed fractions using sodium

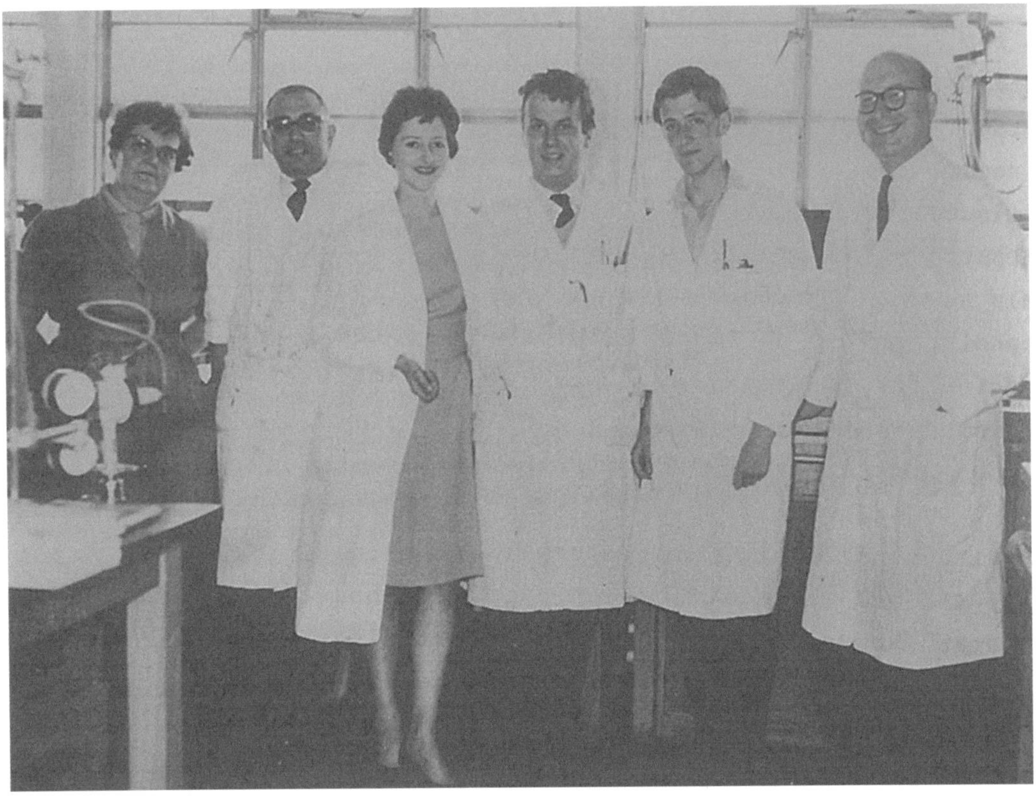

Fig. 2 The proud team that isolated synaptic vesicles from synaptosomes at the Agricultural Research Council Institute of Animal Physiology in Babraham. Right to left: Victor Whittaker, the two technicians Nick Fincham and Gordon Dowe, Jeanette Kirkland, Arthur Michaelson, and Flo Coad (cleaner). From Whittaker VP (1988) The cellular basis of synaptic transmission: an overview. In: Zimmermann H (ed) Cellular and molecular basis of synaptic transmission NATO ASI Series H: Cell Biology, vol. 21, Springer-Verlag, Berlin, Heidelberg, New, York, pp. 1–23. With permission

phosphotungstate was applied for structural characterization. This method was introduced to Whittaker by Bob Horne from the Babraham institute [36]. Notably the material sedimenting on top of the 0.4 M sucrose solution (fraction D) revealed beautifully preserved synaptic vesicles of approx. 50 nm diameter whereas another fraction, sedimenting at the 1.0–1.2 M interface (fraction H), contained seemingly incompletely disrupted synaptosomes. Acetylcholine was bimodally distributed in fractions D and H. But fraction D had the highest acetylcholine activity and specific activity. In order to exclude that smashing up tissue artificially creates synaptic vesicle-like structures, the authors even went to fractionate liver, providing evidence that particles such as synaptic vesicles isolated from brain do not arise spontaneously from lipoprotein membranes [34, 35]. An important gap was now filled. The neurotransmitter is stored inside synaptic vesicles.

It is worth noting how acetylcholine had to be determined these days. No biochemical assay was available and a bioassay had to be applied. In the synaptosome work this was the frog rectus abdominis muscle [37] and in the synaptic vesicle study the more sensitive thin slip of the dorsal muscle of the leech. The contractions of these preparations mounted in an organ bath would be proportional to the acetylcholine concentration within a certain concentration range. These bioassays normally were done immediately after collecting the fractions.

But would the number of acetylcholine molecules stored inside a vesicle derived from the central nervous system correspond to the number of acetylcholine molecules required to induce a muscle mEPSP as observed by Katz and coworkers? Whittaker tackled this problem together with Mike Sheridan, an American postdoc, by counting synaptic vesicles using a tagging method with polystyrene latex beads intermixed with the vesicle fraction [38]. They came up with a number of approximately 2000 molecules acetylcholine per vesicle. This was roughly in the range of previous estimates made on the basis of chemical and biological assays and also fits to estimates made by later electrophysiological experiments at the neuromuscular junction (1000–2000 molecules per quantum) [39, 40]. Subsequently, synaptosomes and synaptic vesicles were isolated by Whittaker's group and others from many other species and tissues and other neurotransmitters were shown to be stored inside synaptic vesicles. Altogether the first half of the 1960s was a very productive period and Whittaker and his associates published a considerable number of additional papers, amongst others, on the synaptosomal distribution of serotonin and noradrenaline or of lipids, distribution of substance P in brain extracts, lactate dehydrogenase as a cytoplasmic marker in the brain, or studies on the yield of synaptosomes from cerebral cortex. In order to prove their contents in neurotransmitter substances, Whittaker and Kris Krnjević, an electrophysiologist and senior member of the Physiology Department, investigated the excitatory and inhibitory effects of soluble components of synaptosomes and synaptic vesicles when squirted by electro-osmosis from multibarrelled micropipettes onto single neurons in the cortex of guinea pigs [41].

The discovery of synaptosomes and the isolation of vesicles from brain have opened many new avenues for the cellular and molecular analysis of the nervous system which cannot be detailed here. These studies showed that synaptosomes maintain a normal membrane potential, contain presynaptic receptors, translocate metabolites and ions, and, when depolarized, release multiple neurotransmitters in a Ca^{2+}-dependent manner. Synaptosomes became also a source of postsynaptic densities or of the presynaptic active zone. Sub-proteomes of isolated synaptosomes are now studied by proteomic techniques revealing their molecular inventory.

The history of the isolation of synaptosomes and synaptic vesicles would be incomplete without reference to the work of Eduardo de Robertis (1913–1988) from the Instituto de Biología Celular, Medical Faculty of the University of Buenos Aires, and the controversy which arose concerning the priority of Whittaker's work. De Robertis started off as an electron microscopist and was one of the pioneers in the study of the fine structure of the synapse [15, 42]. De Robertis and Bennett can be credited with introducing, in 1954, the term "synaptic vesicle" [14]). As a result of that De Robertis worked on the isolation of synaptic vesicles from mammalian brain at about the same time as Whittaker, using fine structural methods only to characterize the subcellular fractions. This resulted in a severe dispute about the priority of the work—as it unfortunately often occurs in science. They both participated in 1960 in Italy (12–17 June) in the "Varenna Symposium on Regional Chemistry, Physiology and Pharmacology of the Nervous System". According to Whittaker, they had some discussions before their presentations when De Robertis mentioned that he had found numerous synaptic vesicles in the mitochondrial fraction. His evidence was based on differential centrifugation and structural data. He apparently had missed the detached nerve terminals. Whittaker, who had performed biochemical analyses in parallel, further reported that he, as a young and naïve scientist, pointed out the synaptosomal profiles in De Robertis' fractions, explained the fractionation properties of these detached nerve terminals, and told him that he now planned to further fractionate them. Next day they both presented their lectures—reporting the isolation of detached nerve terminals. Whittaker reports on this incident also in his review published in 1965 [23]. The Varenna proceedings appeared in print only 2 years after the conference [43, 44]. There are obvious differences in the quality of material presented in the two communications. Since Whittaker anticipated that it would take a long time for the meeting reports to be printed, he presented his data also at the meeting of the Physiological Society which happened to take place in Babraham on July 9th, 1960—to be on the safe side and to have an Abstract printed [31]. After his return, De Robertis published several papers reporting the isolation of nerve endings and synaptic vesicles from rat brain [45, 46] and reported his novel findings at several occasions in the USA. When he felt to be unfairly treated Whittaker would defend himself. Eminent neuroscientists present at the Varenna meeting, particularly Sanford Palay of Harvard University and Frank O. Schmidt of MIT, supported Whittaker which helped to establish his priority. Similarly, Whittaker had concerns about the purity of fractions obtained in other papers of De Robertis [47, 48] where synaptic vesicles were isolated from osmotically lysed synaptosomes by differential centrifugation steps rather than by density gradient centrifugation—the standard method used in all future studies using Whittaker's method [49, 50]. Choline acetyltransferase was claimed by the De Robertis group to be tightly associated with synaptic vesicles. The Whittaker group had found it in the soluble fraction and suggested that the presence of the enzyme in De Robertis' vesicle fraction was due to contamination with undisrupted synaptosomes. Later it was shown by Frode Fonnum, a Norwegian PhD student in Whittaker's laboratory in Babraham, that pH and ionic strength as well as species differences affect the binding of the enzyme to membranes [51, 52]. In later work De Robertis also applied sucrose density gradient centrifugation for fractionating brain tissue and biochemical analysis [53]. Both laboratories carried on producing neurochemical research on synaptosomes. Whittaker later placed his focus on the study of synaptic vesicles and acetylcholine metabolism and release whereas De Robertis turned to the characterization of receptor proteins.

One disadvantage of the central nervous system as starting tissue for vesicle isolation was the heterogeneity of the vesicle fraction regarding transmitter types. No methods were available to specifically isolate cholinergic synaptic vesicles. Richard Keynes from the Institute of Animal Physiology in Cambridge, who worked on the thermodynamics of the electric organ discharge of strongly electric fish, suggested to Whittaker the use of the electric organs of the electric ray *Torpedo* for the isolation of synaptic vesicles. These organs are homologous to skeletal muscle and carry a very rich cholinergic innervation, more than a 100 times that of brain. Whittaker obtained the fish from the Stazione Zoologica in Naples. The first study on the characterization and subcellular fractionation of electric organ was performed in 1966 with Mike Sheridan and Maurice Israël, who had come to Babraham from Paris [54]. This opened up a new avenue for the study of presynaptic cholinergic mechanisms and synaptic vesicle function and biochemistry.

When the author joined Whittaker's laboratory in the Department of Biochemistry at Cambridge University in January 1972 work was ongoing on both mammalian brain and the *Torpedo* electric organ. Using a large zonal rotor cholinergic vesicle from the *Torpedo* electric organ could be isolated at high yield and purity for biochemical analysis. Mike Dowdall who had already worked with Whittaker in New York and Alan Boyne, an American postdoc, had just shown that cholinergic synaptic vesicles from this organ stored ATP in addition to acetylcholine [55]. Size-exclusion chromatography using columns of porous glass beads was introduced by Stephen Morris, an American biophysicist, as an additional tool to produce highly purified synaptic vesicles [56] (Fig. 3).

Morphological and physiological evidence as well as demonstration of vesicular acetylcholine storage strongly supported the hypothesis that acetylcholine is released from synaptic vesicles. But

Fig. 3 Whittaker's team in front of the Department of Biochemistry, University of Cambridge in 1973. Top from right to left: Colin Denston (technician), Mike Dowdall (staff), Victor Whittaker, Gordon Dowe (technician), Stephen Morris (postdoc). Bottom from right to left: Bill Edwards (electron microscopist), Herbert Zimmermann (postdoc), Mia Talerman (secretary), Gerd Ulmar (postdoc), Gail Vickers (PhD student), Fiona Henderson (technician)

it still had to be shown that synaptic vesicles are the actual source of the released neurotransmitter. Answering this question led to many controversial and hot debates. This is discussed here very briefly because Whittaker's laboratory was highly involved in this and because it further reflects Whittaker's stamina when engaging in scientific controversy.

It had been recognized from labeling experiments with radioactive choline that released acetylcholine had a higher specific radioactivity than the acetylcholine pool isolated in the synaptic vesicle fraction [57]. This suggested that newly synthesized acetylcholine was preferentially released and that the more recently synthesized acetylcholine comes from a store different from the bulk vesicular acetylcholine fraction. Was there a sub-pool of vesicles with high acetylcholine turnover? An alternative interpretation would have been that the newly synthesized acetylcholine is in fact directly released from a cytosolic acetylcholine pool. This pool would exist in addition to the vesicular pool and synaptic vesicles would rather function as a sluggish store for the neurotransmitter which feeds into the cytosolic pool. Stimulating the electric organ and analyzing the turnover of acetylcholine and of synaptic vesicles as well as the released radiolabeled acetylcholine would have been a way to tackle this question. This was one of the approaches of the author of this article during the years coming.

After his return to Paris, Israël had started his own laboratory at the Laboratoire de Cytologie, Faculté des Sciences, Hôpital de la Salpêtrière. He also purified synaptic vesicles from the *Torpedo* electric organ and studied acetylcholine release. Unfortunately, the outcome of results and their interpretation differed between the groups which led to many controversies at scientific meetings and also in publications. This continued after Whittaker's move to Göttingen in 1973. The data of the Whittaker group supported an exocytotic mechanism whereas Israël and his associates favored the hypothesis that acetylcholine is released directly from a cytosolic pool through membrane pores or channels of the plasma membrane (mediatophore). As a result of this—for many years—many neuroscientists considered the question undecided.

This controversy cannot be detailed here but the issue is well reflected by a series of ten papers which were published in the years 1978 and 1979 in the newly founded journal Trends in Neurosciences (TINS). Its editors may have hoped to attract readership this way. It started with a paper of John Heuser, then Associate Professor at the Department of Physiology School of Medicine, University of California, San Francisco in the first volume of the journal (TINS, 1, 80–82, 1978). Heuser had performed pioneering work using quick-freeze deep-etch electron microscopy, supporting exocytotic acetylcholine release at the frog neuromuscular junction ("Quick-freezing evidence in favor of the vesicular hypothesis"). The editors felt that they had to simultaneously invite an alternative view and Roger Marchbanks (London), a former coworker of Whittaker in Cambridge, replied with the article "The vesicular hypothesis questioned" where he equally criticized the experimental work of Heuser and of the Whittaker group (TINS, 1, 83–84, 1978). This was answered by Whittaker and Zimmermann where the point was taken—based on their evidence—that the synaptic vesicle pool is in fact heterogeneous and that a recycled vesicle pool rich in newly synthesized acetylcholine can be physically separated from the remainder of the vesicles and that this could be the pool causing preferentially release of the newly synthesized acetylcholine (TINS, 1, 100, 1078). Marchbanks' reply was prompt and highly critical (TINS, 1, 168, 1978). Whittaker then answered his previous pupil in reproachfully aggressive tone (TINS, 2, 55–56, 1979). This was followed by another rather personal reply of Marchbanks (TINS, 2, 56, 1979). Then followed articles of Yves Dunant and Israël (Geneva/Paris) (TINS, 2, 130–132, 1979), Zimmermann (Göttingen) (TINS, 2, 282–285, 1979), Peter Baker und Derek Knight, King's College, London (TINS, 2, 288–291, 1979), and finally Brian Collier, Department of Pharmacology and Therapeutics, McGill University, Montreal (TINS, 2, 285–287, 1979). Collier concluded his article "Synaptic vesicles and quanta" in a diplomatic way: "There is no proof or disproof of the vesicle hypothesis of ACh release." More controversy followed. In an extensive review article on "vesicle recycling and transmitter release" Zimmermann reflected the subject matter against the background of the various methodological approaches [58]. Today it may sound surprising that such heated and mutually damaging debate would at all have been published in a scientific journal. But it is also noteworthy that progress in science eventually helps to overcome scientific controversies.

In Göttingen, Whittaker's group grew in size, studied synaptosomal choline and acetylcholine metabolism, produced the first antibody to recognize cholinergic nerve terminals, and made major contributions to understanding the physical and functional properties of synaptic vesicles from both mammalian brain and *Torpedo* electric organ, including the characterization of synaptic vesicle proteins and lipids. In addition Whittaker became interested in the development of the *Torpedo* electromotor system. Since the transport of the electric rays to Göttingen from the Marine Biological Station in Arcachon in southern France at the Atlantic coast near Bordeaux and the subsequent entertainment of the fish in aquariums were rather tedious he set up an outstation in Arcachon. He himself was a regular visitor there, stationing his small sailing boat in the Bay of Arcachon. Whittaker's dedication to this model system for the study of the cholinergic synapse is reflected in his book "The Cholinergic Neuron and its Target" published after his retirement in 1992 [59]. Altogether these studies laid the foundations for the present-day molecular biological analysis not only of synaptic vesicles but also of all other functional compartments of the synapse. The 2013 Nobel Prize in Physiology or Medicine to James E. Rothman, Randy W. Schekman, and Whittaker's former PhD student Thomas C. Südhof very much rests also on Whittaker's pioneering work.

Whittaker always took clear positions. He offered great and warm support to his coworkers and kept—together with his wife Margaret—an open house for members of his group. He put a lot of effort in training young scientists from all over the world in neurochemistry, in his laboratory and at international training courses. He was gifted with a superb memory, a true polymath with deep interest in all aspects of culture, including music. He loved travelling and to immerse himself in the history and culture of other countries. With Italian, French, and German he managed three additional languages. The final chapter of his four-volume memoirs [6], written at the age of 95, is entitled "Die Tat ist alles, nichts der Ruhm," a metaphor from Goethe's famous dramatic work "Faust". It may be translated as "The deed is all, and not the glory".

References

1. Whittaker VP (1993) Thirty years of synaptosome research. J Neurocytol 22(9):735–742
2. Nicholls DG (1989) Release of glutamate, aspartate, and gamma-aminobutyric acid from isolated nerve terminals. J Neurochem 52(2):331–341
3. Zimmermann H, Fonnum F (2016) Victor P. Whittaker (1919-2016). J Neurochem 139(2):333–335. https://doi.org/10.1111/jnc.13778
4. Whittaker VP (1963) Identification of acetylcholine and related esters of biological origin. In: Koelle GB (ed) Handbuch der experimentellen Pharmakologie. Ergänzungswerk XV. Cholinesterases and anticholinesterase agents. Springer, Berlin, pp 1–39
5. Whittaker VP (2010) Flashbacks: an autobiography. My Scientific Odyssey 1951–1973, vol. 3
6. Whittaker VP (2014) Flashbacks: an autobiography. An Englishman in Germany 1973–1992, vol. 4
7. Dale HH, Feldberg W, Vogt M (1936) Release of acetylcholine at voluntary motor nerve endings. J Physiol Lond 86(4):353–380
8. Fatt P, Katz B (1950) Some observations on biological noise. Nature 166(4223):597–598
9. Fatt P, Katz B (1951) An analysis of the endplate potential recorded with an intracellular electrode. J Physiol Lond 115(3):320–370
10. Fatt P, Katz B (1952) Spontaneous subthreshold activity at motor nerve endings. J Physiol Lond 117(1):109–128
11. Del Castillo J, Katz B (1956) Biophysical aspects of neuro-muscular transmission. Prog Biophys Biophys Chem 6:121–170
12. Palade GE (1954) American association of anatomists. Sixty-seventh annual session. Anat Rec 118(2):335–336. https://doi.org/10.1002/ar.1091180211
13. Sjöstrand FS (1953) The ultrastructure of the retinal rod synapses of the Guinea pig eye. J Appl Phys 24(11):1422. https://doi.org/10.1063/1.1721193
14. De Robertis EDP, Bennett HS (1954) Submicroscopic vesicular component in the synapse. Fed Proc 13:35
15. De Robertis EDP, Bennett HS (1955) Some features of the submicroscopic morphology of synapses in frog and earthworm. J Biophys Biochem Cytol 1(1):47–58
16. Robertson JD (1956) The ultrastructure of a reptilian myoneural junction. J Biophys Biochem Cytol 2(4):381–394
17. Palay SL (1956) Synapses in the central nervous system. J Biophys Biochem Cytol 2(4 Suppl):193–202
18. Claude A (1946) Fractionation of mammalian liver cells by differential centrifugation. I. Problems, methods, and preparation of extract. J Exp Med 84(1):51–59
19. Claude A (1946) Fractionation of mammalian liver cells by differential centrifugation. II. Experimental procedures and results. J Exp Med 84(1):61–89
20. Blaschko H, Welch AD (1953) Localization of adrenaline in cytoplasmic particles of the bovine adrenal medulla. Naunyn Schmiedebergs Arch Exp Pathol Pharmakol 219(1–2):17–22
21. Hillarp NA, Lagerstedt S, Nilson B (1953) The isolation of a granular fraction from the suprarenal medulla, containing the sympathomimetic catechol amines. Acta Physiol Scand 29(2–3):251–263. https://doi.org/10.1111/j.1748-1716.1953.tb01022.x
22. De Duve C, Pressman BC, Gianetto R et al (1955) Tissue fractionation studies. 6. Intracellular distribution patterns of enzymes in rat-liver tissue. Biochem J 60(4):604–617
23. Whittaker VP (1965) The application of subcellular fractionation techniques to the study of brain function. Prog Biophys Mol Biol 15:39–96
24. Barker LA (1976) Subcellular aspects of acetylcholine metabolism. In: Goldberg AM, Hanin I (eds) Biology of cholinergic function. Raven Press, New York
25. Hebb C (1972) Biosynthesis of acetylcholine in nervous tissue. Physiol Rev 52(4):918–957

26. Hebb CO, Smallman BN (1956) Intracellular distribution of choline acetylase. J Physiol Lond 134(2):385–392
27. Hebb CO, Whittaker VP (1958) Intracellular distributions of acetylcholine and choline acetylase. J Physiol Lond 142(1):187–196
28. Whittaker VP (1959) The isolation and characterization of acetylcholine-containing particles from brain. Biochem J 72:694–706
29. Gray EG (1969) Electron microscopy of excitatory and inhibitory synapses: a brief review. Prog Brain Res 31:141–155. https://doi.org/10.1016/S0079-6123(08)63235-5
30. Guillery RW (2000) Early electron microscopic observations of synaptic structures in the cerebral cortex: a view of the contributions made by George Gray (1924-1999). Trends Neurosci 23(12):594–598
31. Gray EG, Whittaker VP (1960) The isolation of synaptic vesicles from the central nervous system. J Physiol (London) 153(Suppl):35P–37P
32. Gray EG, Whittaker VP (1962) The isolation of nerve endings from brain: an electron-microscopic study of cell fragments derived by homogenization and centrifugation. J Anat 96:79–88
33. Michaelson IA, Whittaker VP, Laverty R et al (1963) Localization of acetylcholine, 5-hydroxytryptamine and noradrenaline within subcellular particles derived from Guinea-pig subcortical brain tissue. Biochem Pharmacol 12:1450–1453
34. Whittaker VP, Michaelson IA, Kirkland RJ (1964) The separation of synaptic vesicles from nerve-ending particles ('synaptosomes'). Biochem J 90(2):293–303
35. Whittaker VP, Michaelson IA, Kirkland RJ (1963) The separation of synaptic vesicles from disrupted nervending particles. Biochem Pharmacol 12:300–302
36. Horne RW, Whittaker VP (1962) The use of the negative staining method for the electron-microscopic study of subcellular particles from animal tissues. Z Zellforsch Mikrosk Anat 58:1–16
37. Whittaker VP, Dowe GH (1964) The identification of the acetylcholine-like substance in synaptosomes derived from Guinea-pig brain as acetylcholine itself. Int J Neuropharmacol 3:593–597
38. Whittaker VP, Sheridan MN (1965) The morphology and acetylcholine content of isolated cerebral cortical synaptic vesicles. J Neurochem 12:363–372
39. Katz B, Miledi R (1972) The statistical nature of the acetycholine potential and its molecular components. J Physiol Lond 224(3):665–699
40. Anderson CR, Stevens CF (1973) Voltage clamp analysis of acetylcholine produced end-plate current fluctuations at frog neuromuscular junction. J Physiol Lond 235(3):655–691
41. Krnjević K, Whittaker VP (1965) Excitation and depression of cortical neurones by brain fractions released from micropipettes. J Physiol Lond 179(2):298–322
42. De Robertis E, Franchi CM (1956) Electron microscope observations on synaptic vesicles in synapses of the retinal rods and cones. J Biophys Biochem Cytol 2(3):307–318
43. De Robertis ED (1962) Morphological bases of synaptic processes and neurosecretion. In: Kety S, Elkes J (eds) Proc IV Internat. Soc. Symposium, Varenna, 1960, regional chemistry, physiology and pharmacology of the nervous system. Pergamon Press, Oxford, pp 248–258
44. Whittaker VP 1962 The binding of neurohormones by subcellular particles of brain tissue. In: Kety S, Elkes J Proc IV Internat. Soc. Symposium, Varenna, 1960, regional chemistry, physiology and pharmacology of the nervous system. Pergamon Press, Oxford, 259–263
45. De Robertis E, Pellegrino de Iraldi A, Rodriguez G et al (1961) On the isolation of nerve endings and synaptic vesicles. J Biophys Biochem Cytol 9:229–235
46. De Robertis E, Pellegrino de Iraldi A, Rodriguez de Lores Garnaiz G et al (1962) Cholinergic and non-cholinergic nerve endings in rat brain. I. Isolation and subcellular distribution of acetylcholine and acetylcholinesterase. J Neurochem 9:23–35
47. De Robertis E, Rodriguez de Lores Garnaiz G, Salganicoff L et al (1963) Isolation of synaptic vesicles and structural organization of the acetycholine system within brain nerve endings. J Neurochem 10:225–235
48. De Robertis E, Salganicoff L, Zieher LM et al (1963) Acetylcholine and cholinacetylase content of synaptic vesicles. Science 140(3564):300–301
49. Marchbanks RM, Whittaker VP (1969) The biochemistry of synaptosomes. In: Bittar EE, Bittaker N (eds) The biological basis of medicine, vol 5. Academic Press, London, pp 33–76
50. Whittaker VP (1969) The synaptosome. In: Lajtha A (ed) Handbook of neurochemistry. Springer, Boston, MA, pp 327–364
51. Fonnum F (1967) The 'compartmentation' of choline acetyltransferase within the synaptosome. Biochem J 103(1):262–270

52. Fonnum F (1970) Surface charge of choline acetyltransferase from different species. J Neurochem 17(7):1095–1100
53. De Robertis E (1967) Ultrastructure and cytochemistry of the synaptic region. The macromolecular components involved in nerve transmission are being studied. Science 156(3777):907–914
54. Sheridan MN, Whittaker VP, Israël M (1966) The subcellular fractionation of the electric organ of Torpedo. Z Zellforsch Mikrosk Anat 74(3):293–307
55. Dowdall MJ, Boyne AF, Whittaker VP (1974) Adenosine triphosphate. A constituent of cholinergic synaptic vesicles. Biochem J 140(1):1–12
56. Morris SJ (1973) Removal of residual amounts of acetylcholinesterase and membrane contamination from synaptic vesicles isolated from the electric organ of Torpedo. J Neurochem 21(3):713–715
57. Chakrin LW, Marchbanks RM, Mitchell JF et al (1972) The origin of the acetylcholine released from the surface of the cortex. J Neurochem 19(12):2727–2736
58. Zimmermann H (1979) Vesicle recycling and transmitter release. Neuroscience 4(12):1773–1804
59. Whittaker VP (1992) The cholinergic neuron and its target: the electromotor innervation of the electric ray "Torpedo" as a model. Birkhäuser, Boston

Chapter 3

Synaptosome Preparations: Which Procedure Should I Use?

Peter R. Dunkley and Phillip J. Robinson

Abstract

One of the most extensively used model systems to investigate the functions and the chemical control of the nerve terminal has been the synaptosome. Synaptosomes are pinched-off nerve endings that form by shearing forces during homogenization of neuronal tissue. Depending on the aim of the study, synaptosomes can be used once they are formed within the whole neuronal tissue homogenate, or they can be separated from other subcellular organelles and enriched to various extents, depending on the fractionation procedures adopted. Each procedure varies in the time that it takes and provides synaptosomes with different levels of homogeneity and viability. The major contaminants of synaptosomes that remain after fractionation include neuronal and glial plasma membranes, attached postsynaptic membranes and densities, microsomes, synaptic vesicles, and extra-synaptosomal mitochondria. This chapter documents the most commonly used procedures for making synaptosomes, indicates the most likely contaminants present after each procedure, and assesses the viability of the resulting synaptosomes. This provides researchers with a decision tool to determine which synaptosome preparation procedure best suits the aims of their study.

Key words Synaptosomes, Formation, Fractionation, Homogeneity, Viability

1 Introduction

Since the early 1950s it has been known that synaptic vesicles were present in nerve endings and it was postulated that these vesicles contained chemical neurotransmitters, such as acetylcholine, and that they were somehow involved in the quantized neurotransmitter release observed by De Castillo and Katz [1]. While attempting to isolate these synaptic vesicles from neuronal tissue a subcellular fraction was obtained from brain homogenates that was enriched for the presynaptic markers acetylcholine and choline acetyltransferase and that was different to microsomes, mitochondria, and nuclei [2]. This subcellular fraction was found to contain particles that were much larger than synaptic vesicles and they were eventually identified as intact nerve endings which enclose synaptic vesicle clusters and mitochondria [3, 4]. Nerve endings come in different configurations and thus this preparation

Kathryn M. Murphy (ed.), *Synaptosomes*, Neuromethods, vol. 141,
https://doi.org/10.1007/978-1-4939-8739-9_3, © Springer Science+Business Media, LLC, part of Springer Nature 2018

includes both nerve endings at the terminus of axons as well as *en passant* synaptic boutons [5]. These nerve endings were formed during homogenization of the neuronal tissue in isotonic sucrose solutions by shearing the plasma membranes near to the junction of the axon and the nerve terminal, followed by resealing of the nerve terminal membranes [4]. A few years after their discovery, these pinched-off nerve endings were named "synaptosomes" in the title of a paper, since their physical properties were analogous to those of other subcellular organelles [6].

A Scopus search of PubMed lists almost 14,000 original articles, indicating that synaptosomes are a widely used model system for investigating nerve ending function. This is because synaptosomes contain the same complement of subcellular organelles and proteins, cytoplasm, cytoskeleton, synaptic vesicles, and mitochondria normally found in nerve endings in vivo. The use of synaptosomes began in the early 1960s and rose rapidly in the 1970s and peaked in the 1980s, where they were primarily used to investigate nerve terminal functions, such as neurotransmitter release and endocytosis, neurotransmitter synthesis and degradation, calcium and energy metabolism, as well as control of these functions by ion channels and neuroactive drugs. Synaptosomes are now used for more detailed studies on the biochemical analysis of protein function and mechanisms, as well as molecular aspects of neuroscience, neuropharmacology, and medicine. There are still over 100 papers published per year on synaptosomes.

It is thought that total number of synapses in the human brain could be in the range of 10^{13}–10^{15} [7], and 1 mm^3 of the cerebral cortex with about 90,000 neurons has almost a billion nerve endings [8]. The average synaptosome diameter from mammalian brain is in the range from 0.5 to 0.9 μM which further varies depending on the brain region [9, 10]. This diameter is comparable to the average diameter of a mitochondrion (0.2–0.7 μM), and roughly the size of a bacterium. Up to 60% of cerebral cortical synaptosomes contain mitochondria and up to 15% contain attached postsynaptic densities [11]. The average volume of cerebral cortex synaptosomes is 0.37 μm^3 and they enclose an average of 383 synaptic vesicles, each with a volume of 0.015 μm^3, and 0.12 μm^3 of mitochondria [12].

The original methods for making synaptosomes [3, 4] both included the following critical steps:

1. Tissue Disruption
   (a) Use of fresh whole brain tissue
   (b) Disruption of the tissue in 0.32 M sucrose using an homogenizer
2. Homogenate Fractionation
   (a) Centrifugation to provide a S1 supernatant, which removes nuclei

(b) Centrifugation to provide a P2 pellet, which removes most of the synaptic vesicles, endoplasmic reticulum, and soluble material

(c) Resuspension of the P2 pellet in 0.32 M sucrose, and layering it on to a discontinuous sucrose density gradient and centrifugation, which removes most of the myelin and extra-synaptosomal mitochondria

3. Preparation for Analysis

(a) Collection, dilution, and centrifugation to provide a synaptosome pellet, which contained contaminating neuronal and glial membranes

(b) Resuspension of synaptosomes for analysis

Since these original methods for making synaptosomes, a wide range of procedures have been developed for each of these three critical steps and each has its own advantages in terms of the homogeneity and viability of the final synaptosome preparation and the time it takes to complete the procedure. Each of these critical steps will be discussed so that researchers who wish to use synaptosomes can choose the preparation procedure that is most appropriate for their specific purpose. The focus of the following sections will be primarily on synaptosomes isolated from mammalian brain tissue and not from ganglia or specialized neuronal tissues. It is also important to note that synaptosomes are not synaptoneurosomes (see Box 1).

Box 1: Synaptosomes Versus Synaptoneurosomes

It is important to point out that synaptosomes are not synaptoneurosomes. Synaptoneurosomes are a preparation in which many pinched-off nerve terminals contain attached resealed postsynaptic entities and since membranes from both sides form closed vesicles they appear snowman-shaped using electron microscopy [13]. They were first isolated in 1984 [14]. The synaptoneurosome name is derived from the two elements: the presynaptic synaptosome (synapto), attached to a resealed postsynaptic vesicle (neurosome). The diameter of the synaptoneurosome is about 1.1–1.6 μm, with the enclosed vesicle on either side each being a diameter of about 0.6 μm; accordingly the synaptoneurosome volume of about 0.60 μm^3 is about double that of the average synaptosome [13]. The procedures for preparation of these two subcellular tissue fractions are also quite different. Synaptosomes are formed during brain homogenization in isotonic sucrose using Teflon/glass homogenizers, while synaptoneurosomes are formed during brain homogenization in isotonic Krebs-like buffers using a Dounce glass/glass homogenizer. High-salt buffers must be avoided for synaptosome isolation procedures. The methods used for enriching synaptosomes and synaptoneurosomes are also different, with synaptosomes being isolated using differential and density gradient centrifugation, while synaptoneurosomes are isolated using pressure filtration through nylon mesh and then Millipore filters. Some authors have confused synaptosomes and synaptoneurosomes and this is probably because there is electron microscopic evidence that approximately 10–15% of synaptosomes contain attached postsynaptic densities [11]. However, all synaptosome preparations that have been investigated by electron microscopy contain very few snowman-shaped entities and most of the postsynaptic densities that are present do not have substantial amounts of postsynaptic membrane beyond the density itself, and these membranes are rarely closed off as vesicles. On the other hand synaptoneurosome preparations contain many synaptosomes; for example in 100 electron microscopic fields there were 252 synaptosomes, 67 neurosomes, and 141 synaptoneurosomes [13].

2 Tissue Disruption

2.1 Choice of Tissue

Every nerve terminal has its own shape and size and relationship to the surrounding cells and extracellular matrix, so that when homogenization of the tissue occurs the likelihood of shearing and resealing to form synaptosomes differs widely between these geometries. Consideration therefore needs to be made as to the species; the age and the sex of the animal used; the neuronal tissue chosen; the sacrifice procedure; and whether synaptosomes are prepared from fresh or frozen tissues (Fig. 1). Unfortunately few studies have been specifically designed to systematically investigate the size, contents, homogeneity, viability, or recovery of synaptosomes prepared by varying these parameters.

The original procedures for synaptosome preparation were established using fresh whole rat brains [3], or whole guinea pig brains [4]. Subsequent studies used very similar procedures to prepare synaptosomes from a range of brain tissues taken from many other species including man, monkey, sheep, cat, dog, coypu, rabbit, mouse, pigeon, frog, and cephalopod (see [2] and Table 1 in [15]). Differences between mammalian and fish brain synaptosomes were mostly accounted for by the higher level of myelin in the fish brain [15]. Synaptosome preparations were also obtained from chicken brain tissue using procedures that were substantially modified from the original procedures [16], but these chicken brain synaptosomes did not differ substantially from rat brain synaptosomes prepared using very similar modifications [17, 18]. The original procedures used small amounts of brain tissue and when larger preparations, or use of tissue from large animals, were required a zonal rotor procedure was developed to prepare synaptosomes from rat [19], or bovine [20], brains.

A small number of studies have claimed to have prepared synaptosomes from cultured neurons [21–24]. However major caution is required here, since detailed characterization of the resultant

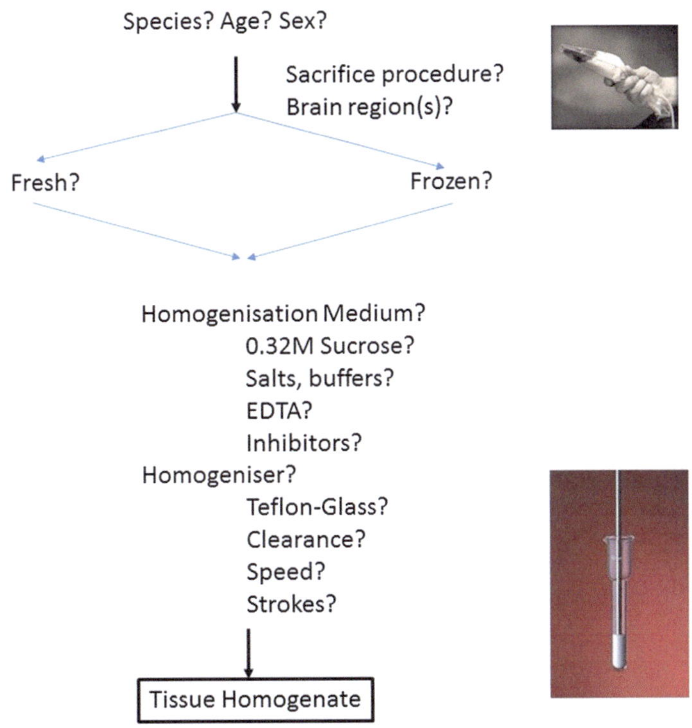

Fig. 1 Tissue disruption. In order to prepare synaptosomes brain tissue must first be disrupted and a homogenate prepared. There are a number of decisions that need to be made including which animal to use, how should it be sacrificed, which brain regions are required, will fresh tissue be used, what homogenization media and what homogenization instrument should be used

synaptosomes has not been reported, as it is particularly important to know whether they maintain the appropriate morphology by at least electron microscopy criteria.

The age of the animal used for making synaptosomes from brain tissue is important, as synaptic development needs to be complete. Nerve endings are formed before birth, but their composition and thus susceptibility to mechanical shear during homogenization are uncertain until maturity. Neonatal animals have been used to prepare "synaptosomes," but there is usually no morphological evidence that the preparations actually contain nerve endings (except see [25]), and certainly they will not contain the adult complement of nerve-ending proteins and postsynaptic elements. The original procedure [3] used 150 g rats that were approximately 5 weeks old, but for rats 30–50-day-old animals were optimal. We recommend that rats should be at least 30 days old. Both male and female animals have been used to prepare synaptosomes from brain tissue. Synaptosomes have also been prepared from a number of specific rat brain regions including the cerebral cortex [9], hippocampus [26], striatum [27], cerebellum [28], and brain stem [29].

Table 1
Criteria used for determining synaptosome homogeneity and viability

Method	Reference	Homogeneity	Viability
Discontinuous sucrose	[3]	EM, protein, AChE, SDH, ACh	
Discontinuous sucrose	[4]	EM, ACh	
Sucrose steps (compared to discontinuous sucrose)	[53]	EM, protein, CA, GPD, SDH, ACE, CAT, CNP	K^+ content, respiratory rate, amino acid release with veratridine
Discontinuous Ficoll	[55]	EM, protein, ACE, CO, NaK-ATPase	
Ficoll floatation	[63]	EM, protein, NADPHCCR, AChE, LDH, cytochromes	Oxygen uptake on depolarization with different substrates
Discontinuous Percoll	[9, 11, 26, 30, 67]	EM, protein, LDH, CO, pMBP, pSynapsin 1, PDH, NA/DA, CAT, muscarine and nicotine binding	Depolarization-dependent Ca uptake, Synapsin 1 phosphorylation and NA release, NA and choline uptake

EM electron microscopy. **Cytosol:** *GPD* glycerol 3-phosphate dehydrogenase; *LDH* lactate dehydrogenase. ***ER:*** *NADPHCCR* NADPH cytochrome C reductase. **Myelin:** *CNP* 2′,3′-cyclic nucleotide-phosphohydrolase; *pMBP* phosphorylated myelin basic protein. **Membranes:** *NaK-ATPase* AChE-acetylcholinesterase (postsynaptic membranes); *CA* carbonic anhydrase (glial membranes); muscarine binding (synaptic plasma membranes). **Synaptosomes:** *ACh/NA/DA* acetylcholine, noradrenaline, and dopamine content (in synaptic vesicles); *CAT* choline acetyltransferase; *pS1* phosphorylated synapsin 1; NA/DA/choline uptake; nicotine binding. **Mitochondria:** *SDH* succinate dehydrogenase; *CO* cytochrome oxidase; *NADHCCR* NADH cytochrome C reductase; cytochrome content; *pPDH* phosphorylated pyruvate dehydrogenase

Many different procedures have been used to sacrifice the animals prior to the preparation of synaptosomes. De Robertis used both anesthetics and decapitation [3] and there was no evidence that the sacrifice procedure substantially changed the morphology of the preparation of synaptosomes, as long as the brain was rapidly removed and cooled to 0–4 °C as soon as possible and all subsequent steps maintained the tissue at this temperature. However, as synaptosomes are used for a wide range of biochemical and physiological studies it is likely that the results of particular investigations could be dependent on the sacrifice procedure used. For example we found (unpublished data) that the phosphorylation profile of synaptosomal proteins differed if anesthetics rather than decapitation without anesthesia was used. Presumably, this was because there were differences in the rate of loss of cellular ATP levels after cell death and because protein kinases require ATP while protein phosphatases do not. This idea was supported by the fact that phosphorylation differences were reduced when synaptosomes were subsequently incubated in a physiological solution to reestablish synaptosomal ATP levels. It is also not possible to use anesthetics for sacrifice if the aim of the study is to investigate the ion channels and transmitter release, or drugs that block synaptic function, as some anesthetic will inevitably carry through with the synaptosomes (see [30]).

We favor decapitation without anesthesia as the means of sacrifice for all studies on synaptosomes. However, individual institutional ethics committees may appropriately ask how this can be done safely and with minimum pain and stress to the animal. We originally used stunning and decapitation [9]. Stunning involves cervical dislocation and takes ~2–3 s. The most rapid method for the subsequent decapitation is to use a guillotine, which takes about 10 s. However, to avoid the stunning step, we have now switched to using a DecapiCone® (see Fig. 1—http://www.braintreesci.com/Products/restrainer.asp) to restrain the animal prior to decapitation. We found that the rats usually enter the cone readily, but we have also used the plastic cone canister holder made of cardboard with an aluminum base as a darkened tunnel entrance to coax the rat to go in the smaller breathing hole of the DecapiCone®, and this has worked well to reduce apparent animal stress.

It is strongly recommended that fresh brain tissue always be used to prepare synaptosomes with minimal time spent between decapitation and homogenization in sucrose (it is possible for a skilled person to achieve this in about 1 min). Frozen tissue has been used for human and rat brain [31–35] and optimal conditions for freezing tissue have been investigated [32]; overtime freezing at −20 °C caused greater loss of activity than at −70 °C and freezing tissue in the absence of fluid was optimal for rat brain.

There are a large number of neurotransmitters present in the nervous system and synaptosomes contain all of these in various proportions, depending on the original brain region used. Attempts to enrich synaptosome preparations for particular neurotransmitters have been undertaken by selecting the starting tissue; acetylcholine is enriched in superior cervical ganglion [36], squid [37], lobster [38], electric organ of Torpedo (an electric ray) [39], and aplysia (sea slug) abdominal ganglia [40], while noradrenaline is enriched in the rat vas deferens and cat spleen [41, 42]. When ganglia or electric organs rather than brain were used, specific methods were required for synaptosome preparation that differed significantly from the original brain procedures. In part this was due to the differences in isotonic conditions for the neuronal tissues [38, 43]. It was also partly due to the high levels of connective tissue, where enzymes such as collagenase were required, and only tissue taken from younger animals was optimal for ganglia [36], while special tissue mincers were required for electric organs [39]. Preparation of slices from the starting tissue and preincubation in

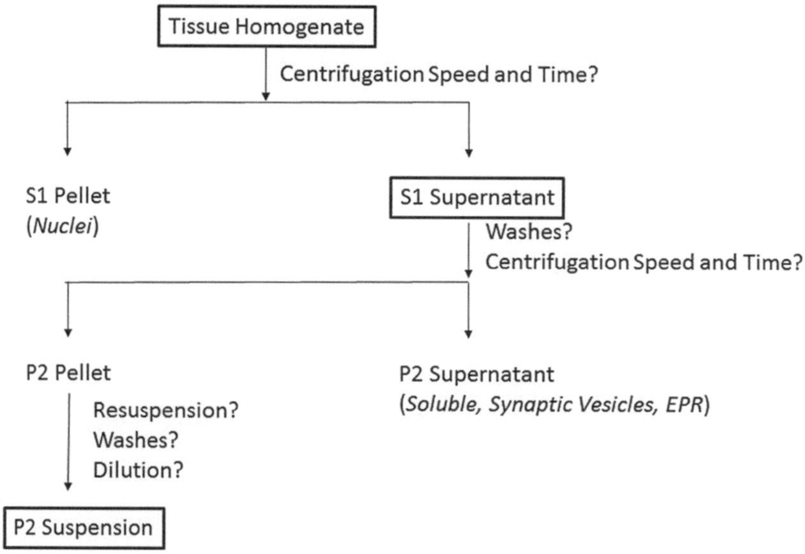

Fig. 2 Homogenate fractionation. Once a homogenate is available it is necessary to determine whether an S1 or a P2 fraction is required and what centrifugation conditions should be used for the preparation of each fraction. It is also necessary to determine how many washes the S1 and P2 fraction needs to undergo in order to ensure the homogeneity of the final synaptosome preparation

the presence of eserine (physostigmine—a reversible acetylcholinesterase activity inhibitor) to preserve the levels of acetylcholine [36], or pargyline (an irreversible monoamine oxidase-B inhibitor) to maintain noradrenaline levels [44], were investigated to improve the neurotransmitter content of synaptosomes, but synaptosome recovery was significantly decreased when compared to starting with fresh brains [36]. Despite this, secretion of neurotransmitters from synaptosomes could still be measured after preincubation of slices and their release modulated by agents that act presynaptically [45].

2.2 Homogenization Procedures

When making synaptosomes from brain tissue the standard procedure is to cut the brain tissue into small pieces, add them to ice-cold 0.32 M sucrose as the homogenization solution, and then disrupt the tissue using a glass homogenizer (Fig. 2). The original procedures used 10 g of brain tissue per 100 mL of 0.32 M sucrose, as either the ratio before homogenization [4] or the final ratio after dilution of the homogenate [3]. The homogenization solution used in most synaptosome preparations from mammalian brains developed after 1962 also contained 0.32 M sucrose as the primary constituent and homogenization occurs at a ratio of between 10 and 20% (w/v). We recommend using 10%. The homogenate must be diluted to ~10%, or less, for optimal centrifugation. Originally Hebb had used 0.25 M sucrose [46], as had Abdel-Latif [47], to homogenize brain tissue, but subsequently switched to 0.32 M sucrose [2]. Gray and Whittaker found that homogenization of brain tissue in hypotonic 0.2 M sucrose provided nerve-ending particles in a P2 fraction that looked very much like nerve endings in intact tissue when examined by electron microscopy [4]. The reason(s) for finally choosing 0.32 M sucrose rather than sucrose of a lower concentration is unclear, but higher concentration of sucrose causes shrinkage of synaptosomes, disrupts neurofilaments, and releases particle-bound acetylcholine [4, 29, 47, 48]. One study examined the effect of hypertonicity on the fine structure of synaptosomes by electron microscopy and showed that shrinking occurs as rapidly as 3 s after exposure to hypertonic buffers [49]. Such drastic changes to the synaptosomes must be considered if sucrose density gradients are used for their purification (see below).

Isotonic 0.15 M NaCl has also been used for homogenization of brain tissue, but many problems have been encountered [4]. The first problem with using saline for homogenization was that the content of acetylcholine was substantially decreased in both the whole homogenate and in the P2 fraction when compared to 0.32 M sucrose. Secondly, the nerve-ending particles coacervated (formed colloidal-like aggregates) and clumped in saline; thus they did not fractionate in the same way as after homogenization in 0.32 M sucrose. The particles were mainly found in the P1 pellet rather than the P2 pellet. Nerve-ending particles acted as if they were negatively charged. When brain tissue was homogenized in 0.32 M sucrose, addition of even low concentrations of a range of salts led to the nerve-ending particles again coacervating and clumping, leading to a lower recovery of nerve-ending particles in the P2 pellet [4]. This suggests that salt-based buffers should be avoided in the homogenization stage.

A number of synaptosome isolation procedures added buffered EDTA [9, 16, 18, 30, 50] to the homogenizing solution to decrease the effects of released metal ions on synaptosome coacervation and clumping. This was acceptable as the salts were at a very low concentration relative to the isotonic saline used above. However, Gurd et al. [18] found that inclusion of buffered EDTA altered the composition and distribution of subcellular fractions derived from their synaptosome preparation.

When acetylcholine was being investigated eserine was often added to the 0.32 M sucrose [3, 4]. Similarly, noradrenaline levels can be maintained by addition of pargyline to the sucrose [44]. The glutamate concentration of synaptosomes can be maintained as long as inorganic ions are not added to the homogenizing and centrifugation media [51].

Originally Hebb and Smallman [46] used an ice-cooled Potter-Elvehjem all-glass homogenizer (often termed glass/glass, referring to the pestle (or plunger) and mortar, respectively) for 1–2 min for preparation of brain homogenates. In contrast, Gray and Whittaker used a Perspex/glass homogenizer [4] with a Perspex pestle with a clearance of 0.25 mm and which was rotating at 840 rpm [4, 52]. The other homogenization conditions can be traced to the procedures described by Hebb and Smallman [46]. The Teflon/glass homogenizer used by de Robertis [3] for 2 min was the first introduction of a Teflon pestle that was "specifically determined to produce a minimal breakage of nerve terminals," as it was found that mild homogenization increased the synaptosome content in the P2 fraction. However, these descriptions of the homogenization procedure were incomplete. A comparison between four different homogenization instruments found that a hand-driven glass homogenizer of the Potter-Elvehjem type with a rubber pestle (rubber/glass) was most useful for maintaining radiolabelled noradrenaline-containing synaptosomes in the 0.8/1.2 M sucrose interface [44]. However, it must be noted that in this procedure brain slices were initially prepared and then incubated prior to their homogenization, rather than starting from brain tissue directly. Subsequent synaptosome isolation procedures have almost universally used a Teflon/glass homogenizer with a Teflon pestle with a clearance of 0.1–0.15 mm [30, 50], 0.2 mm [18], 0.25 mm [44], or 0.5 mm [53] and a homogenizer speed of 500–1000 rpm [30, 50, 53]; the number of up and down strokes varies between groups from 10 [30] to 12 [53] or 16 [50]. While this all indicates some degree of tolerance in these relatively gentle homogenization procedures, it also shows that it is important for each lab to standardize these steps for maximum reproducibility and synaptosome viability.

3 Homogenate Fractionation

When making synaptosomes from brain tissue the standard procedure is to first fractionate the brain homogenate using a low-speed centrifugation step to prepare an S1 supernatant and then to use this fraction in a higher speed centrifugation step to obtain a P2 pellet. Many investigators used this P2 pellet as a crude synaptosomal preparation, while others went on to isolate a more homogeneous synaptosomal fraction, mainly using discontinuous density gradient fractionation procedures. The major issues discussed in this section are the range of fractionation procedures used, the time that they take to perform, and the homogeneity of the resultant synaptosomal fractions.

3.1 Preparation of the S1 Supernatant and the P2 Pellet Fractions

The original procedures for making synaptosomes used the 10% brain homogenate and centrifuged at either 900 × *g* [3] or 1000 × *g* [4], for 10 min to provide a P1 pellet (Fig. 2). All subsequent synaptosome preparation procedures used 1000 × *g* for this centrifugation step, but the time varied among 5 min [54], 7.5 min [18], 10 min [9, 50, 53], and 15 min [47]. The P1 pellet was washed twice by suspension in 0.32 M sucrose, homogenization, and centrifugation [3, 4] and the three supernatant fractions were combined to become the S1 supernatant. Most subsequent synaptosome preparation procedures used only the first S1 supernatant directly and did not wash the P1 pellet [9, 47, 50, 53, 54], or they washed it only once [18]. We routinely wash the P1 once and then pool the two supernatants to make an S1 sample. This lack of a P1 washing step was not evaluated, but presumably any decrease in the yield of synaptosomes that occurred was compensated for by saving time and avoiding any possible disruption of synaptosomes during the resuspension and rehomogenization steps.

The S1 supernatant was used directly for fractionating synaptosomes in the procedures developed by Dodd et al. [53], Nagy et al. [50] (for small amounts of tissue only), and Dunkley et al. [11, 30] (see below). This saved time by avoiding the preparation of the P2 pellet and had the added advantage that damage to the synaptosomes was minimized, as they were not pelleted and then resuspended prior to gradient fractionation.

The original procedures for making synaptosomes took the S1 supernatant and centrifuged it at either 11,000 × *g* for 25 min [3] or 17,000 × *g* for 55 min [4], to provide a P2 pellet. A range of other centrifugation conditions were subsequently used to prepare the P2 pellet including 10,000 × *g* for 20 min [18], 12,000 × *g* for 20 min [50], 14,500 × *g* for 15 min [29], 15,000 × *g* for 30 min [9], 17,000 × *g* for 10 min [54], and 17,300 × *g* for 17 min [16]; no investigation of the differences in subcellular content of the resulting P2 pellets has been undertaken. Given the fragility of synaptosomes, logic suggests employing the shortest time at the lowest centrifugation speed to minimize damage, but this may decrease yields.

In one of the other original procedures for making synaptosomes the P2 pellet was washed once in 0.32 M sucrose prior to sucrose gradient fractionation [3]. Washing the P2 pellet up to three times substantially reduced the microsome contamination in the final synaptosome preparations and these extra washing steps were essential when synaptic plasma membranes were prepared from the synaptosomes [16–18], as otherwise the plasma membranes were heavily contaminated. However, washing added to the time taken for preparation of the synaptosomes and potentially increased the chance of damaging them, so these P2 washing steps were not generally adopted.

The P2 pellet was then resuspended in a range of solutions depending on whether it was to be used directly as the source of synaptosomes, or whether further fractionation was to be undertaken, with each fractionation procedure requiring a different resuspension solution. If no further fractionation occurred, the P2 pellet was generally resuspended in a buffered physiological solution that depended on the particular analytical procedure to be undertaken. In one of the original procedures for making synaptosomes the P2 pellet was resuspended in 0.32 M sucrose prior to sucrose density gradient fractionation [4]. In subsequent procedures the P2 pellet was resuspended in 0.25 M sucrose [47], or 10% (~0.29 M) sucrose [55], or 14% Ficoll in 0.32 M sucrose buffered with 5 mM Hepes at pH 7.6 [18], prior to Ficoll gradient fractionation. In more recent procedures the P2 pellet is resuspended in 0.32 M sucrose buffered with 5 mM Hepes at pH 7.5 and 0.1 mM EDTA [50], or in 0.32 M sucrose buffered with 5 mM Tris at pH 7.4 and 1 mM EDTA plus 0.25 mM DTT [9], prior to Percoll gradient fractionation.

The volume used for resuspension of the P2 pellet depends on the analysis procedure if the P2 fraction is used directly as the synaptosome preparation. However, if further fractionation of the P2 suspension is required to provide more homogeneous synaptosomes, then the volume is generally determined by the number of centrifuge tubes used and the volume available in the tubes after the gradient has been generated (see below).

3.2 Discontinuous Sucrose Gradient Procedures

The original procedures for making synaptosomes layered the resuspended P2 pellet (~1 volume) on top of a discontinuous gradient of 0.8 M sucrose (2 volumes) and 1.2 M sucrose (2 volumes) and centrifuged the gradient in a swing-out rotor at 90,400 × *g* for 2 h [4], or layered the resuspended P2 pellet on top of a discontinuous gradient of 0.8 M, 1.0 M, 1.2 M, and 1.4 M sucrose and centrifuged the gradient at 50,000 × *g* for 2 h [3]. Gray and Whittaker [4] collected three fractions from the two interfaces and the pellet and these were identified by electron microscopy as containing mainly myelin fragments, nerve-ending particles (synaptosomes), and free mitochondria, respectively (Fig. 3). De Robertis et al. collected five fractions from the interfaces and the pellet and these were identified by electron microscopy as containing mainly myelin; a mixture of myelin, microsomes, synaptic and glial plasma membranes, and small synaptosomes; synaptosomes with few mitochondria inside and unidentified membranes; synaptosomes with more mitochondria inside; and free mitochondria.

The original procedures for making synaptosomes provided an enormous number of papers and substantially advanced our knowledge of nerve terminals, but they also had a number of disadvantages. The centrifugation step was only suitable for small amounts of tissue and took a long time (2 h). Exposure of the synaptosomes to the hypertonic sucrose in the gradient (0.8–1.4 M sucrose) caused them to shrink [49], leading to increased density of the vesicles within the synaptosomes [4, 56], a decreased content of

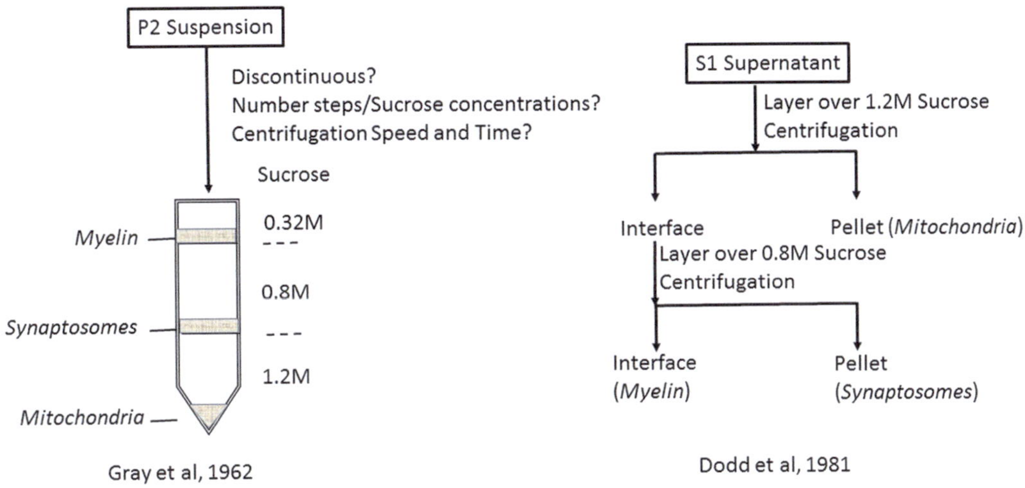

Fig. 3 Sucrose gradients. Synaptosomes can be prepared on sucrose density gradients starting with either the resuspended P2 pellet or the S1 fraction. Each has advantages, but the major advantage of the S1 procedure is that it takes substantially less time without compromising on the homogeneity and viability of the final synaptosome preparation

noradrenaline [44], as well as increased proportion of synaptosomes in the mitochondrial fraction [29]. Exposure of synaptosomes to hypertonic sucrose also led to a reduced capacity to lyse them and hence smaller yields of synaptic plasma membranes were obtained [55]. Cotman et al. also found that glial plasma membranes had a similar isopycnic banding (differences in density between the components) density to synaptosomes on a sucrose gradient [57], suggesting that some of the unidentified membrane contaminants in the synaptosome preparations were likely to be of glial origin.

A range of adaptations of the discontinuous sucrose density gradients were then developed to overcome some of the disadvantages of the original synaptosome isolation procedures. Whittaker used a continuous sucrose gradient and found myelin and free mitochondrial free synaptosomes in the middle of the gradient with contamination only at the top and bottom of the gradient [58]. This was likely to be due to the fact that synaptosomes at the top of the gradient were devoid of mitochondria, while those at the bottom of the gradient were rich in mitochondria. Similar results were seen on the discontinuous gradients of De Robertis [3]. Using a linear sucrose gradient for fractionation of rat striatum suggested that dopaminergic synaptosomes were heavier and moved further into the gradient than the bulk of the striatal synaptosomes [59], presumably as they contained more mitochondria.

Kornguth et al. [60] added $MgCl_2$ to the homogenization solution and the discontinuous sucrose gradient used by De Robertis and found that most of the synaptosomes (referred to as synaptic complexes) were found at the 1.0–1.2 M sucrose interface. This was presumably due to the synaptosomes coacervating due to the salts. When the synaptosomes were then fractionated on a $CsCl_2$ gradient, band 4 was enriched in synaptic complexes and there was no myelin or mitochondrial contamination. However, the $CsCl_2$ gradient required a 72-h centrifugation.

Hajos used 0.3 M sucrose for homogenization of rat cerebral cortex and preparation of a P2 pellet that was dispersed in 0.3 M sucrose [56]. The P2 suspension was layered over 0.8 M sucrose and centrifuged at 9000 × *g* for 30 min, which was substantially less centrifugal force and time than that required for the original sucrose gradients. In fact this was a nonequilibrium centrifugation step, as myelin stayed at the 0.3 M/0.8 M sucrose interface and free mitochondria formed a pellet, but the synaptosomes were collected from the 0.8 M sucrose layer, as the force was insufficient for them to pellet with the free mitochondria. However, the synaptosomes still had sufficient time during centrifugation to shrink in the hypertonic sucrose [56] and had to be re-equilibrated in isotonic sucrose before use.

Dodd et al. used rat cerebral cortex and prepared an S1 supernatant [53] that was directly layered onto 1.2 M sucrose and centrifuged at 256,000 × *g* for 15 min (Fig. 3). The interfacial material was collected, diluted with 0.32 M sucrose, and layered over 0.8 M sucrose before centrifugation at 256,000 × *g* for 15 min. The synaptosomes were found in the pellet fraction and they were resuspended in isotonic sucrose or other buffer before use. This method took only 1–1.5 h and saved time by avoiding the preparation of the P2 pellet and the pelleting of the synaptosomes after collection from the sucrose density gradient. It also was able to be adapted to alternative fixed-angle centrifuges so that large volumes could be handled for bulk preparations of synaptosomes. The method was used for a range of tissues, including human postmortem tissue, and the resulting material compared favorably with synaptosomes prepared by traditional sucrose density gradient procedures and the Hajos procedure in terms of morphology defined by electron microscopy, enzyme analysis, and functional studies [53].

3.3 Discontinuous Ficoll Gradient Procedure

The sucrose density gradient procedures overcame the long preparation times and the need to use small amounts of tissue, but in all cases shrinkage of synaptosomes in hypertonic sucrose occurred. This was overcome by the use of isotonic sucrose containing increasing concentrations of Ficoll [47, 48, 54, 61] for the density gradient centrifugation step. Ficoll is an uncharged highly branched, high-mass copolymer of sucrose and epichlorohydrin and it has very low tonicity when compared to sucrose. Ficoll provides density without increasing osmotic pressure. Cotman and Mathews [55] used 0.32 M sucrose for homogenization of rat

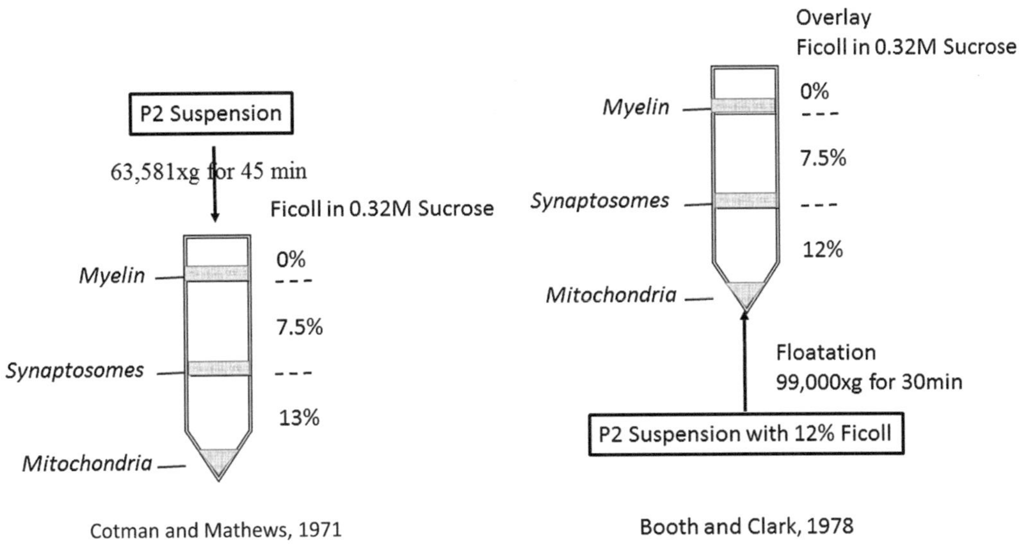

Fig. 4 Ficoll gradients. Synaptosomes can be prepared on Ficoll density gradients starting with the resuspended P2 pellet fraction added to the top of the gradient or to the bottom of the gradient. Each has advantages, but the major advantage of the later procedure is that it takes substantially less time without compromising on the homogeneity and viability of the final synaptosome preparation

forebrain and preparation of a P2 suspension, which was layered onto a gradient of 7.5% and 13% Ficoll, both in 0.32 M sucrose; after centrifugation at 63,581 × *g* for 45 min synaptosomes were identified at the 7.5%/13% Ficoll interface (Fig. 4). Unlike synaptosomes prepared with sucrose density gradients, those prepared using isotonic Ficoll density gradients did not shrink [62] and they were easily lysed when alkaline hypotonic solutions were used [55]. Furthermore, the resulting synaptosomes were metabolically more active than those prepared on sucrose density gradients, but the contamination of the synaptosomes with myelin, mitochondria, and unidentified membranes was not improved [63, 64].

In order to reduce the free mitochondrial contamination of the synaptosomes a floatation method was developed for the Ficoll density gradient, where the P2 pellet was resuspended in either 14% Ficoll (made up in 0.32 M sucrose buffered with 5 mM Hepes pH 7.6) [18] or 12% Ficoll (made up in 0.32 M sucrose buffered with 50 μM potassium EDTA pH 7.4) [63]. These suspensions were then overlaid with 7.5% Ficoll in buffered sucrose [18], or with both 7.5% Ficoll in buffered sucrose and buffered sucrose [63]. Centrifugation was then undertaken and the synaptosomes floated to the 7.5%/14%, or 7.5%/12%, Ficoll interface and mitochondrial contamination of the synaptosomes was reduced to less than 5% (Fig. 4). The Booth and Clark method [63] also reduced the total synaptosome preparation time to less than 2 h, as their density gradient centrifugation was at 99,000 × *g* for only 30 min. The Gurd et al. procedure [18] took a substantially longer time as they washed their P2 fraction to reduce microsomal contamination and their gradient centrifugation was for 3 h.

There are some disadvantages of the Ficoll density gradient procedures. Commercial Ficoll solutions are not always uniform and vary from batch to batch. Ficoll also attaches to biological membranes [50, 65, 66], potentially compromising some studies.

3.4 Discontinuous Percoll Gradient Procedures

A second isotonic procedure for preparation of synaptosomes from a P2 fraction used colloidal silica gradients instead of Ficoll gradients, again to provide density but not extra osmotic pressure. The first attempt used colloidal silica together with the polymer polyethylene glycol to form a continuous gradient [44]. The synaptosomes isolated using this procedure did not shrink and the recovery of noradrenaline was higher than when compared to synaptosomes isolated on sucrose gradients. However, this procedure had the disadvantages that silica gradients formed during the centrifugation due to the differences in size of the colloidal silica particles and the polymer was difficult to remove from the synaptosomes.

Improvements in colloidal silica preparations led to the development of Percoll, which is uniform colloidal silica particles covered with a polyvinylpyrrolidone monolayer. Percoll can be used to form gradients without the need for a stabilizing polymer and Percoll can easily be separated from subcellular organelles. Percoll gradients were then formed using isotonic sucrose and they were used to prepare synaptosomes from a P2 fraction [9, 50]. Analysis of the synaptosomes indicated that the synaptosomes were very homogeneous, containing very low levels of microsomes, myelin, and free mitochondria, and they were metabolically viable and able to take up calcium and release neurotransmitters [9, 50]. The major advantage of these Percoll procedures was that the synaptosomes could be prepared in approximately 1 h as the density gradient step took only 20 min at 15,000 × g [50], or 5 min at 35,000 × g [9]. Other advantages were that it was not necessary to have an ultracentrifuge to prepare the synaptosomes, nor was it necessary to separate the Percoll from the synaptosomes as Percoll does not interfere with biological assays. A further potential advantage was that the synaptosomes present in fractions 3 and 4 differed in terms of their content of serotonin and dopamine, suggesting that nerve endings enriched for particular neurotransmitters could be prepared [10].

Nagy et al. were able to prepare synaptosomes by directly applying an S1 supernatant from small tissue samples to a discontinuous Percoll gradient [50], but no analysis of the resulting synaptosomes was presented. Dunkley et al. [11] applied an S1 fraction obtained from rat cerebral cortex directly to a step density

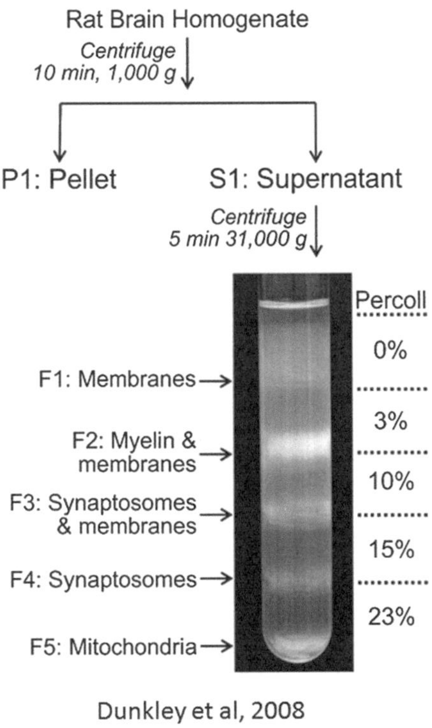

Fig. 5 Percoll gradients. Synaptosomes can be prepared on Percoll density gradients starting with either the resuspended P2 pellet or the S1 fraction. Each has advantages, but the major advantage of the S1 procedure is that it takes substantially less time without compromising on the homogeneity and viability of the final synaptosome preparation. This figure was first published in Nature Protocols [30]

gradient consisting of 3%, 10%, 15%, and 23% Percoll in 0.32 M sucrose (containing 1 mM EDTA, 0.25 mM dithiothreitol, adjusted to pH 7.4) and obtained five subcellular fractions that were analyzed by electron microscopy, enzyme analysis, and phosphoprotein analysis (Fig. 5). Fractions 1 and 2 contained most of the myelin and unidentified membranes, while fractions 3 and 4 contained most of the synaptosomes and fraction 5 contained most of the free mitochondria [11]. These fractions were also analyzed for their viability [67] by determining the depolarization-dependent uptake of ^{32}Pi and protein phosphorylation, the content and uptake of noradrenaline, as well as the depolarization-dependent release of noradrenaline; fraction 4 synaptosomes were the most viable.

The S1 Percoll procedure had all of the advantages of the P2 Percoll procedure, but the major additional advantage that it reduced the preparation time for the synaptosomes to well under an hour, as it required only one 10-min and one 5-min centrifugation step, making it the fastest synaptosome preparation procedure available. It also increased the yield of synaptosomes 1.4-fold and avoided the potentially damaging resuspension of the P2 fraction required by many other procedures. Initial analysis suggested that the S1 Percoll procedure also had the substantial advantage that glial and neuronal membrane contamination was markedly reduced compared to other synaptosome isolation procedures, as most of the unidentified membranes were found in fractions 1 and 2 [11]. This was confirmed for synaptosomes prepared from an S1 fraction of rat hippocampus, when the presynaptic markers choline acetyltransferase, choline uptake, and nicotine binding were found mainly in fraction 4, while the postsynaptic muscarine binding site was found mainly in fractions 1 and 2 [26]. A full description of the laboratory methods for the S1 Percoll gradient procedure has been published [30].

Commercial Percoll has the disadvantage that there is small batch-to-batch variation and every new bottle of Percoll needs to be centrifuged, or course-filtered, to remove particulate material before it is used [30].

4 Other Synaptosome Preparation Procedures

Synaptosomes prepared by sucrose density gradient techniques were subsequently fractionated into GABA or acetylcholine-enriched synaptosomes [68]. This was achieved by incubation of the synaptosomes with antibodies to either glutamate decarboxylase or choline acetyltransferase and then incubation with protein A coupled to magnetic microspheres. Protein A bound to the antibodies and the synaptosomes could then be isolated using a magnet to attract the magnetic microspheres. A generally similar magnetic isolation approach was used by Bowman et al. [69, 70] to isolate a subfraction of synaptosomes expressing omega-CgTx GVIA-sensitive, N-type voltage-sensitive calcium channels by using a biotinylated derivative of omega-CgTx GVIA attached to streptavidin-coated magnetic beads. They found that chicken forebrain synaptosomes all had this calcium channel, while rat cortex synaptosomes did not. It was not possible to release the synaptosomes from the beads for further investigation.

A resuspended P2 pellet was fractionated using a phase partition and centrifugation procedure to prepare synaptosomes in less than an hour [71]; the phases were composed of dextran T500 and polyethylene glycol 4000. The synaptosomes were relatively homogeneous with microsomes, free mitochondria, and myelin constituting only 4.4%, 3.2%, and 0.1%, respectively. An isopycnic gradient of dextran and polyethylene glycol was also used to rapidly prepare synaptosomes [72].

5 Preparation for Analysis

All of the procedures outlined above had as their primary purpose to prepare homogeneous and viable synaptosomes at high yield (see Table 1). Therefore most of the papers undertook some form of investigation of these parameters. The most frequently used procedure for analysis of homogeneity is electron microscopy, as this is able to identify myelin, mitochondrial, and membrane contamination. In addition, electron microscopy is able to characterize the synaptosomes in terms of their size, mitochondrial content, and postsynaptic attachments. Whittaker and Gray suggested four distinguishing morphological criteria for a synaptosome: "the presence within (1) a sealed plasma membrane of (2) one or more small mitochondria, (3) a local concentration of vesicles resembling in size and number the synaptic vesicles seen in nerve terminals and varicosities in situ and (4) the frequent presence of lengths of characteristically electron-dense membrane resembling postsynaptic membranes adhering by means of some poorly-staining material to the periphery of the synaptosome" [4, 73]. Of course not all of these elements are present in every section through a synaptosome, as the plane of sectioning may miss the presence of either mitochondria or postsynaptic membranes. In practice, we defined a synaptosome as a clearly membrane-bound body containing three or more vesicles of 40–60 nm diameter (i.e., the typical diameter of synaptic vesicles) and synaptosome-like structures were not included as synaptosomes if their plasma membrane was not intact (i.e., sealed) [9].

The next most frequent procedure for analysis of homogeneity was analysis of specific enzymes or proteins that are specifically localized in particular subcellular organelles. These generally included analysis of myelin, mitochondria, and endoplasmic reticulum (ER) proteins, but sometimes also postsynaptic neuronal and/or glial membranes.

Many of the studies investigated the viability of the synaptosomes, and the most frequently used analyses focussed on maintenance of membrane potentials, metabolic activity, mitochondrial respiratory function, metabolism or uptake of calcium or specific neurotransmitters, or release of neurotransmitters on depolarization. We have found that measurement of depolarization-dependent protein phosphorylation (e.g., Synapsin I), or dephosphorylation (e.g., Dynamin I) is a particularly helpful criteria as it requires metabolically viable mitochondria to produce the ATP for phosphorylation and depolarization-dependent increases in calcium to drive the phosphorylation and dephosphorylation events [67, 74]. Most other studies using synaptosomes did not undertake any analysis of their homogeneity or viability and have relied entirely on the data in these original published procedures.

When synaptosomes are required for analysis they need to be prepared appropriately and the exact procedure depends on what was to be analyzed. The crudest synaptosome preparation available to be used directly is the S1 supernatant, which is in 0.32 M sucrose and contains soluble proteins, synaptic vesicles, microsomes, ER, synaptic and glial membranes, myelin, and free mitochondria. The S1 supernatant can only be used when what is being analyzed is specific to the nerve terminal and is not interfered with by the presence of any of the contaminants or the sucrose. The P2 pellet has frequently been used as a crude synaptosome preparation and it contains contaminating myelin, membranes, and free mitochondria. The synaptosomes prepared using the Dodd et al. procedure (see Fig. 3) are also prepared as a pellet. It is therefore necessary to resuspend either the P2 pellet or the Dodd et al. pellet, into a buffer suitable for the required analysis and, as this requires mechanical action, it is best performed gently in order to avoid potential damage to the synaptosomes. If it is necessary for analysis to remove the sucrose this is performed by resuspension and recentrifugation of the P2 pellet prior to resuspension in the analysis buffer. The P2 pellet can only be used when what is being analyzed is specific to the nerve terminal and is not interfered with by the presence of any of the extra-synaptosomal contaminants or the sucrose. For example, based on the premise that only the synaptosomes are sealed and metabolically viable and have membrane potentials, it is reasonable to use them for synaptic vesicle exocytosis, transmitter uptake, or stimulus-dependent phosphorylation analyses.

All of the other synaptosome preparations require removal of the synaptosomes from an interface within a sucrose, Ficoll, or Percoll gradient. This also has to be performed gently and various methods have been employed. We recommend the procedure used in the paper by Dunkley et al. [30], which involves careful removal of the gradient material with a suction pump followed by removal of the fractions with a Pasteur pipette. These synaptosomes need to be centrifuged to form a pellet and possibly washed if the gradient material interferes with the planned analysis; for example Percoll only needs to be removed if electron microscopy is planned, as the silica damages the cutting knives and can show up as small electron-dense particles in the images. This washing is always undertaken using resuspension and recentrifugation.

Preparation of synaptosomes can take from 1 to 5 h or more depending on the exact procedure chosen. Analysis can also take a substantial time and therefore it is frequently necessary to freeze the synaptosomes immediately after preparation and undertake the analysis at a later time. This may not matter for some analyses such as protein or some enzyme activities, but freezing and thawing may substantially compromise some analyses especially those requiring physiologically active and fully intact synaptosomes, such as neurotransmitter release or recycling of synaptic vesicles. However, synaptosomes can be stored frozen and thawed and their viability can be retained, depending on the exact procedures used (see [30] for details).

6 Preparation of Synaptosomal Subpopulations

Several groups have successfully purified synaptosomal subpopulations by adapting flow cytometry or a fluorescence-activated cell sorter to the organelles by methods sometimes called fluorescence-activated synaptosome sorting (FASS). Flow cytometry has the ability to definitively identify each individual particle or synaptosome as being positive or not to signals from specific fluorescent markers, as well as sorting by particle size, where typically particles with sizes between 0.5 and 1.5 μm are collected. Typically such studies are applied to synaptosomes already purified using Percoll or Ficoll methods. The early studies focussed on internal dyes labelling synaptosomes rather than contaminants (primarily mitochondria), such as voltage-sensitive dyes [75] or the viable cytoplasmic dye calcein blue AM [76]. Other studies focussed on specific surface antibodies for sorting specific subpopulations of synaptosomes, such as tyrosine hydroxylase [77] for dopaminergic nerve endings. In addition to transmitter synthesis markers, they can be sorted based on the presence of other antigens like Aβ, phospho-tau, and SNAP-25 in Alzheimer's disease [78–81], or other pathology markers [82]. A more recent innovative application of this technique purified specific synaptosomal types from knock-in mice with fluorescent glutamatergic synapses, producing synaptosomes which were almost devoid of other synapse types or glial contaminants [83, 84]. This approach has allowed analysis of the proportion of glutamatergic synaptosomes that contain PSDs and it was concluded that most contain the post-synapse, but not the full spine [83]. This differs from the generally accepted 15% of synaptosomes containing PSDs; the difference may be due to the brain region used, absence of synaptosomes containing other (especially inhibitory) neurotransmitters, particular nature of glutamatergic synapses, or details of the analysis procedures. Future adaptation of these methods has considerable potential for isolation of other classes of synapses.

7 Limitations of Synaptosomes

Synaptosomes have proven to be an excellent model system for the study of nerve endings and synaptic transmission, but it is important to also acknowledge that synaptosomes have a number of limitations. Synaptosomes are only viable for a few hours; thus they must be used on the same day of preparation (although storage frozen is possible [85]). The small size of synaptosomes makes them not ideal for fluorescence imaging (although this is possible [83, 86], especially when super-resolution microscopy is used [12]). Synaptosomes cannot be used for most molecular biology approaches like transfections of tagged genes or RNAi knockdown of message and proteins. Thus, it is not easy to manipulate the levels of any protein they contain (apart from freeze/thaw [85] and cell-penetrating tags [87]). Being sealed structures, the proteins and vesicles within synaptosomes are unable to interact with their axon, preventing the normal exchange of proteins [88], or synaptic vesicles [89], that may leave or depart synapses during or after stimulation. Although they maintain a membrane potential, synaptosomes have lost the appropriate sodium and potassium channels of the axon that propagate action potentials. Thus, they need to be depolarized primarily by chemical means and are not very responsive to field stimulation (our unpublished observations). A number of earlier studies demonstrated modest effects on metabolism and transmitter release from electrically stimulated synaptosomes, but the stimulus time was typically tens of minutes and the responses were typically modest.

8 Conclusions

There are dozens of papers describing procedures for the preparation of synaptosomes from brain tissue. Although there are many sources for the tissues most of the procedures were developed using fresh whole rat brain or rat cerebral cortex. There are also many procedures for homogenization of the brain tissues, but most are now based on the use of 10% tissue in 0.32 M sucrose and a Teflon/glass homogenizer. All synaptosome preparation procedures involve fractionation of the homogenate to form an S1 supernatant and most now involve subsequent preparation of a P2 pellet. Many studies use the P2 pellet as the source of "crude" synaptosomes. Thereafter, many methodological differences exist, depending on the purpose of the experiment, and this will determine the quantity of synaptosomes required and the need for homogeneous (free from contamination by other subcellular organelles) and/or viable (optimally functional) synaptosomes.

Many of the early procedures involved multiple washes of either the P1 or the P2 pellets and use of discontinuous or continuous sucrose density gradients. The gold standard criteria for homogeneity is electron microscopy; analysis of the presence of enzymes or proteins specific to contaminating subcellular organelles was also extensively used. The gold standard for viability is metabolic integrity and/or nerve terminal functions such as neurotransmitter uptake or release. These procedures took many hours and required an ultracentrifuge, always led to nerve terminal shrinkage due to the synaptosomes exposure to hypertonic sucrose, and frequently provided only limited yields of homogeneous and viable synaptosomes. It was clear from the electron microscopic studies that synaptosomes are not uniform in terms of their diameter, their vesicle and mitochondrial content, and their postsynaptic membrane association. The synaptosome preparations always contained some neuronal and glial membrane contamination and also contained microsomal or free mitochondrial contamination depending on the number of washes of the P1 and P2 fractions and which part of the sucrose gradients they were taken from. The synaptosome preparations were also not uniform in terms of their metabolic viability, presumably due to the differences in mitochondrial content, and they also differed in neurotransmitter content and capacity for uptake and release of specific neurotransmitters.

More recently the procedures for synaptosome preparation involved fewer washes of the P1 and/or P2 pellets, direct application of the S1 supernatant rather than the P2 pellet to the gradients, and the use of isotonic Ficoll or Percoll gradients rather than sucrose gradients. These procedures substantially reduce the time taken to prepare synaptosomes and the potential damage to synaptosomes due to the number of resuspension/washing steps and the shrinkage due to hypertonic conditions. The homogeneity and viability of the synaptosomes still depended on the exact part of the gradient that they are taken from and the yield of synaptosomes is not substantially improved over the sucrose gradient procedures. However, with the Percoll gradient procedure there is also the additional advantages that no ultracentrifuge was required and there is a substantial reduction in the neuronal and glial membrane contamination.

It is clear from this review that the procedures for making synaptosomes have improved substantially over time, mainly due to the advent of Ficoll and Percoll. However, because of the wide range of procedures available it is now even more critical to carefully determine the choice of synaptosome preparation method and the specific fractions to be used for analysis based on the exact purpose of the experiment.

Acknowledgments

We would like to thank all of our colleagues who have helped to develop the Percoll gradient procedure for preparation of synaptosomes [9, 11, 26, 30, 67] for their hard work, camaraderie, and intellectual input. The NHMRC of Australia is thanked for past funding of projects arising from the use of synaptosomes.

References

1. Del Castillo J, Katz B (1956) Biophysical aspects of neuro-muscular transmission. Prog Biophys Biophys Chem 6:121–170
2. Hebb CO, Whittaker VP (1958) Intracellular distributions of acetylcholine and choline acetylase. J Physiol 142(1):187–196
3. Derobertis E et al (1962) Cholinergic and non-cholinergic nerve endings in rat brain .1. Isolation and subcellular distribution of acetylcholine and acetylcholinesterase. J Neurochem 9:23–35
4. Gray EG, Whittaker VP (1962) Isolation of nerve endings from brain - an electron-microscopic study of cell fragments derived by homogenization and centrifugation. J Anat 96:79–88
5. Whittaker VP, Gray EG (1962) Synapse - biology and morphology. Br Med Bull 18(3):223–228
6. Whittaker VP, Michaelson IA, Kirkland RJ (1964) Separation of synaptic vesicles from nerve-ending particles (synaptosomes). Biochem J 90(2):293–303
7. Rockland KS (2002) Non-uniformity of extrinsic connections and columnar organization. J Neurocytol 31(3–5):247–253
8. Faisal AA, White JA, Laughlin SB (2005) Ion-channel noise places limits on the miniaturization of the brain's wiring. Curr Biol 15(12):1143–1149
9. Dunkley PR et al (1986) A rapid method for isolation of synaptosomes on Percoll gradients. Brain Res 372(1):115–129
10. Robinson PJ, Lovenberg W (1986) Dopamine and serotonin in two populations of synaptosomes isolated by percoll gradient centrifugation. Neurochem Int 9(3):455–458
11. Dunkley PR et al (1988) A rapid Percoll gradient procedure for isolation of synaptosomes directly from an S1 fraction: homogeneity and morphology of subcellular fractions. Brain Res 441(1–2):59–71
12. Wilhelm BG et al (2014) Composition of isolated synaptic boutons reveals the amounts of vesicle trafficking proteins. Science 344(6187):1023–1028
13. Hollingsworth EB et al (1985) Biochemical characterization of a filtered synaptoneurosome preparation from Guinea pig cerebral cortex: cyclic adenosine 3′:5′-monophosphate-generating systems, receptors, and enzymes. J Neurosci 5(8):2240–2253
14. Schwartz RD et al (1984) Barbiturate and picrotoxin-sensitive chloride efflux in rat cerebral cortical synaptoneurosomes. FEBS Lett 175(1):193–196
15. Whittaker VP, Greengard P (1971) Isolation of synaptosomes from brain of a teleost fish, centriopristes-striatus. J Neurochem 18(2):173–176
16. Babitch JA et al (1976) Preparation of chick brain synaptosomes and synaptosomal membranes. Biochim Biophys Acta 433(1):75–89
17. Morgan IG et al (1971) Isolation of plasma membranes from rat brain. Biochim Biophys Acta 241(3):737–751
18. Gurd JW et al (1974) Isolation and partial characterization of rat brain synaptic plasma membranes. J Neurochem 22(2):281–290
19. Cotman C, Mahler HR, Anderson NG (1968) Isolation of a membrane fraction enriched in nerve-end membranes from rat brain by zonal cetrifugation. Biochim Biophys Acta 163(2):272–275
20. Leskawa KC et al (1979) Large-scale preparation of synaptosomes from bovine brain using a zonal rotor technique. Neurochem Res 4(4):483–504
21. Kishi M et al (1991) Pharmacological characteristics of choline transport system in mouse cerebral cortical neurons in primary culture. Jpn J Pharmacol 55(2):223–232
22. Restituito S et al (2011) Synaptic autoregulation by metalloproteases and gamma-secretase. J Neurosci 31(34):12083–12093
23. Bate C, Williams A (2012) Neurodegeneration induced by clustering of sialylated glycosylphosphatidylinositols of prion proteins. J Biol Chem 287(11):7935–7944
24. Leshchyns'ka I et al (2015) Abeta-dependent reduction of NCAM2-mediated synaptic adhesion contributes to synapse loss in Alzheimer's disease. Nat Commun 6:8836
25. Faundez V et al (1992) Epidermal growth factor receptor in synaptic fractions of the rat central nervous system. J Biol Chem 267(28):20363–20370
26. Thorne B, Wonnacott S, Dunkley PR (1991) Isolation of hippocampal synaptosomes on Percoll gradients: cholinergic markers and ligand binding sites. J Neurochem 56(2):479–484
27. De Belleroche JA, Bradford HF (1975) The release of endogenous 3,4-dihydroxyphenethylamine from synaptosomes isolated from corpus striatum. Biochem Soc Trans 3(1):99–101
28. Clark M, Dar MS (1989) Release of endogenous glutamate from rat cerebellar synap-

tosomes: interactions with adenosine and ethanol. Life Sci 44(22):1625–1635

29. Tamir H, Rapport MM, Roizin L (1974) Preparation of synaptosomes and vesicles with sodium diatrizoate. J Neurochem 23(5):943–949
30. Dunkley PR, Jarvie PE, Robinson PJ (2008) A rapid Percoll gradient procedure for preparation of synaptosomes. Nat Protoc 3(11):1718–1728
31. Hardy JA et al (1983) Metabolically active synaptosomes can be prepared from frozen rat and human brain. J Neurochem 40(3):608–614
32. Dodd PR et al (1986) Optimization of freezing, storage, and thawing conditions for the preparation of metabolically active synaptosomes from frozen rat and human brain. Neurochem Pathol 4(3):177–198
33. Dodd PR et al (1989) Uptake of gamma-aminobutyric acid and L-glutamic acid by synaptosomes from postmortem human cerebral cortex: multiple sites, sodium dependence and effect of tissue preparation. Brain Res 490(2):320–331
34. Mazzo F et al (2016) Reconstitution of synaptic ion channels from rodent and human brain in Xenopus oocytes: a biochemical and electrophysiological characterization. J Neurochem 138(3):384–396
35. Franklin W, Taglialatela G (2016) A method to determine insulin responsiveness in synaptosomes isolated from frozen brain tissue. J Neurosci Methods 261:128–134
36. Wilson WS, Cooper JR (1972) Preparation of cholinergic synaptosomes from bovine superior cervical ganglia. J Neurochem 19(12):2779–2790
37. Dowdall MJ, Whittaker VP (1973) Comparative studies in synaptosome formation: preparation of synaptosomes from head ganglion of squid, Loligo-Pealii. J Neurochem 20(4):921–935
38. Newkirk RF et al (1976) Comparative studies in synaptosome formation: preparation of synaptosomes from the ventral nerve cord of the lobster (Homarus americanus). Brain Res 101(1):103–111
39. Israel M et al (1985) Large-scale purification of Torpedo electric organ synaptosomes. J Neurochem 44(4):1107–1110
40. Chin GJ et al (1989) Aplysia synaptosomes. I. Preparation and biochemical and morphological characterization of subcellular membrane fractions. J Neurosci 9(1):38–48
41. Bisby MA, Fillenz M (1969) Isolation of peripheral synaptosomes from a sympathetically innervated tissue. J Physiol 204(2):105P+
42. Bisby MA, Fillenz M (1971) The storage of endogenous noradrenaline in sympathetic nerve terminals. J Physiol 215(1):163–179
43. Simon EJ et al (1976) Comparative studies on synaptosomes - applicability of rapid method for preparing synaptosomes to elasmobranch brain. Neurochem Res 1(1):83–92
44. Lagercrantz H, Pertoft H (1972) Separation of catecholamine storing synaptosomes in colloidal silica density gradients. J Neurochem 19(3):811–823
45. Chin GJ, Shapiro E, Schwartz JH (1989) Aplysia synaptosomes. II. Release of transmitters. J Neurosci 9(1):49–55
46. Hebb CO, Smallman BN (1956) Intracellular distribution of choline acetylase. J Physiol 134(2):385–392
47. Abdel-Latif AA (1966) A simple method for isolation of nerve-ending particles from rat brain. Biochim Biophys Acta 121(2):403–406
48. Kurokawa M, Sakamoto T, Kato M (1965) A rapid isolation of nerve-ending particles from brain. Biochim Biophys Acta 94(1):307–309
49. Ashton AC, Ushkaryov YA (2005) Properties of synaptic vesicle pools in mature central nerve terminals. J Biol Chem 280(44):37278–37288
50. Nagy A, Delgado-Escueta AV (1984) Rapid preparation of synaptosomes from mammalian brain using nontoxic isoosmotic gradient material (Percoll). J Neurochem 43(4):1114–1123
51. Geddes JW, Newstead JD, Wood JD (1980) Stability of the glutamate content of synaptosomes during their preparation. Neurochem Res 5(10):1107–1116
52. Whittaker VP (1959) The isolation and characterization of acetylcholine-containing particles from brain. Biochem J 72:694–706
53. Dodd PR et al (1981) Rapid preparation of nerve ending particles (synaptosomes) from rat-brain - comparison with 2 standard methods. J Anat 132:462
54. Cotman C et al (1970) Analytical differential centrifugation: an analysis of the sedimentation properties of synaptosomes, mitochondria and lysosomes from rat brain homogenates. Arch Biochem Biophys 136(2):436–447
55. Cotman CW, Matthews DA (1971) Synaptic plasma membranes from rat brain synaptosomes: isolation and partial characterization. Biochim Biophys Acta 249(2):380–394
56. Hajos F (1975) An improved method for the preparation of synaptosomal fractions in high purity. Brain Res 93(3):485–489
57. Cotman C, Herschman H, Taylor D (1971) Subcellular fractionation of cultured glial cells. J Neurobiol 2(2):169–180

58. Whittaker VP (1968) The morphology of fractions of rat forebrain synaptosomes separated on continuous sucrose density gradients. Biochem J 106(2):412–417
59. Van der Krogt JA, Koot-Gronsveld E, Van den Berg CJ (1983) Subcellular fractionation of striatum: sedimentation properties of dopaminergic synaptosomes. Life Sci 33(7):605–613
60. Kornguth SE, Anderson JW, Scott G (1969) Isolation of synaptic complexes in a caesium chloride density gradient: electron microscopic and immunohistochemical studies. J Neurochem 16(3):1017–1024
61. Autilio LA et al (1968) Biochemical studies of synapses in vitro. I. Protein synthesis. Biochemistry 7(7):2615–2622
62. Day ED et al (1971) Zonal centrifuge profiles of rat brain homogenates - instability in sucrose, stability in Iso-Osomotic Ficoll-sucrose. Anal Biochem 39(1):29–45
63. Booth RF, Clark JB (1978) A rapid method for the preparation of relatively pure metabolically competent synaptosomes from rat brain. Biochem J 176(2):365–370
64. Verity MA (1972) Cation modulation of synaptosomal respiration. J Neurochem 19(5):1305–1317
65. Jansen GJ, Vaneerten MTW, Forrester IT (1982) Biochemical events associated with Ficoll washing of ram spermatozoa. Proc N Z Soc Anim Prod 42:95–97
66. Asgeirsson D et al (2006) Increased glomerular permeability to negatively charged Ficoll relative to neutral Ficoll in rats. Am J Physiol Renal Physiol 291(5):F1083–F1089
67. Harrison SM, Jarvie PE, Dunkley PR (1988) A rapid Percoll gradient procedure for isolation of synaptosomes directly from an S1 fraction: viability of subcellular fractions. Brain Res 441(1–2):72–80
68. Docherty M, Bradford HF, Wu JY (1987) The preparation of highly purified GABAergic and cholinergic synaptosomes from mammalian brain. Neurosci Lett 81(1–2):232–238
69. Bowman D, Smith W, Mccormack A (1995) Affinity purification of rat cortical and chicken forebrain synaptosomes using a biotinylated derivative of omega-Cgtx Gvia. Neuropharmacology 34(7):743–752
70. Bowman D et al (1994) Affinity purification of synaptosomes using a biotinylated derivative of Omega-Conotoxin Gvia. Br J Pharmacol 112:U176–U176
71. Enriquez JA et al (1990) Rat brain synaptosomes prepared by phase partition. J Neurochem 55(6):1841–1849
72. Muino-Blanco T et al (1993) Use of a resolving density gradient created with dextran and poly(ethylene glycol) to purify brain synaptosomes. J Biochem Biophys Methods 27(1):1–10
73. Whittaker VP (1993) Thirty years of synaptosome research. J Neurocytol 22(9):735–742
74. Robinson PJ et al (1994) Phosphorylation of dynamin-I and synaptic-vesicle recycling. Trends Neurosci 17(8):348–353
75. Wolf ME, Kapatos G (1989) Flow cytometric analysis of rat striatal nerve terminals. J Neurosci 9(1):94–105
76. Gylys KH, Fein JA, Cole GM (2000) Quantitative characterization of crude synaptosomal fraction (P-2) components by flow cytometry. J Neurosci Res 61(2):186–192
77. Wolf ME, Zigmond MJ, Kapatos G (1989) Tyrosine hydroxylase content of residual striatal dopamine nerve terminals following 6-hydroxydopamine administration: a flow cytometric study. J Neurochem 53(3):879–885
78. Sokolow S et al (2012) Isolation of synaptic terminals from Alzheimer's disease cortex. Cytometry A 81(3):248–254
79. Wang DS et al (2005) Decreased neprilysin immunoreactivity in Alzheimer disease, but not in pathological aging. J Neuropathol Exp Neurol 64(5):378–385
80. Sokolow S et al (2015) Pre-synaptic C-terminal truncated tau is released from cortical synapses in Alzheimer's disease. J Neurochem 133(3):368–379
81. Gylys KH, Bilousova T (2017) Flow cytometry analysis and quantitative characterization of tau in synaptosomes from Alzheimer's disease brains. Methods Mol Biol 1523:273–284
82. Postupna NO et al (2014) Flow cytometry analysis of synaptosomes from post-mortem human brain reveals changes specific to Lewy body and Alzheimer's disease. Lab Investig 94(10):1161–1172
83. Biesemann C et al (2014) Proteomic screening of glutamatergic mouse brain synaptosomes isolated by fluorescence activated sorting. EMBO J 33(2):157–170
84. Luquet E et al (2017) Purification of synaptosome populations using fluorescence-activated synaptosome sorting. Methods Mol Biol 1538:121–134
85. Daniel JA et al (2012) Analysis of synaptic vesicle endocytosis in synaptosomes by high-content screening. Nat Protoc 7(8):1439–1455
86. Choi SW, Gerencser AA, Nicholls DG (2009) Bioenergetic analysis of isolated cerebrocortical nerve terminals on a microgram

scale: spare respiratory capacity and stochastic mitochondrial failure. J Neurochem 109(4):1179–1191

87. Marcelli S et al (2016) Targeting SUMO-1ylation contrasts synaptic dysfunction in a mouse model of Alzheimer's disease. Mol Neurobiol 54(8):6609–6623

88. Chi P, Greengard P, Ryan TA (2003) Synaptic vesicle mobilization is regulated by distinct synapsin I phosphorylation pathways at different frequencies. Neuron 38(1):69–78

89. Darcy KJ et al (2006) Constitutive sharing of recycling synaptic vesicles between presynaptic boutons. Nat Neurosci 9(3):315–321

Part III

Making Synaptosomes

Chapter 4

Protocol for a High-Throughput Semiautomated Preparation for Filtered Synaptoneurosomes

Justin L. Balsor and Kathryn M. Murphy

Abstract

This book outlines various techniques for the isolation and application of synaptosomes for neuroscience research. One of the simplest techniques involves fractionation of synaptoneurosomes through filtration and low-speed centrifugation (Hollingsworth J Neurosci 5:2240–2253, 1985). This approach however is limited by two essential stages of the synaptoneurosome tissue preparation: the manual homogenization and filtrations are both laborious and slow. We have updated this traditional technique to include modern benchtop homogenizers and centrifugal filter units to simplify these labor-intensive stages, to make each stage faster and to reduce the variability between samples. Here we outline our protocol to produce filtered synaptoneurosomes that reduce sample preparation time, increase the amount of tissue recovered, and, most importantly, increase protein enrichment.

Key words Synaptoneurosomes, Synaptic proteins, High-throughput, Centrifugal filtration

1 Introduction

1.1 General Introduction

Synaptosome preparations enrich brain tissue samples for synaptic proteins, making it an important tool for studying proteins involved in synaptic development and plasticity [1–7]. Two types of protocols are used to enrich the synaptic fraction: the use of a sucrose or Percoll gradient to make *synaptosomes* [8, 9], or filtration procedures to make *synaptoneurosomes* [10, 11]. What makes these synaptosome or synaptoneurosome preparations useful for studying synaptic development and plasticity is the approximate twofold enrichment of synaptic proteins [10, 12]. That enrichment improves the signal-to-noise so that synaptic proteins can be accurately quantified even when the synaptosome or synaptoneurosome is prepared from a small brain sample (e.g., ~30 mg). Furthermore, low-abundance synaptic proteins (e.g., PICK and

Kathryn M. Murphy (ed.), *Synaptosomes*, Neuromethods, vol. 141,
https://doi.org/10.1007/978-1-4939-8739-9_4,

GRIP) that are difficult to reliably quantify in homogenate samples can be routinely studied with synaptoneurosomes [13].

Our research uses filtered synaptoneurosomes to study synaptic development and plasticity in visual cortex of animals and humans. We have also developed a high-throughput, semiautomated protocol for preparing synaptoneurosomes because many aspects of synaptic function can be studied with this preparation. For example, we have examined both pre- and postsynaptic proteins involved in establishing nascent synapses and mediating experience-dependent strengthening or weakening of connections [13–21]. Our studies, as well as ones from other labs, have found that using a synaptoneurosome preparation can uncover synapse-specific aspects of neurodevelopment and plasticity [12, 14]. Indeed, our lab has used synaptoneurosomes to highlight similarities and differences between the total pool of synaptic proteins found in homogenate samples and the amount concentrated at the synapse by the synaptoneurosome [14, 19, 21].

This chapter focuses on describing the protocol we developed for high-throughput and semiautomated preparation of synaptoneurosomes that is ideal to process large numbers of samples quickly, inexpensively, and with high reliability [18].

1.2 Synaptoneurosome Steps

All filtered synaptoneurosome preparations are made using two steps: tissue homogenization and fractionation (Dunkley & Robinson, Chap. 3). The simple protocol described by Hollingsworth et al. [10] helped to popularize the use of filtered synaptoneurosomes. That method describes manual homogenization with a glass-glass Dounce tissue grinder, followed by 5 μm filtration and low-speed centrifugations. Two challenges emerge when trying to scale up any of the manual protocols for making synaptoneurosomes. First, these protocols are very slow because each sample is prepared by hand, one at a time. Second, substantial sample-to-sample variability can be introduced because both homogenization and filtration steps are done by hand. Many groups have tried to improve the quality of the synaptoneurosome preparation by introducing new methods for tissue lysing [22–24], fractionation [25], or adjusting the filter pore size [11, 25]. Yet, all of those protocols still use hand processing for tissue lysing and/or filtering, so they continue to have sample-to-sample variability and are not easily scaled up for large projects. We have addressed these problems by developing a new high-throughput, semiautomated protocol to prepare filtered synaptoneurosomes [18].

1.3 Tissue Homogenization

Synaptoneurosome protocols with manual steps for tissue homogenization (e.g., glass-glass Dounce tissue grinders) use qualitative descriptions to explain how to lyse the sample. In practice, these protocols are difficult to reproduce since only the number of strokes is quantified, and most other aspects of the tissue lysing, such as how fast or how much pressure to use with each stroke, are difficult to describe. New protocols however use modern benchtop homogenizers to grind, homogenize, and lyse tissue samples [18]. These homogenizers can prepare multiple samples at a time, using built-in programmable steps that lead to highly reproducible preparation of many samples.

1.4 Sample Fractionation

Tissue fractionation protocols typically include syringe filtration steps to remove larger extra-synaptic components from the homogenate. These are followed by low-speed centrifugations to further separate heavier synaptic fractions from the lighter supernatant. The manual syringe filtrations vary according to the number of filtrations, size of the filter pores, and qualitative description of how to carry out the steps. In addition, syringe filters such as Swinnex holders have about 1 mL of dead space, so at least 1.5 mL of sample has to be loaded to recover enough filtrate to continue the protocol. This loss of sample is problematic when using small, valuable samples that do not generate enough homogenate for subsequent filtration. We eliminated these problems by replacing the manual filtration steps with centrifugal filtration units, where even a small volume of homogenate is reliably filtered. Each centrifugal filtration unit is loaded with up to 500 μL of homogenate, and then spun with a known *g*-force in a centrifuge to collect the filtered sample needed to prepare the synaptoneurosome. There is no dead space where sample volume can be lost, and many samples can be filtered at once, leading to reliable high-throughput filtration.

Here we detail the steps to prepare filtered synaptoneurosomes using our new protocol [18]. The protocol explains how to implement our semiautomated, high-throughput procedure and compares samples prepared with this new protocol versus older manual protocols to illustrate the optimizations that lead to improved reliability, enhanced protein yield, greater synaptic enrichment, and faster preparation of synaptoneurosomes.

2 Materials and Equipment List

2.1 Laboratory Equipment

1. Peristaltic pump.
2. Biosafety cabinet (equipped with drain).
3. 1.5 mL Microcentrifuge vial (7× tissue piece):
   (a) 1 on crushed dry ice
   (b) 6 on wet ice
4. −80 °C Freezer.
5. Weigh boats.
6. Scale.

7. Scoopula.
8. 2 mL Lysing Matrix-D tube (1× tissue piece, MP Biomedicals, Cat. #116913050).
9. Benchtop Homogenizer (FastPrep 24 Tissue and Cell homogenizer, MP Biomedicals, Cat. #116004500).
10. 2 mL Ultrafree-MC SV Centrifugal Filter Units with Microporous Membrane (1× tissue piece, Millipore, Cat # UFC30SV00).
11. Centrifuge (e.g., Sorvall legend Micro 21R).
12. 96-Well sample assay plate.
13. Pipettor (volumes between 0.5 and 200 μL).
14. Biological incubator (capable of 45 °C).
15. Absorbance reader (e.g., iMark Plate reader, BioRad).
16. Heating block.

2.2 Chemicals and Reagents for Synaptoneurosome Tissue Preparation

2.2.1 Animal Perfusion and Tissue Collection

1. PO_4 solution (0.2 M): Make stock ahead of time:
   (a) 4 L dH_2O.
   (b) 30.8 g NaOH.
   (c) 117.1 g Monobasic NaH_2PO_4.
2. PBS: Make fresh, store at 4 °C:
   (a) 50 mL 10× saline.
   (b) 450 mL dH_2O.
   (c) 500 mL PO_4 solution (0.2 M).
3. Euthanyl (sodium pentobarbital) (150 mg/kg rat, 165 mg/kg cat).

2.2.2 Synaptoneurosome Preparation

1. Homogenization buffer recipe yields 50 mL (3 mL × tissue piece):
   (a) 5 mL HEPES (10 mM).
   (b) 5 mL EDTA (2 mM).
   (c) 5 mL EGTA (2 mM).
   (d) 125 μL DTT (0.5 mM).
   (e) 500 μL Leupeptin (10 mg/L).
   (f) 500 μL Soybean trypsin inhibitor (50 mg/L).
   (g) 50 μL Microcystin (100 nM).
   (h) 33.825 mL H_2O.
      - Check the pH level of the combined ingredients, and adjust to pH 7.5.
      - Add 50 μL PMSF.

- Put solution on wet ice until added to frozen tissue. Can be frozen at −20 °C for later use.

2. Stock HEPES (10 mM):
   (a) 100 mL dH_2O.
   (b) 2.38 g HEPES powder (N-2-hydroxyethylpiperazine-N′-2-ethanesulfonic acid, Bishop, Cat #Hep 001).
   (c) Stir until powder is dissolved. Aqueous HEPES can be aliquoted and frozen at −20 °C, and thawed for use in homogenization buffer.
3. Stock EDTA (2 mM):
   (a) 100 mL dH_2O.
   (b) 0.5844 g EDTA powder (ethylenediaminetetraacetic acid disodium salt dihydrate, Sigma, Cat# E-1644).
   (c) Stir until powder is dissolved. Aqueous EDTA can be aliquoted and frozen at −20 °C, and thawed for use in homogenization buffer.
4. Stock EGTA (2 mM):
   (a) 90 mL dH_2O.
   (b) 10 mL NaOH.
   (c) Stir until powder is dissolved.
   (d) To the above solution, add 0.768 g EGTA powder (ethylene-bis(oxyethylenenitrilo)tetraacetic acid, Sigma, Cat #E4378). Stir until powder is dissolved. Aqueous EGTA can be aliquoted and frozen at −20 °C, and thawed for use in homogenization buffer.
5. Stock SDS solution (10%) (20 μL × tissue piece):
   (a) 10 g of SDS powder (sodium dodecyl sulfate, Bishop, Cat#SDS001).
   (b) 100 mL dH_2O.
6. Stock SDS solution (1%) (boiling, 100 μL × tissue piece):
   (a) 100 mL of stock SDS solution (10%) (see above).
   (b) Add 900 mL of dH_2O.

2.2.3 BCA Assay and Preparing Tissue for Western Blotting

1. Albumin Standards:
   (a) Dilute stock albumin (2 μg/mL) to the following concentrations using saline/dH_2O:
   (b) 1 μg/mL.
   (c) 0.5 μg/mL.
   (d) 0.25 μg/mL.
2. BCA working solution

(a) 50 mL Reagent A (Pierce™ BCA protein Assay Reagent A, Thermo Fisher Scientific, Cat# 23223).

(b) 1 mL Reagent B (Pierce™ BCA protein Assay Reagent B, Thermo Fisher Scientific, Cat#23224).

3. Sample loading buffer (M260 NextGel® Sample loading buffer 4×, Amresco LLC, Solon, OH, USA).
4. Laemmli buffer (Cayman Chemical Company, Ann Arbor, MI, USA).

3 Methods

3.1 Animal Perfusion and Tissue Collection

Here we describe the perfusion and tissue collection protocols that we use for our studies of visual cortex. If perfusion is not a suitable option for your experiment we provide an alternative protocol with modified filtration steps in a subsequent chapter (Siu et al., Chap. 15).

3.1.1 Animal Perfusion

1. Euthanize the animal using the protocol approved by your institutional animal ethics committee.
2. Use transcardial perfusion to clear blood from the animal. Insert the needle from the peristaltic pump into the left ventricle of the heart, cut the right atrium with scissors, and begin the perfusion with cold 0.1 M phosphate buffered saline (PBS) at a rate of 50–100 mL/min, for about 4 min, or until circulating fluid becomes clear.
3. Quickly remove the brain from the skull and immerse in cold PBS.

3.1.2 Tissue Collection

1. Cut samples from the desired location with sharp iris scissors or a surgical blade. For example, the samples we cut from rat cortex are typically 3 × 3 × 1 mm to yield ~500 mg of tissue that is ready for synaptoneurosome preparation.
2. Quickly transfer tissue pieces to precooled centrifuge vials. Label the vial with the needed identifiers, and place vial back on dry ice to rapidly freeze the sample.
3. Then store the frozen samples at −80 °C, or begin with the synaptoneurosome preparation.

3.2 Synaptoneurosome Preparation

In this chapter, we describe our published protocol [18] for the high-throughput and semiautomated preparation of filtered synaptoneurosomes, and compare this protocol with the manual hand preparation popularized by Hollingsworth et al., 1985 [10] (Fig. 1). The primary steps needed to make filtered synaptoneurosomes, homogenization, and filtration, as well as a set of low-speed centrifugations, are outlined in Fig. 1. An approximation of how long it takes to prepare a single sample at each of these three steps

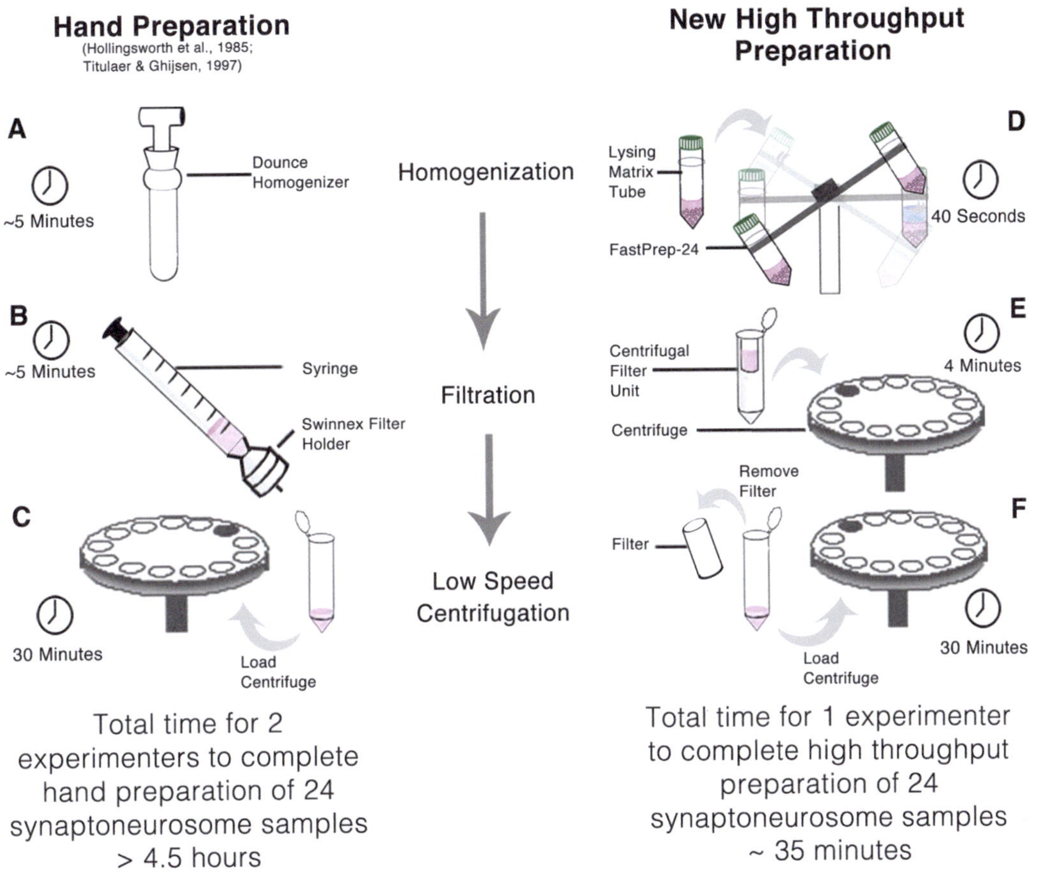

Fig. 1 Overview of process for preparing synaptoneurosomes using either the manual hand or new high-throughput, semiautomated protocol, including a comparison of the primary steps in the processes from homogenization to filtration, and low-speed centrifugation. (**a**) The hand preparation began with two experimenters working together to homogenize samples one at a time using glass-glass Dounce homogenizers. (**b**) The resultant homogenate was then filtered by applying hand pressure to a syringe, and forcing homogenate through hydrophobic membranes held in place by the Swinnex filter holder. (**d**) The high-throughput preparation began with one experimenter placing samples in lysing matrix tubes, and running samples in the FastPrep®-24 Tissue and Cell Homogenizer for 40 s. (**e**) The resultant homogenate was then transferred to centrifugal filter units, and centrifuged for 4 min. The filter component of the centrifugal filter unit was removed before low-speed centrifugation. (**c**, **f**) In both the hand and high-throughput preparations, samples were centrifuged at low speed for 30 min. To prepare 24 samples with the hand preparation, the total time require was 4.5 h and two experimenters were required. To prepare 24 samples with the high-throughput preparation, the total time require was 35 min, and only one experimenter was required. Reprinted from Murphy et al. [18] with permission from Elsevier

is also outlined in Fig. 1. Homogenization and filtration following the manual hand preparation each took about 5 min to prepare just one sample, and when scaling to 24 samples just these two steps add a minimum of 4 h to a tissue preparation protocol. Furthermore, the manual hand preparation requires at least two experimenters, because while one person is homogenizing and filtering tissue the other is needed to clean the glass-glass Dounce homogenizers, as well as clean and assemble the Swinnex filter holders between samples. In contrast, one experimenter can easily perform the new high-throughput, semiautomated steps in under 10 min and the single experimenter is able to homogenize and filter all 24 samples at once.

3.2.1 Synaptoneurosome Preparation: Preliminary Steps

Our protocol significantly speeds up the process and decreases the number of experimenters required to make synaptoneurosomes from fresh frozen tissue, but it is still important to work quickly to prevent tissue thawing. Therefore, we include these steps to reduce the likelihood of tissue thawing before samples are introduced to homogenization buffer (*see* **Note 1**).

1. Thaw homogenization buffer (2.2.2 Synaptoneurosome Preparation) on ice. Homogenization buffer can be made ahead of time, and it can now be taken out to thaw, but homogenization buffer must remain cold when adding to tissue pieces.
2. Turn on the centrifuge ahead of time, and allow sufficient time to cool to operating temperature of 4 °C.
3. Label all 1.5 mL centrifuge vials. Three vials are required for each sample, with separate vials designated for "Homogenate," "Supernatant," and "Synaptoneurosome," respectively (e.g., for sample R123, one vial each is labeled as R123 HOM, R123 SUP, and R123 SYN). Place these on ice.
4. Pipette 10 μL of 10% SDS solution into each "Homogenate" and "Supernatant" vial. Place these vials on ice (*see* **Note 2**).
5. Label 1 Lysing Matrix-D tube for each sample (e.g., for sample R123, write "R123" clearly on both tube side and cap). Place on ice.
6. Label 1 Ultrafree-MC SV Centrifugal Filter Unit for each sample.
7. Obtain either freshly sectioned tissue samples or fresh frozen tissue samples, and place vials containing the tissue samples on crushed dry ice.

3.2.2 Synaptoneurosome Preparation: Homogenization

Modern benchtop homogenizers lyse tissue samples using built-in programmable steps. For example, in this protocol, tissues are homogenized at an angular velocity of 6 m/s for a duration of 40 s. Furthermore, the Lysing Matrix-D described in this protocol is specifically designed by MP Biomedicals for use with soft brain tissues. Controlling the lysing matrix and the lysing intensity results in a highly reproducible preparation of many samples.

1. Remove the tissue section from the original vial, weigh the sample on a balance, and record the weight. Quickly transfer the sample to the labeled Lysing Matrix-D tube and add cold

homogenization buffer to the Lysing Matrix-D tubes. The amount of homogenization buffer required in each vial depends on sample size. Use a ratio of 1 mL of homogenization buffer to 0.05 g of tissue to determine the volume of homogenization buffer required for each sample. Record the amount of homogenization buffer added. Cap tubes, and place on ice (*see* **Note 3**).

2. Secure Lysing Matrix-D tubes in FastPrep-24 Tissue and Cell Homogenizer. Set the FastPrep-24 to a speed of 6 m/s, set the duration to 40 s, and run the FastPrep-24.
3. Inspect vials to ensure that samples are completely homogenized. Check for visible chunks of tissue resuspended in the homogenization buffer. If the tissue is not fully homogenized, run all samples for a second time at 6 m/s for a duration of 40 s.
4. Remove 90 μL of homogenized tissue from lysing matrix tubes and transfer it to the labeled "Homogenate" vial. Gently vortex the sample, and place vial on ice.

3.2.3 Synaptoneurosome Preparation: Micro-MC Spin Tube Filtration

The syringe filtration component of tissue fractionation is replaced with centrifugal filter units in our protocol. The product description for Ultrafree-MC SV Centrifugal Filter Units outlines a range of allowable parameters that can be used with these filter units. First, the recommended speed of 12,000 × *g* is a qualitative description of the force applied to pass homogenate tissue through the microporous membranes. We optimized the filtration time by measuring the amount of filtrate collected across the range of minutes recommended by Millipore (1–4 min). We determined that 4 min yields a significantly higher recovery 334 μL (SEM 22 μL, $n = 4$) (Fig. 2) of the total original homogenate (500 μL) than 1 (average 183 μL, SEM 62 μL, $n = 4$) or 2 min (average 180 μL, SEM 59 μL, $n = 4$) ($p < 0.05$). Here we describe our protocol for filtering the homogenate tissue through these filter units.

1. Using a 3 mL syringe equipped with an 18 5/8 G needle tip, take up the remaining homogenate from the Lysing Matrix D-tube and transfer to Ultrafree-MC SV Centrifugal Filter Units (*see* **Note 4**).
2. Place the newly filled filter spin tubes in the centrifuge. Set the centrifuge speed to 12,000 × *g*, the duration for 4 min, and run the centrifuge.
3. When complete, remove and inspect the filters to ensure that no visible tears occurred. Be sure to note any relevant details pertaining to the condition of the filters. Dispose of empty filter units.
4. Gently vortex all samples.
5. Place the filter spin tubes (now without filters) in the centrifuge. Set the centrifuge speed to 1000 × *g*, set the duration for 20 min, and run the centrifuge.

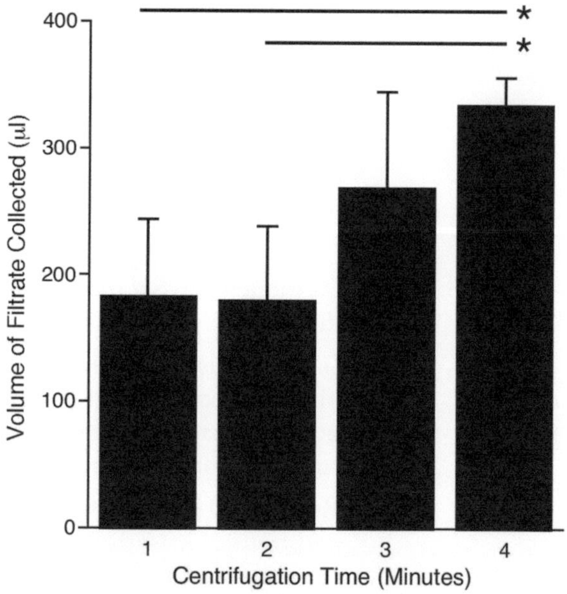

Fig. 2 Volume of filtered homogenate tissue collected using Ultrafree-MC SV centrifuge tubes. Identical amounts of homogenate tissue (500 μL) were loaded into filter units of the centrifuge filter units (Millipore) and the percent volume of filtrate was compared across 1-min filtration intervals (1–4 min). Centrifugal filtration for 4 min yielded greater sample recovery (average 335 μL, SEM 22 μL, $n = 23$) than either 1 (average 183 μL, SEM 62 μL, $n = 4$) ($p < 0.05$) or 2 min (average 180 μL, SEM 59 μL, $n = 4$) ($p < 0.05$). Reprinted from Murphy et al. [18] with permission from Elsevier

3.2.4 Synaptoneurosome Preparation: Supernatant

The following sections describe the set of low-speed centrifugations that separate heavier synaptic fractions from the lighter supernatant. For the following steps it is imperative that the remaining tissue in the centrifuged vials is not disturbed since this concentrated pellet now contains all resealed synaptoneurosome entities.

1. Transfer 90 μL of supernatant tissue from the top of the centrifuged filter vial into the respective "Supernatant" vial. Vortex the supernatant vial and place the sample on ice.
2. Carefully take up the remaining supernatant material surrounding the pellet with a pipette and expel it into a waste beaker.

3.2.5 Synaptoneurosome Preparation: Synaptoneurosome

1. Turn on the heating block, place vial containing 25 mL of SDS solution (1%) (2.2.2 Synaptoneurosome Preparation) inside the heating block, and bring to a boil.
2. Add 1 mL cold homogenization buffer to "Synaptoneurosome" vial to resuspend each pellet and vortex each sample until the pellet has dissolved entirely. Place the vials on ice until ready to proceed to the next step.

3. Place "Synaptoneurosome" vials in the centrifuge in a counterbalanced fashion. Set the centrifuge speed to 1000 × *g*, set the duration for 10 min, and run the centrifuge.
4. A pellet should once again be collected in the bottom of the vial. Use a pipette to discard the remaining supernatant, being careful not to disturb the pellet.
5. The "Synaptoneurosome" vials should contain little more than a pellet at this point. Using a pipette, add 100 μL of boiling SDS solution (1%) to each of the "Synaptoneurosome" vials, and gently vortex until the pellet is resuspended.
6. "Homogenate," "Supernatant," and "Synaptoneurosome" samples are now prepped. Place all vials in the heating block for a duration of 10 min, and then store at −20 °C.

3.3 BCA Assay and Preparing Tissue for Western Blotting

The total protein concentration from both homogenate and synaptoneurosome tissue can be obtained following bicinchoninic acid (BCA) assay guidelines (Pierce, Rockford, IL, USA). This provides a highly accurate measurement of the total protein available in a tissue preparation before investing in Western blotting. We measured the total protein by performing BCA assays of whole homogenates and synaptoneurosomes prepared from rat cortex. We compared total protein concentration between samples prepared following traditional hand preparation using glass-glass Dounce homogenizers or with the new high-throughput FastPrep-24 Tissue and cell homogenizer. We found similar total protein concentrations in whole homogenates prepared by hand (average = 3.34 μg/mL, SEM 0.16 μg/mL, $n = 50$) as those prepared using the high-throughput preparation (3.43 μg/mL, SEM 0.24 μg/mL, $n = 24$) (Fig. 3a). The total protein concentration was greater ($p < 0.05$) when synaptoneurosomes were centrifuged using the Ultrafree-MC centrifugal filtration units (2.37 μg/mL, SEM 0.18 μg/mL, $n = 23$) than when samples were pressure filtered by hand (1.88 μg/mL, SEM 0.13 μg/mL, $n = 88$) (Fig. 3b).

We briefly describe here the BCA protocol to compare total protein concentration between the semiautomated and manual tissue preparations at either the whole homogenate stage or after creating synaptoneurosome.

1. Leaving one column of wells blank, pipette 3 μL each of the four different albumin protein standards (e.g., 0.25, 0.5, 1, 2) in order to establish the profile for experimenter pipetting variability. Do this three times for each standard. Since multiple samples are loaded multiple times, create a map of your assay plate by assigning samples to specific wells.
2. Gently vortex samples once more before pipetting 3 μL each of "Homogenate," "Supernatant," and "Synaptoneurosome" for each sample. Do this three times for each sample.

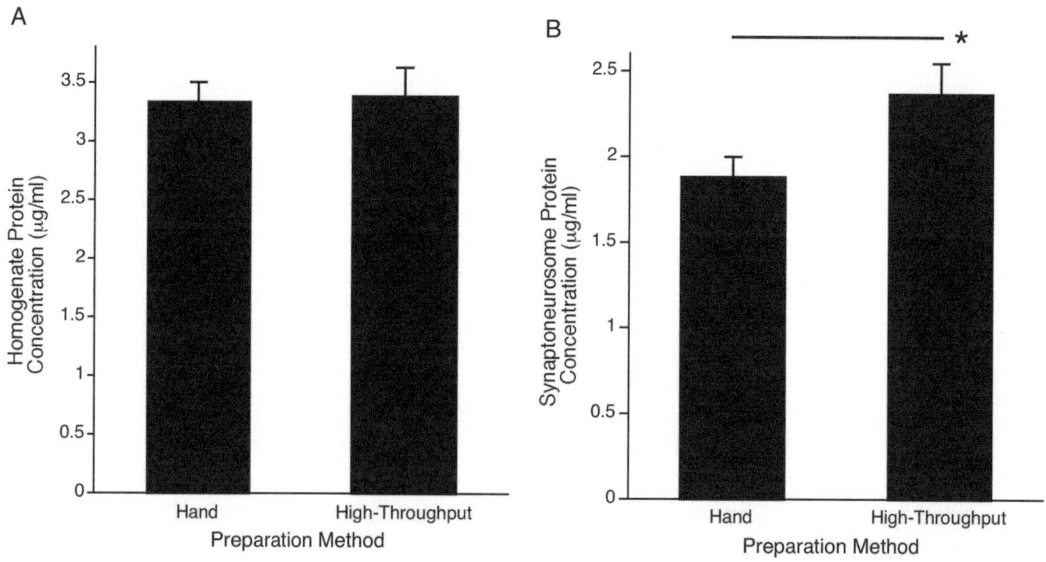

Fig. 3 Total protein concentration for the hand or high-throughput techniques measured in (**a**) homogenate and (**b**) synaptoneurosome tissue fractions prepared from cat and rat cortical tissues. (**a**) There was no difference between total protein concentration measured by BCA assay in the homogenate samples prepared using either the hand (3.34 μg/mL, SEM 0.16 μg/mL, $n = 50$) or the high throughput technique (3.43 μg/mL, SEM 0.24 μg/mL,+ $n = 24$). (**b**) There was greater total protein concentration in the synaptoneurosome samples prepared using the high-throughput technique (2.4 μg/mL, SEM 0.18 μg/mL, $n = 23$) than in the hand preparation (1.88 μg/mL, SEM 0.13 μg/mL, $n = 88$; $p < 0.05$). Reprinted from Murphy et al. [18] with permission from Elsevier

3. In all wells, add 300 μL of the BCA solution. Place the lid on the protein assay plate, and place the plate in the incubator at 45 °C for 45 min.
4. Quantify the colorimetric change in the samples by scanning the plate in the iMark Microplate Absorbance Reader (Bio-Rad Laboratories, Hercules, CA, USA). Use the information obtained from this assay to determine total protein concentration in each sample. In an Excel spreadsheet, first plot the protein concentration of the four pipetted albumin standard (i.e., 0.25, 0.5, 1, and 2 μg/mL) against the net light absorbance at the designated wavelength of your spectrophotometer (e.g., 562 nm). Plotting a line of best fit will determine the strength of the relationship between protein concentration and absorbance with the function $y = mx + b$. We have stringent guidelines for this step in the protocol and require a minimum correlation of $r^2 = 0.99$ between the values for these two variables for all protein standards. If $r^2 = 0.99$ is not achieved, the assay must be redone.
5. Next, obtain the average absorbance (nm) from the three runs of each sample, and identify total protein concentration (μg/mL) by dividing this average by the slope of the function, and adding any linear shift in the function (y-intercept).

6. In this protocol an ideal final concentration for diluted tissue samples is 1 μg/μL. Use the information about the average protein concentration obtained from the plate reader to dilute samples down to 1 μg/μL. It helps to start with a common volume of tissue (e.g., 80 μL) from each "Homogenate," "Supernatant," and "Synaptoneurosome" vial, and transfer this to new vials for use with Western blotting. Dilute samples with sample loading buffer (4×) and then with Laemmli buffer (1×) to weigh samples down for loading into the wells on SDS-PAGE gels, and to equate protein volume across samples.

3.4 Check for Synaptic Enrichment

We quantified the expression of a specific postsynaptic anchoring protein for glutamatergic receptors, postsynaptic density95 (PSD95), in both whole homogenate and synaptoneurosomes for each sample. We then performed a typical enrichment check by dividing the expression of PSD95 in the synaptoneurosome fraction by the PSD95 expression in the homogenized fraction (synaptoneurosome/homogenate). We calculated the amount of enrichment after 1, 2, 3, and 4 min of centrifugal filtration, and found the best enrichment after 4 min (Fig. 4a). Since 4 min of centrifugation produced the greatest amount of synaptic filtrate (Fig. 2a) as well as the greatest enrichment, we compared PSD95 enrichment in this subset of samples versus those prepared through traditional manual preparations. We found less than 2-fold enrichment (Fig. 2b, average 1.6, SEM 0.2, $n = 24$) in the manual preparation but an average of 2.7-fold enrichment (SEM 0.3, $n = 23$) using the new high-throughput preparation, with some samples having up to five times greater PSD95 expression in the synaptoneurosome than the homogenate. Importantly, these high enrichment values were obtained using very small tissue samples (25 mg) that were orders of magnitude smaller than tissue samples required for the traditional preparation (1000 mg [10]).

The synaptoneurosome fraction is dominated by synaptic proteins, since cytosolic proteins are removed as supernatant at each successive stage of the tissue fractionation protocol. Therefore it is imperative that protein expression is *not* normalized against classic housekeeping proteins typically used as loading controls in Western blotting experiments, since these proteins are expected to be depleted by the synaptoneurosome preparation. To demonstrate this, we compared the expression of a classic housekeeping protein, β-tubulin, against PSD95 expression in a subset of homogenate and synaptoneurosome samples prepared through both the hand and high-throughput techniques (Fig. 5). We found greater average β-tubulin expression in the homogenate fraction (1.12) than the synaptoneurosome (0.06; $p < 0.001$ Fig. 5a light gray), while the opposite is true for PSD95 expression, which again demonstrates higher protein expression in the synaptoneurosome (0.155) versus the homogenate (0.11; $p < 0.05$ Fig. 5a dark grey). This

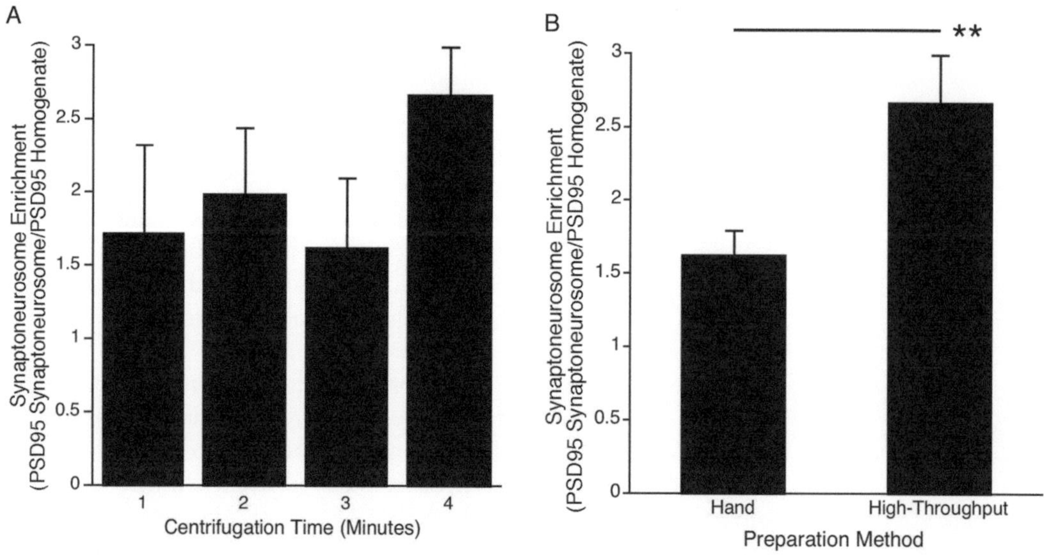

Fig. 4 Synaptoneurosome enrichment of PSD95 expression comparing across (**a**) different lengths of filtration or (**b**) hand versus new high-throughput preparation. (**a**) Centrifugal filtration for 4 min led to better enrichment for PSD95 than 1, 2, or 3 min (average for 1–3 min was 1.8 X's enrichment, SEM 0.3, $n = 12$; average for 4 min was 2.7 X's enrichment, SEM 0.3, $n = 23$) ($p < 0.05$). (**b**) There was significantly greater enrichment of the synaptoneurosomes prepared using the high-throughput (2.7 X's enrichment, SEM 0.3, $n = 23$) than the hand preparation (1.6 X's enrichment, SEM 0.2, $n = 24$) ($p < 0.01$). Reprinted from Murphy et al. [18] with permission from Elsevier

confirms the low expression of β-tubulin in homogenate versus synaptoneurosome reported by others [12]. In fact, an enrichment check performed for β-tubulin produces an anti-enrichment for β-tubulin, while standard enrichment is demonstrated for PSD95 (Fig. 5b). It is vital that loading controls are used properly when analyzing a synaptoneurosome preparation, since normalizing protein expression against a housekeeping protein that demonstrates anti-enrichment (i.e., enrichment values less than 1) will lead to artificially inflated measurements of expression for proteins of interest.

4 Notes

1. One purpose of the homogenization buffer is to slow the process of protein denaturing. There are a variety of homogenization buffers available for purchase, but here we include our in-house design since this combination of calcium chelators and protease inhibitors has proven successful for our Western blotting purposes. Be sure to modify this recipe to suit your own needs.
2. After applying heat, SDS solution coats proteins with an even electric charge, so that electrophoretic separation of proteins is

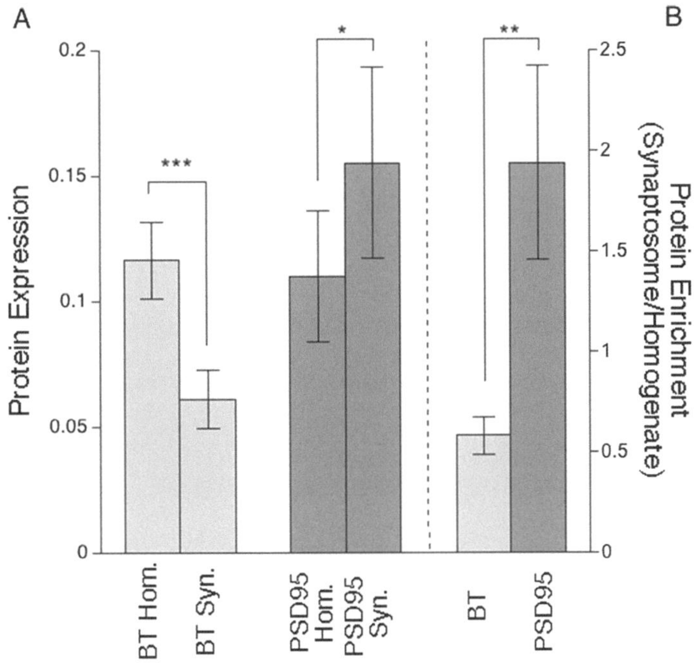

Fig. 5 Comparison of β-tubulin (BT) and PSD95 expression in whole homogenate and synaptoneurosomes (**a**) with matching enrichment (**b**). (**a**) There was less β-tubulin expression in the synaptoneurosome (0.06, SEM = 0.01) than in the whole homogenate sample (0.12, SEM = 0.02; $p < 0.001$), while PSD95 had greater expression in the synaptoneurosome (0.16, SEM = 0.04) than in the homogenate (0.11, SEM = 0.03) sample. This led to protein enrichment (>1×) for PSD95 (1.9×, SEM = 0.48) and protein depletion of β-tubulin (0.58×, SEM = 0.10, $p < 0.05$) in the synaptoneurosome fractions

based primarily on the size of unfolded proteins. If this is undesirable for your synaptoneurosome preparation, skip this step.

3. Be aware of the total volume of homogenization buffer being added to each Lysing Matrix-D Tube. The current protocol uses smaller pieces of tissue, and the maximum volume these tubes accommodate is 2 mL. If you are sectioning larger pieces that require more homogenization buffer, MP Biomedicals makes larger lysing matrix tubes that can be used with a larger shaker unit for the Fast-Prep 24 Tissue and cell homogenizer. Ensure that you have the proper lysing matrix (Lysing Matrix-D) that is specifically designed for use on soft tissues like cortical tissue, since all parameters in this protocol were designed for use with soft tissue.
4. We advise having extra filter spin tubes ready in case there is excess homogenized tissue. Again be aware of the total volume that these filtration units can accommodate to avoid overfilling vials. These particular units accommodate 500 μL in the filter unit.

Acknowledgements

This work was supported by NSERC Grant RGPIN-2015-06215 awarded to KM, and an OGS awarded to JB.

References

1. Suzdak PD et al (1986) A selective imidazobenzodiazepine antagonist of ethanol in the rat. Science 234:1243–1247
2. Suzdak PD, Schwartz RD, Skolnick P, Paul SM (1986) Ethanol stimulates gamma-aminobutyric acid receptor-mediated chloride transport in rat brain synaptoneurosomes. Proc Natl Acad Sci U S A 83:4071–4075
3. Morrow AL, Pace JR, Purdy RH, Paul SM (1990) Characterization of steroid interactions with gamma-aminobutyric acid receptor-gated chloride ion channels: evidence for multiple steroid recognition sites. Mol Pharmacol 37:263–270
4. Weiler IJ et al (1997) Fragile X mental retardation protein is translated near synapses in response to neurotransmitter activation. Proc Natl Acad Sci U S A 94:5395–5400
5. Wu L et al (1998) CPEB-mediated cytoplasmic polyadenylation and the regulation of experience-dependent translation of α-CaMKII mRNA at synapses. Neuron 21:1129–1139
6. Quinlan EM, Olstein DH, Bear MF (1999) Bidirectional, experience-dependent regulation of N-methyl-D-aspartate receptor subunit composition in the rat visual cortex during postnatal development. Proc Natl Acad Sci U S A 96:12876–12880
7. Quinlan EM, Philpot BD, Huganir RL, Bear MF (1999) Rapid, experience-dependent expression of synaptic NMDA receptors in visual cortex in vivo. Nat Neurosci 2:352–357
8. Hebb CO, Whittaker VP (1958) Intracellular distributions of acetylcholine and choline acetylase. J Physiol 142:187–196
9. Cohen RS, Blomberg F, Berzins K, Siekevitz P (1977) The structure of postsynaptic densities isolated from dog cerebral cortex: I. Overall morphology and protein composition. J Cell Biol 74:181–203
10. Hollingsworth EB et al (1985) Biochemical characterization of a filtered synaptoneurosome preparation from Guinea pig cerebral cortex: cyclic adenosine 3′:5′-monophosphate- generating systems, receptors, and enzymes. J Neurosci 5:2240–2253
11. Weiler IJ, Greenough WT (1991) Potassium ion stimulation triggers protein translation in synaptoneurosomal polyribosomes. Mol Cell Neurosci 2:305
12. Heynen AJ, Quinlan EM, Bae DC, Bear MF (2000) Bidirectional, activity-dependent regulation of glutamate receptors in the adult hippocampus in vivo. Neuron 28:1–10
13. Murphy KM, Tcharnaia L, Beshara SP, Jones DG (2012) Cortical development of AMPA receptor trafficking proteins. Front Mol Neurosci 5:65
14. Murphy KM, Beston BR, Boley PM, Jones DG (2005) Development of human visual cortex: a balance between excitatory and inhibitory plasticity mechanisms. Dev Psychobiol 46:209–221
15. Beston BR, Jones DG, Murphy KM (2010) Experience-dependent changes in excitatory and inhibitory receptor subunit expression in visual cortex. Front Synaptic Neurosci 2:138
16. Williams K, Irwin DA, Jones DG, Murphy KM (2010) Dramatic loss of Ube3A expression during aging of the mammalian cortex. Front Aging Neurosci 2:18
17. Pinto JGA, Hornby KR, Jones DG, Murphy KM (2010) Developmental changes in GABAergic mechanisms in human visual cortex across the lifespan. Front Cell Neurosci 4:16
18. Murphy KM, Balsor J, Beshara S, Siu C, Pinto JGA (2014) A high-throughput semi-automated preparation for filtered synaptoneurosomes. J Neurosci Methods 235:35–40
19. Pinto JGA, Jones DG, Williams CK, Murphy KM (2015) Characterizing synaptic protein development in human visual cortex enables alignment of synaptic age with rat visual cortex. Front Neural Circuits 9:3
20. Beshara S, Beston BR, Pinto JGA, Murphy KM (2015) Effects of fluoxetine and visual experience on glutamatergic and GABAergic synaptic proteins in adult rat visual cortex. eNeuro 2 ENEURO0126–15.2015. https://doi.org/10.1523/ENEURO.0126-15.2015
21. Siu CR, Beshara SP, Jones DG, Murphy KM (2017) Development of glutamatergic proteins in human visual cortex across the lifespan. J Neurosci 37:6031–6042
22. Chandler LJ, Crews FT (1990) Calcium-versus G protein-mediated phosphoinosit-

ide. Hydrolysis in rat cerebral cortical synaptoneurosomes. J Neurochem 55:1022–1030

23. Villasana LE, Klann E, Tejada-Simon MV (2006) Rapid isolation of synaptoneurosomes and postsynaptic densities from adult mouse hippocampus. J Neurosci Methods 158:30–36

24. Chang JW et al (2012) Synaptoneurosome micromethod for fractionation of mouse and human brain, and primary neuronal cultures. J Neurosci Methods 211:289–295

25. Titulaer MN, Ghijsen WE (1997) Synaptoneurosomes. A preparation for studying subhippocampal GAB_A receptor activity. Methods A Mol Biol 72:49–59

Chapter 5

Subcellular Fractionation of Brain Tissue from Small Tissue Explants

Véronique De-Smedt-Peyrusse, Laetitia Darriet, Pierre Trifilieff, Etienne Herzog, and Maria Florencia Angelo

Abstract

For several decades, neurobiologists have used subcellular fractionation methods to analyze the molecular structure and some functional features of the cells in the central nervous system. Indeed, the brain tissue is built through the networking of neuronal, glial, and vascular cells in an intermingled meshwork of micrometer-sized structures. Subcellular fractionation protocols allow for the separation of specific compartments such as synapses (called "synaptosomes"), synaptic plasma membranes, and synaptic vesicles for their analysis at the molecular level. Most current protocols were established to process large amounts of tissue as required in previous experimental paradigms. Here, we provide a protocol to prepare synaptosomes from as little as 10 mg of tissue or a full fractionation to enrich crude synaptic vesicles and synaptic plasma membranes from 20 mg of tissue. This protocol will be useful to anyone aiming at addressing specific questions regarding local microcircuits in combination with connectomics and proteomics approaches.

Key words Subcellular fractionation, Synapse, Synaptosome, Synaptic plasma membrane, Synaptic vesicles

1 Introduction

The brain tissue is an extremely complex structure hosting the circuitry of neuronal, glial, and vascular networks. It is built from the assembly and wiring of billions of cells with specialized shapes that allow for the formation of trillions of contacts. The separation and enrichment of specific compartments of brain tissue have soon been envisioned as a way to understand the features of each component of the networks. First attempts to study subcellular components of brain were done by Brody and Bain in the early 1950s [1–3], followed by the work of Whittaker who achieved a landmark

Etienne Herzog and Maria Florencia Angelo are contributed equally to this work.

Kathryn M. Murphy (ed.), *Synaptosomes*, Neuromethods, vol. 141,
https://doi.org/10.1007/978-1-4939-8739-9_5,

in the subcellular fractionation of brain tissue leading to the isolation of synaptosomes and synaptic vesicles (from torpedo at that time) at the beginning of the 1960s [4–10]. Since then, neuroscientists have been working extensively with this kind of preparations and tried to improve and modify the technique according to their needs [7, 10–15]. Brain tissue is homogenized in at least five volumes of isotonic sucrose buffer in a glass/Teflon potter and the separation of each fraction achieved through differential sedimentations and density gradient centrifugations. Hypo-osmotic lysis of synaptosomes allows the separation of synaptic vesicles and intracellular organelles from synaptic plasma membranes [5, 8, 16]. Further purification protocols include immunoisolation [13, 17–27] and a flow cytometry-based procedure developed to isolate fluorescently labeled synapse subpopulations (fluorescence-activated synaptosome sorting, FASS) [23, 28].

Recent advances in the exploration of brain connectomics call for a refinement of subcellular fractionation and downstream approaches to grasp synaptic specificities at the regional and projection pathway levels [29, 30]. Here we propose a miniaturized subcellular fractionation protocol (see overview, Fig. 1) that allows the enrichment of synaptosomes and sub-synaptic components from down to a single hippocampus or for instance the nuclei accumbens pooled from five mouse brains.

2 Materials

1. Dissection tools, including NIH Style Neuro Punch (Gauge 19—18036-19) and Adult Mouse Brain Slicer Matrix (i.e., Zivic BSMAS001-1).
2. Drill type homogenizer (i.e., Heidolph RZR 2102 control).
3. Glass/Teflon potter with a capacity of 2 cm^3.
4. Glass/Teflon potter with a capacity of 10 cm^3
5. Refrigerated benchtop microcentrifuge (reaching up to 12,500 × *g*).
6. Ultracentrifuge with swinging-bucket and fixed-angle rotors (i.e., Beckman-Coulter Optima MAX-XP ultracentrifuge with TLS-55 rotor [2.2 mL Ultra-Clear™ Tubes (ref 347356)] and TLA-100.3 rotor [3.5 mL Thickwall Polycarbonate Tubes (ref 349622)].
7. Phosphate-buffered saline (PBS): NaCl 154 mM, $Na_2HPO_4{\cdot}2H_2O$ 3 mM, KH_2PO_4 1.06 mM.
8. Isosmolar buffer: 0.32 M Sucrose, 4 mM HEPES pH 7.4 (5 mL per condition).
9. 0.8 M Sucrose solution: 0.8 M Sucrose, 4 mM HEPES pH 7.4 (1 mL per condition).

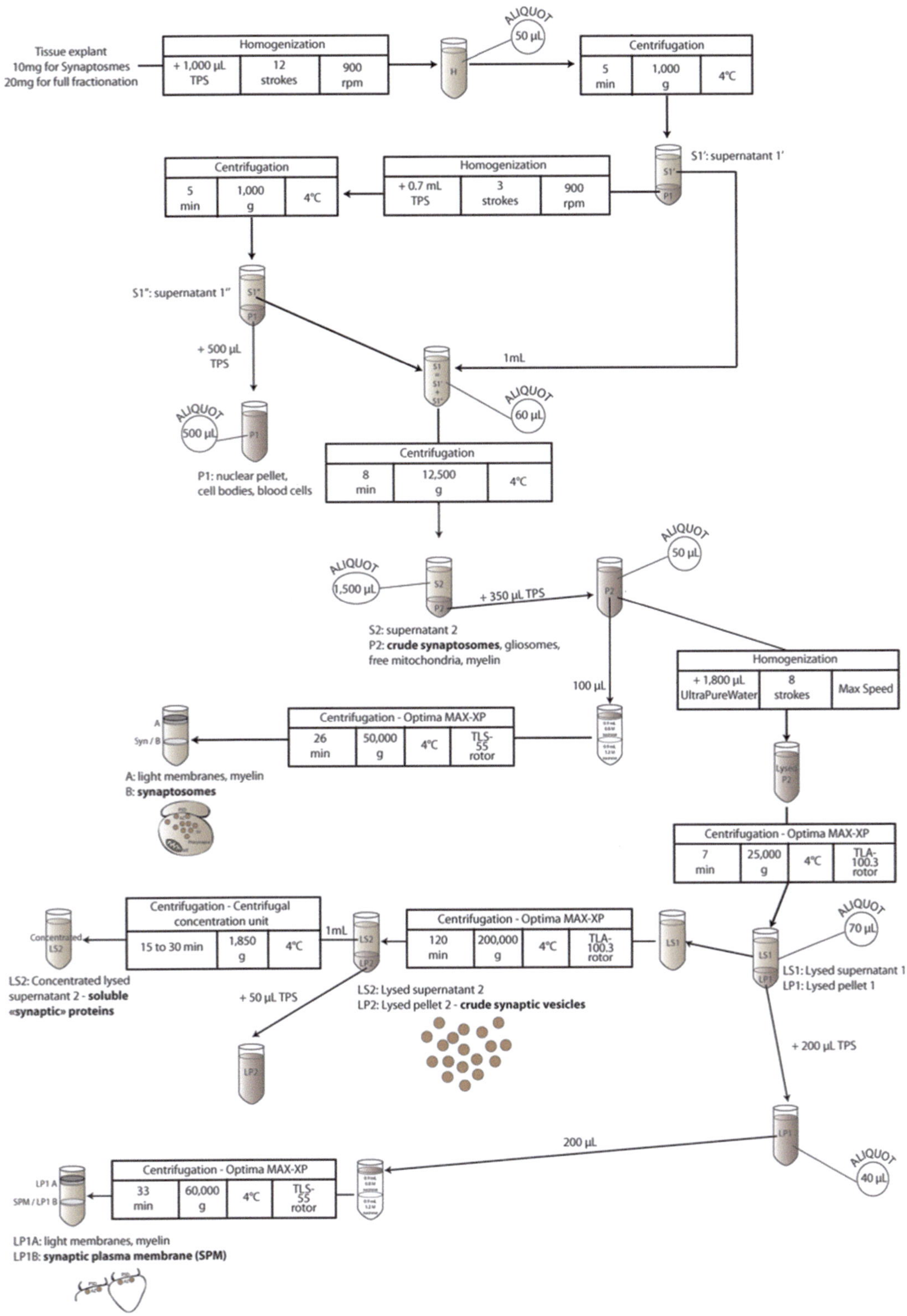

Fig. 1 At-a-glance diagram of the subcellular fractionation protocol. This diagram illustrates our subcellular fractionation procedure for the preparation of crude synaptosomes (P2), gradient-purified synaptosomes (B), synaptic plasma membranes (LP1B), and crude synaptic vesicles (LP2). Synaptosomes, synaptic plasma membranes, and isolated synaptic vesicles are represented schematically

10. 1.2 M Sucrose solution: 1.2 M Sucrose 4 mM HEPES pH 7.4 (1 mL per condition).
11. Protease inhibitors: Protease inhibitor cocktail set III EDTA free (ref 539134 Calbiochem).
12. 2.0 mL Round-bottom microcentrifuge tubes.
13. 1.5 mL Microtubes.
14. Sterilizing filter unit (0.22 μm pore size).
15. Centrifugal concentrator Unit (30 kDa membrane, Ref Millipore UFC 803024).
16. BD Micro-Fine U-100 Insulin Syringe 0.5 mL.
17. Glass Pasteur pipette and rubber bulb.
18. Distilled water.
19. 70% Ethanol.

3 Methods

An overview of the different steps of the protocol is available in Fig. 1. Buffers should be prepared in advance, filtered (0.22 μm pore size filter unit), and stored at 4 °C. The glassware should be very clean (avoid all detergent traces): wash the required glass/Teflon potter(s) by rinsing three times with distilled water, brushing three times with distilled water, brushing three times with 70% ethanol, and rinsing three times with 70% ethanol. Dry in a place protected from dust. Just before use, rinse the potter once with isosmolar buffer and leave it on ice. Add protease inhibitors to isosmolar buffer to obtain a 1:1000 solution.

3.1 Brain Tissue Dissection

Sacrifice the mouse by cervical dislocation (follow the local regulatory guidelines regarding ethics on animal handling). Decapitate and quickly extract the brain from the skull.

1. Clean and cool down the brain by placing it in ice-cold PBS. All subsequent steps must be performed on ice-cold surface.
2. Dissect out the regions of interest.
3. Repeat steps in 3.1 for each mouse. We routinely performed our fractionation with 20 mg of tissue. Yet for synaptosome isolation without further steps one can dissect as little as 10 mg of tissue (see Fig. 1). Pool the tissue samples.

3.2 Tissue Homogenization

1. Weigh the sample in order to measure the yield of your fractionation. Introduce tissue pieces into a 2 cm^3 clean ice-cold glass/Teflon potter. Add 1 mL of ice-cold isosmolar buffer

into the potter. Homogenize tissue with 12 strokes at 900 rpm. Keep the potter as cold as possible during homogenization.

2. Collect the homogenate with a glass Pasteur pipette and transfer it to a 2 mL round-bottom microcentrifuge tube. Rinse glass potter by adding 0.25 mL of ice-cold isosmolar buffer and performing three manual strokes. Pool this wash with the previous homogenate. Further rinse the Teflon potter by running 0.25 mL ice-cold isosmolar buffer and letting it spill into the glass potter and add the wash to the homogenate. The final homogenate volume is approximately of 1.5 mL. Mix the homogenate by inverting the tube three times (avoid vortexing to prevent aggregation). Aliquot 50 μL of homogenate (H fraction) prior to proceeding.

3.3 Synaptosome Fractionation

1. Centrifuge at 1000 × *g* for 5 min at 4 °C in a benchtop microcentrifuge.
2. Collect and transfer 1 mL of the supernatant (S1') to a new 2 mL round-bottom microcentrifuge tube, and mix by inverting the tube three times.
3. Add 0.7 mL of ice-cold isosmolar buffer to the P1 tube, resuspend the pellet, and transfer to the 2 cm^3 ice-cold glass/Teflon potter. Homogenize with three strokes at 900 rpm to increase synaptosome extraction.
4. Transfer the P1 homogenate with a glass Pasteur pipette to a 2 mL round-bottom microcentrifuge tube and centrifuge again at 1000 × *g* for 5 min at 4 °C in a benchtop centrifuge to obtain S1″ and P1.
5. Collect S1″ and add to S1' to generate S1. Mix by inverting the tube three times and aliquot 60 μL. Resuspend P1 in 500 μL of ice-cold isosmolar buffer by several pipetting strokes and keep it as an aliquot.
6. Centrifuge S1 at 12,500 × *g* for 8 min at 4 °C in a benchtop centrifuge.
7. During this centrifugation step, prepare the discontinuous sucrose gradients in ultracentrifuge tubes (we use 11x35 mm Beckman Coulter thin-wall ultraclear 2.2 mL ref 347356). Lay down first 0.9 mL of ice-cold 1.2 M sucrose solution, and cover slowly with 0.9 mL of 0.8 M sucrose solution using a micropipette.
8. Once the centrifugation is complete (Fig. 1), aliquot 1.5 mL of the supernatant (S2) and discard the rest of S2. Resuspend the pellet (P2) in 350 μL of ice-cold isosmolar buffer with several up and down strokes with a micropipette. Aliquot 50 μL of P2.

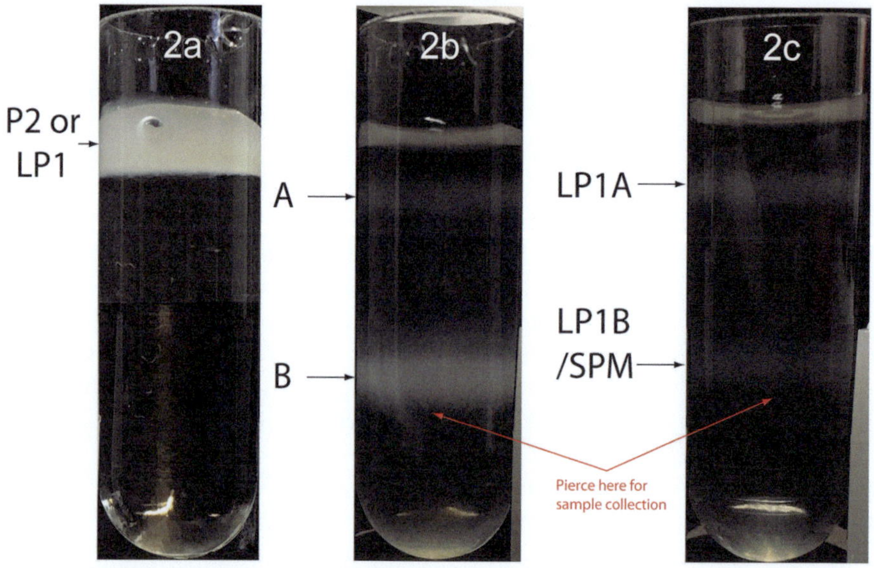

Fig. 2 Discontinuous sucrose density gradient centrifugations. Separation of synaptosomes and synaptic plasma membranes from myelin and denser cellular structures is performed by centrifugation on a sucrose density gradient. (**a**) P2 or LP1 fractions in isosmolar buffer (0.32 M sucrose) are layered on top of a 0.8 M/1.2 M discontinuous sucrose density cushions. (**b**, **c**) After centrifugation, three fractions are separated in the tubes. Fraction A contains mainly myelin; fraction B enriches the synaptic membranes (synaptosomes in **b**, or synaptic plasma membranes in **c**). The pellet contains denser material including free mitochondria as described earlier [4–6]

9. Slowly lay 100 μL of P2 on top of the sucrose gradient with a micropipette. Leave 2 mm between the rim of the tube and the gradient surface (see Fig. 2a).
10. Centrifuge at 50,000 × *g* for 21 min at 4 °C (we use Beckman-Coulter Optima MAX-XP ultracentrifuge with TLS-55 rotor). The centrifuge must be at full speed for 21 min for full separation.
11. After ultracentrifugation, two thin layers (A and B) and a pellet should be visible (see Fig. 2b). Collect fraction A with a micropipette from the top and then fraction B with a 0.5 mL syringe through the tube wall by piercing below the B fraction interface (we use BD Micro-Fine U-100 Insulin Syringe, see Fig. 2b). Collect each fraction in a different microtube and keep them as aliquot.

3.4 Synaptic Membrane Fractionation

1. Transfer 200 μL of P2 to a 10 cm^3 ice-cold glass/Teflon potter. Create an osmotic shock by adding 1800 μL of ice-cold ultrapure water in a quick stroke quickly followed by eight strokes at full speed of the potter (*see* **Note 1**).

2. Transfer the lysate to an ultracentrifuge tube (we use 13 × 51 mm Beckman Coulter thick-wall polycarbonate 1.5 mL ref 349622) and centrifuge at 25,000 × *g* for 7 min at 4 °C (we use Beckman-Coulter Optima MAX-XP ultracentrifuge with TLA-100.3 rotor).
3. During this centrifugation step, prepare a discontinuous sucrose gradient with the same procedure as in Sect. 3.3, step 7.
4. Once the centrifugation is complete, aliquot 70 μL of LS1 and transfer the rest to a new ultracentrifuge tube (we use 13 × 51 mm Beckman Coulter thick-wall polycarbonate ref 349622).
5. Resuspend LP1 in 200 μL of ice-cold isosmolar buffer with several up and down strokes with a micropipette and aliquot 40 μL of LP1.
6. Slowly lay 200 μL of LP1 on top of the discontinuous sucrose gradient with a micropipette.
7. Separate the LP1 on the gradient at 60,000 × *g* for 33 min at 4 °C (we use Beckman-Coulter Optima MAX-XP ultracentrifuge with TLS-55 rotor). It is critical that the centrifuge is at full speed for 33 min for full separation of synaptic plasma membranes.
8. After centrifugation, two thin LP1A and LP1B fractions float between the phases of sucrose densities (see Fig. 2c). Collect the LP1A fraction (upper layer) with a micropipette from the top. The LP1**B** fraction is collected through the tube wall with a 0.5 mL syringe (we use BD Micro-Fine U-100 Insulin Syringe). To that end, pierce below the LP1B fraction interface and aspirate. Collect each fraction in a different microcentrifuge tube and keep them as aliquot.

3.5 Synaptic Vesicle Fractionation

1. Centrifuge LS1 at 200,000 × *g* for 120 min at 4 °C (we use Beckman-Coulter Optima MAX-XP ultracentrifuge with TLA-100.3 rotor).
2. Collect the whole supernatant (LS2) and transfer to a pre-wet centrifugal concentration unit (we use Amicon ultra-4 centrifugal filter units 30 kDa Ref Millipore UFC 803024).
3. Centrifuge at 1850 × *g* at 4 °C for 15–30 min until the volume of concentrated LS2 reached 50 μL (we use a benchtop centrifuge GR412 Jouan).
4. The LP2 pellet containing crude synaptic vesicles appears like a transparent brown film. Resuspend LP2 in 50 μL of ice-cold isosmolar buffer with several up and down strokes with a micropipette and mix using a 0.5 mL syringe to eliminate clumps.

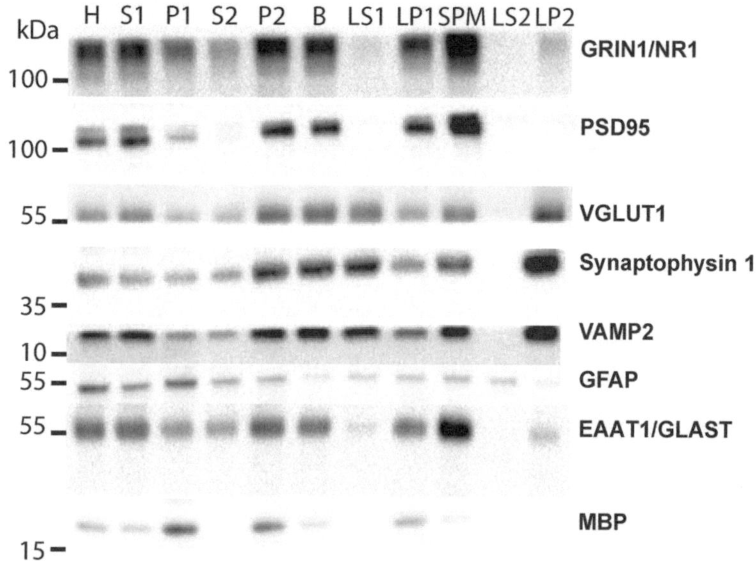

Fig. 3 Western blot analysis of the subcellular fractionation. The quality of the subcellular fractionation is efficiently monitored using a western blot procedure with a set of markers for typical compartments. Here we probe synaptic plasma membranes using the subunit 1 of NMDA receptors (GRIN1/NR1) and the postsynaptic density protein PSD95. Note the enrichment of these in SPM compared to H. An eight- to tenfold enrichment is expected in SPM compared to H for these markers. Markers for synaptic vesicles were probed to estimate the enrichment of SV membranes at LP2. VGLUT1 (the vesicular glutamate transporter), synaptophysin 1 and VAMP2 (the vSNARE) display a typical eight- to tenfold enrichment in LP2 compared to H. To trace astrocyte compartments we probed the glial fibrillary acidic protein (GFAP), and the plasma membrane glutamate transporter (EAAT1/GLAST). GFAP depletes from the B/SYN fraction and following ones compared to H. However, EAAT1/GLAST significantly enriches in SPM. Hence, glial membranes co-enrich with synaptic membranes. Finally, we probed the myelin basic protein (MBP) to follow myelin membranes. MBP is efficiently depleted from B/SYN, SPM, and LP2. Note that the comparison of several fractions for several markers is the only way to prove the quality of your fractionation and the enrichment/depletion of a given protein with a given fraction

3.6 Monitoring of the Subcellular Fractionation Quality

After protein titration, perform a western blot experiment with all fractions side by side using standard procedures. See example in Fig. 3. Several typical markers will help estimate the quality of the fractionation.

4 Notes

1. Osmotic lysis allows for the release of synaptic contents (synaptic vesicles, mitochondria, cytosol, …) and the separation of synaptic plasma membrane and docked synaptic vesicles. It is critical to proceed precisely and quickly for a successful lysis.

Acknowledgments

MFA salary was supported by the Fondation Pour la Recherche Médicale (ING20150532192), and EH received the following funding from the French Agence Nationale de la Recherche: ANR-10-LABX-43 BRAIN and ANR-12-JSV4-0005-01 VGLUT-IQ. PT received the following funding: ANR-10-IDEX-03-02, NARSAD Young investigator grant from the brain and behavior foundation, Région Aquitaine, INRA. Experiments were performed thanks to the equipment of the Bordeaux Neurocampus central facility for biochemistry and biophysics of protein.

References

1. Brody TM, Bain JA (1951) Effect of barbiturates on oxidative phosphorylation. Proc Soc Exp Biol Med 77:50–53
2. Brody TM, Bain JA (1952) A mitochondrial preparation from brain. Biol Chem 195:685
3. Spanner S (1972) Methods of separating the subcellular components of brain tissue. In: Zambotti V, Tettamanti G, Arrigoni M (eds) Glycolipids, glycoproteins, and mucopolysaccharides of the nervous system. Advances in experimental medicine and biology, vol 25. Plenum Press, New York, pp 195–207
4. Whitakker VP (1959) The isolation and characterization of acetylcholine-containing particles from brain. Biochem J 72:694–706
5. Whittaker VP, Michaelson IA, Kirkland RJ (1964) The separation of synaptic vesicles from nerve-ending particles ('synaptosomes'). Biochem J 90:293–303
6. Whittaker VP (1965) The application of subcellular fractionation techniques to the study of brain function. Prog Biophys Mol Biol 15:39–96
7. Whittaker VP (1993) Thirty years of synaptosome research. J Neurocytol 22:735–742
8. Gray EG, Whitakker VP (1962) The isolation of nerve endings from brain: an electron-microscopic study of cell fragments derived by homogenization and centrifugation. J Anat 96:79–88
9. Zimmermann H, Fonnum F (2016) Victor P. Whittaker (1919-2016). J Neurochem 139:333–335. https://doi.org/10.1111/jnc.13778
10. Huttner WB, Schiebler W, Greengard P, De Camilli P (1983) Synapsin I (protein I), a nerve terminal-specific phosphoprotein. III. Its association with synaptic vesicles studied in a highly purified synaptic vesicle preparation. J Cell Biol 96:1374–1388
11. Cotman CW, Matthews DA (1971) Synaptic plasma membranes from rat brain synaptosomes: isolation and partial characterization. Biochim Biophys Acta 249:380–394
12. Jones DH, Matus AI (1974) Isolation of synaptic plasma membrane from brain by combined flotation-sedimentation density gradient centrifugation. Biochim Biophys Acta 356:276–287
13. Boyken J, Grønborg M, Riedel D, Urlaub H, Jahn R, Chua JJ (2013) Molecular profiling of synaptic vesicle docking sites reveals novel proteins but few differences between glutamatergic and GABAergic synapses. Neuron 78:285–297. https://doi.org/10.1016/j.neuron.2013.02.027
14. Weingarten J, Lassek M, Mueller BF, Rohmer M, Lunger I, Baeumlisberger D, Dudek S, Gogesch P, Karas M, Volknandt W (2014) The proteome of the presynaptic active zone from mouse brain. Mol Cell Neurosci 59:106–118. https://doi.org/10.1016/j.mcn.2014.02.003
15. Luquet E, Biesemann C, Munier A, Herzog E (2017) Purification of synaptosome populations using fluorescence-activated synaptosome sorting. Methods Mol Biol 1538:121–134
16. De Robertis E, Rodriguez De Lores Arnaiz G, Pellegrino De Iraldi A (1962) Isolation of synaptic vesicles from nerve endings of the rat brain. Nature 194:794–795
17. Bajjalieh SM, Frantz GD, Weimann JM, McConnell SK, Scheller RH (1994) Differential expression of synaptic vesicle protein 2 (SV2) isoforms. J Neurosci 14:5223–5235
18. Burre J, Beckhaus T, Schagger H, Corvey C, Hofmann S, Karas M, Zimmermann H,

Volknandt W (2006) Analysis of the synaptic vesicle proteome using three gel-based protein separation techniques. Proteomics 6:6250–6262. https://doi.org/10.1002/pmic.200600357

19. Farr CD, Gafken PR, Norbeck AD, Doneanu CE, Stapels MD, Barofsky DF, Minami M, Saugstad JA (2004) Proteomic analysis of native metabotropic glutamate receptor 5 protein complexes reveals novel molecular constituents. J Neurochem 91:438–450. https://doi.org/10.1111/j.1471-4159.2004.02735.x
20. Gronborg M, Pavlos NJ, Brunk I, Chua JJ, Munster-Wandowski A, Riedel D, Ahnert-Hilger G, Urlaub H, Jahn R (2010) Quantitative comparison of glutamatergic and GABAergic synaptic vesicles unveils selectivity for few proteins including MAL2, a novel synaptic vesicle protein. J Neurosci 30:2–12. https://doi.org/10.1523/JNEUROSCI.4074-09.2010
21. Husi H, Ward MA, Choudhary JS, Blackstock WP, Grant SG (2000) Proteomic analysis of NMDA receptor-adhesion protein signalling complexes. Nat Neurosci 3:661–669
22. Lassek M, Weingarten J, Einsfelder U, Brendel P, Muller U, Volknandt W (2013) Amyloid precursor proteins are constituents of the presynaptic active zone. J Neurochem 127:48–56. https://doi.org/10.1111/jnc.12358
23. Lassek M, Weingarten J, Volknandt W (2015) The synaptic proteome. Cell Tissue Res 359:255–265. https://doi.org/10.1007/s00441-014-1943-4
24. Morciano M, Burre J, Corvey C, Karas M, Zimmermann H, Volknandt W (2005) Immunoisolation of two synaptic vesicle pools from synaptosomes: a proteomics analysis. J Neurochem 95:1732–1745. https://doi.org/10.1111/j.1471-4159.2005.03506.x
25. Morciano M, Beckhaus T, Karas M, Zimmermann H, Volknandt W (2009) The proteome of the presynaptic active zone: from docked synaptic vesicles to adhesion molecules and maxi-channels. J Neurochem 108:662–675. https://doi.org/10.1111/j.1471-4159.2008.05824.x
26. Weingarten J, Lassek M, Mueller BF et al (2014) The proteome of the presynaptic active zone from mouse brain. Mol Cell Neurosci 59C:106–118. https://doi.org/10.1016/j.mcn.2014.02.003
27. Yao J, Nowack A, Kensel-Hammes P, Gardner RG, Bajjalieh SM (2010) Cotrafficking of SV2 and synaptotagmin at the synapse. J Neurosci 30:5569–5578. https://doi.org/10.1523/JNEUROSCI.4781-09.2010
28. Biesemann C, GronborgM LE et al (2014) Proteomic screening of glutamatergic mouse brain synaptosomes isolated by fluorescence activated sorting. EMBO J 33:157–170. https://doi.org/10.1002/embj.201386120
29. Tannu NS, Hemby SE (2006) Methods for proteomics in neuroscience. Prog Brain Res 158:41–82. https://doi.org/10.1016/S0079-6123(06)58003-3
30. Schreiner D, Savas JN, Herzog E, Brose N, de Wit J (2017) Synapse biology in the 'circuit-age'-paths toward molecular connectomics. Curr Opin Neurobiol 42:102–110. https://doi.org/10.1016/j.conb.2016.12.004

Chapter 6

Preparation of P2 or Percoll-Purified Synaptosomes from Mammalian Brain Tissue

Jing Xue, Annie Quan, and Phillip J. Robinson

Abstract

Isolated nerve endings have been widely used as a model for studying synaptic functions in the brain. These nerve endings can be separated from the axons and postsynaptic connections when neuronal tissue is homogenized in an isotonic buffer. The plasma membrane of the sheared nerve endings reseals to form pinched-off nerve terminals called synaptosomes. Synaptosomes show similar morphology and contain all the intact molecular machinery found in nerve terminals in vivo, including the proteins, synaptic vesicles, and mitochondria, necessary for regulating synaptic function for the release, uptake, and storage of neurotransmitters upon different stimulations. The methods for making synaptosomes have varied and been modified for different experimental applications. The goal of this chapter is to describe optimized procedures used in our laboratories for isolation of intact and functional synaptosomes from whole mammalian brain tissue, at two levels of purity: the "crude" P2 and the Dunkley S1 method of Percoll-purified fractions, for molecular, biochemical, and functional assays. These represent the shortest experimental methods to date. The protocol involves mild homogenization of the brain tissue in isotonic conditions, with medium centrifugation speeds to minimize the mechanical damage to the synaptosomes and the preparation time.

Key words Percoll-purified, Crude P2 gradient, Functional assay

1 Introduction

Neurons communicate with each other at the highly specialized structure known as the synapse. Ultrastructural analyses (Fig. 1a) show that each synapse is a junction comprised of a presynaptic (axonal) terminal containing synaptic vesicles (SVs) filled with neurotransmitters, mitochondria, and other cell organelles, a synaptic cleft which is a narrow extracellular space approximately 20–30 nm wide containing a plaque of intercellular filamentous material, and a highly electron-dense postsynaptic density (PSD) membrane which is part of a dendritic spine [1–4]. In response to a threshold action potential the SVs filled with neurotransmitters fuse with the presynaptic membrane and release their contents into the synaptic cleft via exo-

Kathryn M. Murphy (ed.), *Synaptosomes*, Neuromethods, vol. 141,
https://doi.org/10.1007/978-1-4939-8739-9_6, © Springer Science+Business Media, LLC, part of Springer Nature 2018

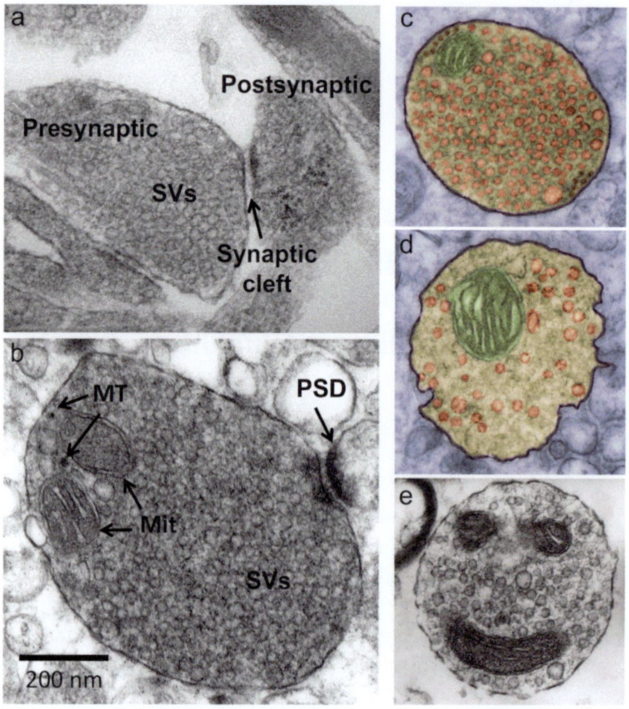

Fig. 1 Gallery of electron micrographs. (**a**) Synapse from cultured rat hippocampal neurons at 20 days in vitro (DIV). The synapse comprises presynaptic nerve terminal filled with synaptic vesicles (SVs), synaptic cleft, and an electron-dense postsynaptic density (PSD) membrane showing the features of presynaptic terminals, synaptic cleft, and postsynaptic terminals with PSD. (**b**, **c**) Intact synaptosomes isolated via discontinuous Percoll gradient fractionation. Resting synaptosomes have sealed plasma membrane, filled with SVs, intact mitochondria forming a synaptic contact with a PSD membrane. (**d**, **e**) Drug-treated synaptosomes to block SV recycling with compromised synaptic function have membrane invaginations, depleted SVs, large endosomes/vacuoles, and intact mitochondrial structures after 10-min treatment with roscovitine (100 μM, panel **d**) or Ryngo-1-22 (100 μM, panel **e**), each followed by depolarization as we previously described [23, 87]. The synaptosomes in panels **c** and **d** are manually colored to highlight vesicles and endosomes (orange), mitochondria (green), and microtubules (brown)

cytosis. The neurotransmitters diffuse across the synaptic cleft, bind to the receptor proteins on the postsynaptic membrane, and activate the postsynaptic terminals to complete the process of synaptic transmission. To maintain the continuity of synaptic transmission the limited number of SVs in presynaptic terminals have to be rapidly retrieved and recycled via endocytosis. The amount of neurotransmitter released from the synapse changes the efficacy of synaptic transmission. The ability of the neurons to change the strength of their synaptic connections is known as synaptic plasticity, and is believed to form a key element of learning and memory mechanisms in the brain. Synaptic transmission and plasticity are crucial for all aspects of nervous system function and necessary for proper development of the central nervous system. Abnormalities in synaptic transmission and plasticity contribute to the pathology of many neurological disorders [5], including schizophrenia [6–8], autism [9, 10], other intellectual disabilities [11, 12], and neurodegenerative diseases, such as Alzheimer's [13] and Parkinson's disease [14].

Our current understanding of synaptic transmission in nerve terminals at the molecular, biochemical, and functional physiological levels has been built upon a combination of electrophysiological studies of in situ synapses combined with molecular studies on isolated nerve terminals [2–4, 15–17]. When nervous system tissues are homogenized in an isotonic solution of appropriate viscosity and with an appropriate shear force, the nerve terminals are torn from the connecting axons, and postsynaptic and glial cells. The plasma membrane of the terminals reseals to form spherical structures first reported in the 1950s [3, 15, 18]. Since they represent an authentic subcellular organelle, they were later termed "synaptosomes" by Whittaker in 1964 [1]. Synaptosomes are generated from both nerve terminals and varicosities by a mild liquid shear being applied to the brain tissue by the homogenizer, when the tissue is suspended in an iso-osmotic sucrose medium. The liquid shear is formed in the annulus between the stationary glass and the rotating Teflon pestle. There remains no evidence that synaptosome formation might favor any one type of synapse or transmitter [4]. The efficiency of formation of synaptosomes has been estimated to be within 70% and 100% [19].

Ultrastructural analysis of viable, healthy intact synaptosomes (Fig. 1b, c) shows strikingly analogous morphology to nerve terminals in brain or cultured cells (Fig. 1a). They exhibit a range of diameters typically from 0.5 to 0.9 μm, although structures up to 3 μm are known, depending on the source of the starting neuronal tissue from which they were prepared [20, 21]. Synaptosomes retain the same complement of subcellular organelles, cytoskeleton, synaptic vesicles, and mitochondria normally found in nerve terminals in vivo, and make synaptic contact with remnant postsynaptic membranes. Accordingly, drug-treated synaptosomes that have deficient synaptic functions show a synaptic vesicle (SV) depletion, larger endosomes, or vacuoles and often have plasma membrane invaginations representing aborted endocytosis (Fig. 1d, e) [22, 23].

Functionally, synaptosomes are able to respire, synthesize ATP, maintain membrane potential and ion homeostasis, transport ions and neurotransmitters, and release and retrieve synaptic vesicles, and they have broad range of transmitter and peptide hormone receptors on their surface membrane [4, 24–31]. However, synaptosomes are only rarely able to synthesize a small number of new proteins, apart from within their mitochondria which have an independent protein synthesis system, as they largely lack mRNA and the DNA replication and translation machinery [24, 32, 33], although very low levels of cytoplasmic synthesis have occasionally been observed [34, 35].

The original methods developed for generation of synaptosomes [3, 36] included a number of variables that have now been standardized between laboratories. The first step is tissue disruption or homogenization, typically performed on ice with pre-chilled tubes and reagents. For the fresh brain tissue this was initially trialled in a variety of different isotonic buffers, but almost all labs now use 0.32 M sucrose. Higher concentrations of sucrose cause shrinkage of synaptosomes, disruption of neurofilaments, and release of particle-bound acetylcholine [3, 37–39]. The volume used to homogenize the brain is typically around 10% (w/v) followed by centrifugation. Homogenization was initially performed in a variety of homogenizers with varying degrees of rotation, but most labs now use commercial Teflon/glass homogenizers and rotation speeds of less than 1000 rpm.

After generation of synaptosomes the second step is to isolate or enrich them from the homogenate. The choice of method for enrichment requires consideration of the subsequent methodological applications or research goal. Here we focus on the fastest two methods that have been used in our laboratories for almost 20 years. The first step is universally a low-speed centrifugation to remove cells, nuclei, and debris, resulting in a supernatant called the "S1" containing the bulk of the synaptosomes in 0.32 M sucrose. While stopping at this point is the fastest method, few subsequent studies are compatible with sucrose. Every centrifugation and resuspension step results in damage to the structures and minimal steps and careful gentle resuspension are essential. The S1 is typically subjected to higher speed centrifugation to pellet the synaptosomes as a "P2" pellet. The P2 pellet can be resuspended in the buffer of choice for the subsequent experiments, typically a saline-based physiological buffer such as Krebs-like buffers. The P2 pellet represents a relatively "crude" synaptosome preparation, with sufficient suitability for many applications. Based on our morphological analysis from electron microscope images, this P2 fraction contains up to ~30% synaptosomes (for a morphological definition of synaptosomes see Box 1). However, the P2 contains high amounts of contaminants including free mitochondria, myelin and glial membrane fragments, and lysosomes, all of which may interfere with the study of dynamic secretory or metabolic processes.

The third step is to purify the synaptosomes. Although multiple methods are available with small differences in strengths and

Box 1: Synaptosome Morphology

Initially described as "nerve-ending particles," the name "synaptosome" was first used by Whittaker [1]. A number of criteria have been suggested to define a synaptosome from adult brain, almost exclusively based on morphology by electron microscopy. In identifying synaptosomes as detached synapses, Gray and Whittaker first suggested four distinguishing morphological criteria: (1) a round sealed plasma membrane, within which is found (2) a concentration of vesicles resembling in size and number of the synaptic vesicles seen in nerve terminals and varicosities in situ and (3) one or more small mitochondrial profiles, and (4) the frequent presence of lengths of electron-dense membrane resembling postsynaptic membranes adhering to the periphery of the synaptosome [3, 4]. Since synaptosomes are essentially round in shape and are typically assessed by thin-section electron microscopy, not all profiles observed by electron microscope images contain mitochondria and fewer show observable PSD profiles since these may be present outside the plane of section, or may not be present at all. The mitochondrial profiles may number more than one in thin sections, yet 3D reconstructions or electron tomography shows that these can also represent a single horseshoe-shaped mitochondrion [40–43]. Some synaptosomes have been confirmed by serial section not to contain mitochondria [44].

The minimal morphological criteria for standard electron microscopy are that the synaptosomes are small (0.5–0.9 μm), with larger structures being rarer; have well-sealed plasma membrane; and contain in their cytoplasm an obvious cluster of synaptic vesicles (typically around 40 nm in diameter) from a few to dozens in the plane of section. Other morphologic features of adult brain synaptosomes are known, but will not be observed within the plane of section in every synaptosome. These include a single microtubule coil, typically located opposite to the synapse, that does not appear to have a cytoskeletal supportive function [45–47] and which is strongly associated with mitochondria [45] (e.g., colored in brown in Fig. 2c). Occasionally a clathrin-coated vesicle will be observed, typically 1 or 2 in 10% of the synaptosomes, and large dense-cored vesicles, typically 1 in less than 5% of the synaptosomes, depending on the tissue source [48]. Many synaptosomes contain larger round endosome-like vesicles, typically from 150 to 200 nm in diameter [22]. These may represent many different types of structures such as endoplasmic reticulum, endosomes, or vacuoles. However, many may represent activity-dependent bulk endosomes, large numbers of which form after strong stimulation and which have a half-life of around 30 min [49, 50]. Note that although synaptosomes may be purified from immature or newborn brain, they have few or no vesicles and PSD and are not suitably described by the criteria here [51, 52].

weaknesses, all employ differential centrifugation through various forms of density gradient material, such as sucrose, Ficoll, or Percoll. Since these original methods for purifying synaptosomes were described, we and others have developed a range of procedures, varying each of the three main steps, each with its own advantages in terms of the homogeneity and viability of the final

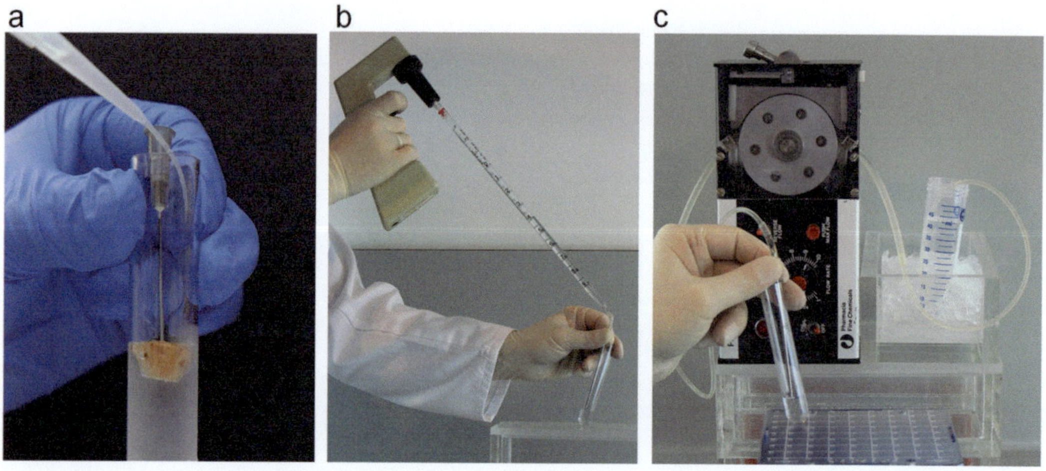

Fig. 2 Methods for generating discontinuous Percoll gradients. (**a**) Gradients generated manually with pipettor and modified floating cork to minimize disruption to gradient layers. (**b**) Gradients generated with a manual pipet aid. (**c**) Gradients generated with the aid of a peristaltic pump. (Panels **b** and **c** are reproduced with permission from Dunkley et. al., (2008), Nat Protoc, 3(11): 1718-28, Copyright © 2006 Nature Publishing Group)

synaptosome preparation and the time it takes to complete the procedure [20, 21, 53–56]. In our hands, the Dunkley S1 Percoll gradient method provides the greatest speed with the least synaptosomal damage and requires the minimum of specialized equipment. It is important to note that there were two Dunkley methods. The original Dunkley P2 method in 1986 started with the P2 pellet being resuspended and applied to the Percoll gradient [16]. Subsequently, in 1988 the Dunkley S1 method was developed, whereby the S1 soluble fraction was directly applied instead [53], saving considerable time by removing one of the longer and potentially mechanically damaging centrifugation steps, and improving the yield by 40%. The final purified synaptosome fractions typically are ~50% synaptosomes [20, 57–59].

The focus of this chapter is on the detailed procedures for isolating synaptosomes from mammalian brain tissue at two different degrees of purity that we have used and applied to a number of synaptic function studies:

1. The "crude P2 synaptosome fraction" is made from differential centrifugation of brain homogenates. The P2 synaptosome fraction has been widely used for studies on synaptic functions in the nerve terminals when purity is not the main consideration, but rather, obtaining a larger quantity of synaptosomes quickly for functional studies [23, 60–63]. This preparation can be particularly useful for protein purification and protein-protein interaction analysis. For metabolic labeling with $^{32}P_i$ this preparation is suitable, as ATP production from the $^{32}P_i$ occurs only in sealed, viable synaptosomes which would contain

mitochondria and maintain a membrane potential, and thus exhibit stimulus-responsive changes in phosphorylation. Thus phosphorylation changes occur only within the synaptosomes. Some example applications include:

(a) Protein-protein interaction studies using in vitro pulldown assays [62, 64]

(b) Metabolic labeling, for example $^{32}P_i$-radioactive labeling for stimulus-dependent phosphoprotein analyses [23, 25, 62, 65–67]

2. Discontinuous Percoll gradient procedure for making highly pure synaptosomes as developed and described by the Dunkley S1 method [20, 53]: Percoll is a colloidal suspension of polyvinylpyrrolidone-coated silica particles of a uniform 17 nm diameter. Percoll has two major advantages: (1) low viscosity, allowing more rapid sedimentation and use of lower centrifugal forces, and (2) it can be prepared in an isotonic solution (such as isotonic sucrose) to maintain osmolarity. Two laboratories independently developed Percoll gradient procedures for preparation of synaptosomes starting with the P2 fraction and applying it to a discontinuous Percoll gradient [16, 20, 68]. Synaptosomes isolated from Percoll gradients are much more homogeneous than the P2 method, containing low levels of microsomes, myelin, and free mitochondria, and they are metabolically viable and able to take up calcium and release neurotransmitters. The procedure developed in our laboratory [20, 53] described here requires less time for centrifugation since it starts with the S1 fraction rather than the P2. The preparation is used for studies requiring low background or very high purity. Specific experimental applications include:

(a) Measuring glutamate release from the nerve terminals [62, 69–71]

(b) Morphological analysis using electron microscopy [22, 23]

2 Materials and Reagents

2.1 Mammalian Brain Tissue Extraction and Homogenization

1. Sprague-Dawley adult rats (≥8 weeks old) (*see* **Note 1**).
2. Homogenization buffer (Tris/sucrose/EDTA buffer, ice cold): 0.32 M Sucrose, 1 mM EDTA, 5 mM Tris, pH 7.4 (adjusted with 0.5 M HCl).
3. Animal dissection and tissue extraction instruments: Plastic film tubes called DecapiCone® (cat. no. DCL-120, Braintree Scientific Inc., MA, USA), small animal rodent guillotine (Kent Scientific, Torrington, CT, USA) with sharp blade for animal decapitation, sharp scissors, and curved forceps or tweezers for harvesting the brain from animal cranial cavity.

4. Wheaton homogenizer: Potter-Elvehjem type Teflon/glass tissue grinder (15 mL (cat. no. 358009) or 30 mL (cat. no. 358011) depending on the amount of tissue used and volume of homogenization buffer). The grinding chamber clearance between glass and Teflon pestle is 0.1–0.15 mm.
5. A suitable variable electric motor to drive the tissue grinder between 500 and 800 rpm like a Wheaton overhead stirrer. Alternatively, a commercial variable low-speed drill from a hardware store mounted on a drill stand can serve as an inexpensive option.

2.2 Homogenate Fractionation

1. HEPES-buffered Krebs—HBK (without calcium): 143 mM NaCl, 4.7 mM KCl, 1.18 mM $MgSO_4$, 20 mM HEPES/Tris, 10 mM D-glucose, pH 7.4.
2. HEPES-buffered Krebs—HBK (with calcium): 143 mM NaCl, 4.7 mM KCl, 1.18 mM MgSO4, 0.1 mM (low) or 1.2 mM (high) $CaCl_2$, 20 mM HEPES/Tris, 10 mM D-glucose, pH 7.4.
3. Carbogen gas (95% oxygen (O_2), 5% carbon dioxide (CO_2), medical grade) and sintered glass gas sparger.
4. High-speed centrifuge (Avanti J-25I centrifuge with rotor JA25.50 from Beckman Coulter, CA, USA).
5. 50 mL Round-bottom polycarbonate centrifuge tubes for centrifugation of the brain homogenates in the rotor JA25.50 (Beckman Coulter, cat. no. 363647).
6. Plastic disposable Pasteur pipettes.

2.3 Percoll Gradient Fractionation

1. Percoll reagent (GE Healthcare, cat. no. 17-0891-01).
2. Gradient buffer (4 × X Tris/sucrose/EDTA buffer): 1.28 M Sucrose, 4 mM EDTA, 20 mM Tris, pH 7.4 (adjust with HCl).
3. 250 mM 1,4-Dithiothreitol (DTT).
4. 12 mL Round-bottom polycarbonate centrifuge tubes (Nalgene, cat. no. 3117-0120).
5. Rubber adapters for round-bottom centrifuge tubes (Beckman Coulter, cat. no. 870329).
6. Peristaltic pump and tubing of a diameter suitable to deliver ≤1 mL/min.
7. Glass pipettes.
8. Vacuum filtration apparatus (Sartorius, Goettingen, Germany, glass vacuum filter holder for 47 mm diameter filters, cat. no. 16309) to prefilter the Percoll reagent.

9. Glass fiber filter with binder, 47 mm, 1 μm pore size (Merck-Millipore, Darmstadt Germany, cat. no. AP1504700) to fit vacuum filtration apparatus.

3 Methods

Experiments involving animals must be carried out by trained personnel in accordance with all appropriate national and institutional regulations. A Material Safety Data Sheet (MSDS) for the reagents and the Standard Operating Procedure (SOP) for the experiment methods should be obtained where available and read before use.

3.1 Preparing Crude P2 Synaptosomes

1. Immobilize and restrain the adult rats in the plastic film tubes commercially called DecapiCone® (larger size >150 g) (*see* **Note 2**).
2. Sacrifice the adult rat(s) by decapitation with guillotine (*see* **Notes 2** and **3**).
3. Rapidly remove brain tissue from cranial cavity with sharp scissors and forceps. Immediately wash and rinse the blood from the tissue in ice-cold homogenization buffer (*see* **Note 4**). Then transfer the rinsed brain tissue to another beaker of ice-cold homogenization buffer to obtain a 10% (v/w) suspension (e.g., 10 mL/1 g of rat brain), and quickly mince with scissors.
4. Immediately homogenize the minced tissue in a 15 mL Teflon/glass tissue grinder (pre-chilled on ice) with seven up and down strokes using a motor-driven pestle at 700 rpm.
5. Centrifuge the brain homogenate in 50 mL centrifuge tubes at 948 × *g* (2800 rpm in JA25.50 rotor) for 10 min at 4 °C.
6. The supernatant is called S1. Transfer the S1 supernatant to a new ice-cold centrifuge tube (leave behind the "whitish" part of the supernatant). Resuspend the pellet in 10 mL of ice-cold homogenization buffer and mix with plastic Pasteur pipet or 1 mL wide-bore pipet tip. Centrifuge the resuspended mixture at 948 × *g* for 10 min at 4 °C.
7. The new supernatant is called S2. Collect the S2 supernatant and combine with the supernatant (S1) from step 6 and centrifuge at 23,036 × *g* (13,800 rpm in JA25.50 rotor) for 30 min at 4 °C to obtain the P2 pellet.
8. Discard the supernatant. Resuspend the P2 pellet in about 20 mL ice-cold HEPES-buffered Krebs (HBK) without calcium, for buffer exchange, and gently mix well with plastic Pasteur pipet (*see* **Note 5**).
9. Recentrifuge the P2 fraction at 23,036 × *g* for 10 min at 4 °C.

10. Discard the supernatant. Resuspend the pellet in 8 mL HBK (usually containing glucose and either 0.1 mM Ca^{2+} which is essential for $^{32}P_i$-radioactive labeling, or 1.2 mM Ca^{2+} for physiological experiments not requiring $^{32}P_i$). Mix the synaptosomes gently and well with a plastic Pasteur pipet or alternatively with 1 mL wide-bore pipet tip. The resuspended synaptosome concentration should be approximately 5 mg/mL protein for ideal working concentration for further assays.
11. Transfer the crude P2 synaptosomes into an appropriate-volume polypropylene sample tube and keep on ice ready for further experimentation.
12. If using crude P2 synaptosomes for protein-protein interaction studies, after relevant intact synaptosome treatments are performed (for example control vs. depolarization), the synaptosomes are typically pelleted and then lysed in total synaptosome lysis buffer (5 mM Tris, 150 mM NaCl, 1 mM EDTA, 1 mM EGTA, 1% Triton X-100, 1 mM phenylmethylsulfonyl fluoride (PMSF), 40 μg/mL leupeptin, complete protease — EDTA-free inhibitor tablets (Roche Life Sciences, Basel, Switzerland), Phosphatase Inhibitor Cocktail (Merck-Millipore)).

3.2 Purifying Synaptosomes Using Discontinuous Percoll Gradient Fractionation

1. The stock Percoll reagent must be adjusted in pH to 7.4 before use due to small batch variations. For each 250 mL of Percoll reagent, adjust the pH to 7.4 with 0.5 M HCl. The Percoll is then filtered directly through a glass fiber filter membrane (Merck-Millipore AP15 prefilter) to remove aggregated, incompletely coated particles which can contaminate subcellular fractions on centrifugation. Each Percoll batch slightly varies in the amount of minor aggregates and they can interfere with a few experimental approaches such as electron microscopy. After filtration, aliquots of the Percoll can be stored indefinitely at −20 °C or at 4 °C for use *within about a week. Filtration should be repeated if stored for longer than 1 week at* 4 °C and for thawed stocks, *especially if the synaptosome preparation is to be used for* $^{32}P_i$ *radioactive labeling or electron microscopy analysis.*
2. Prepare Percoll gradient solutions comprising layers of 3% (v/v), 10% (v/v), 15% (v/v), and 23% (v/v) gradient solutions. Each of the Percoll gradient solutions is made according to Table 1.
3. Generate discontinuous Percoll gradients in the 12 mL round-bottom polycarbonate centrifuge tubes. Keep all tubes cool on ice at all times. There are three alternative ways of making the discontinuous Percoll gradients as previously described and herein [20] (*see* **Note 6**). Method (a) is our preferred method:

Table 1
Preparation of Percoll gradient solutions

Final concentration (v/v)		Percoll stock (pH 7.4) (mL)	4× Gradient buffer (mL)	Water (mL)	250 mM DTT (μL)
(a)	3%	3	25	72	100
(b)	10%	10	25	65	100
(c)	15%	15	25	60	100
(d)	23%	23	25	52	100

(a) Create the gradient layers manually with a pipettor (for example, Gilson P1000 Pipetman (Middleton, WI, USA)) and with the aid of a cork (manually cut into shape and inserted into the 12 mL polycarbonate round-bottom centrifuge tube with a needle attached for later cork retrieval (see Fig. 2a)). The cork distributes the liquid and minimizes layer disturbance. Firstly, add 2 mL of 23% (v/v) Percoll gradient solution into the empty centrifuge tube and then insert the cork (with a fine-gauge needle attached to its top) onto the 23% (v/v) layer. Then overlay the 15% (v/v) Percoll gradient solution using a 1 mL pipettor with a gel-loading tip inserted on a standard 1 mL pipettor tip. Hold the tube at an angle and allow the gel-loading tip to touch the inside surface of the tube. Dispense the Percoll gradient solution with a steady speed without disturbing the underlying layer. Repeat the procedure with the 10% (v/v) layer next and the 3% (v/v) layer last.

(b) Create the gradient layers manually with a pipet aid (see Fig. 2b). The gradient is made by carefully adding 2 mL of 23% (v/v) Percoll gradient solution into the centrifuge tube, and then overlaying the 15% (v/v), 10% (v/v), and finally the 3% (v/v) sequentially on top with a manual pipet aid. The pipet aid is filled with the gradient solution and depressing the lower button empties the pipet. The flow rate must be very slow and steady, and it is usually proportional to the amount the button is depressed.

(c) Create the discontinuous Percoll gradients with the aid of a peristaltic pump (see Fig. 2c). The gradient is made by carefully adding 2 mL of 23% (v/v) Percoll gradient solution into the centrifuge tube, and then overlaying the 15% (v/v), 10% (v/v), and 3% (v/v) sequentially on top using a constant low flow rate of <1 mL/min.

4. Sacrifice the animal by decapitation and collect the required brain tissue as described above in Sec. 2.1, steps 1–2, and associated notes.
5. Homogenize the tissue and centrifuge as described above in Sec. 2.1, steps 3–6. The volume of the combined supernatants, S1 and S2, is approximately 16 mL from one rat brain. Bring the total volume to 17 mL with homogenization buffer, ensuring that the total protein concentration of synaptosomes is less than 5 mg/mL, and add DTT to the mixture to a final concentration of 0.25 mM.
6. Slowly layer 2 mL of combined S1 and S2 mixture directly onto the top layer (3% (v/v)) of the discontinuous Percoll gradients using a manual pipettor. The cork is not required at this stage.
7. Carefully insert the 12 mL round-bottom polycarbonate centrifuge tubes with the layered synaptosome/Percoll gradient solutions into the rubber centrifuge adapters. Centrifuge the tubes at 31,400 × *g* (16,112 rpm on the rotor JA25.50) for 5 min at 4 °C with slow acceleration and deceleration centrifugation rates (*see* **Note 7**). It is critical to control the centrifugation timing, as variations will affect the homogeneity and purity of the different fractions (*see* **Note 7**). Fig. 3 shows that centrifugation of the discontinuous Percoll gradient generates five major fractions (F1–F5) as evidenced by electron microscopy (Fig. 4) analyses of each of the fractions previously described [20]: F1 (above the 3% (v/v) Percoll layer) contains small unidentified membranous material; F2 (between 3% (v/v) and 10% (v/v) Percoll layers) mainly has myelin, membranes, and membrane vesicles; and F3 (in between 10% (v/v) and 15% (v/v) Percoll layers) has synaptosomes and some membrane vesicles. The synaptosomes are primarily enriched in F4 fraction (between 15% (v/v) and 23% (v/v) Percoll layers), and F5 (underneath the 23% (v/v) Percoll layer) contains extra-synaptosomal mitochondria.
8. Harvest the F3 and F4 synaptosome layers (about 2 mL from each tube) to an ice-cold glass beaker using a pipettor with attached needle tip. We typically pool these two fractions, although they can be individually investigated if desired [58]. For the (slightly) highest purity synaptosomes possible, F4 can be collected alone. Dilute the total harvested F3 and F4 synaptosome layers with ice-cold homogenization buffer to a total volume of 80 mL, i.e., the Percoll/synaptosome layers will be diluted at least fourfold. This wash step is to maximally remove the Percoll solution as well as to recover the synaptosomes.
9. Split the mixture into 2 × 50 mL polycarbonate centrifuge tubes. Then centrifuge at 31,400 × *g* (16,112 rpm on rotor

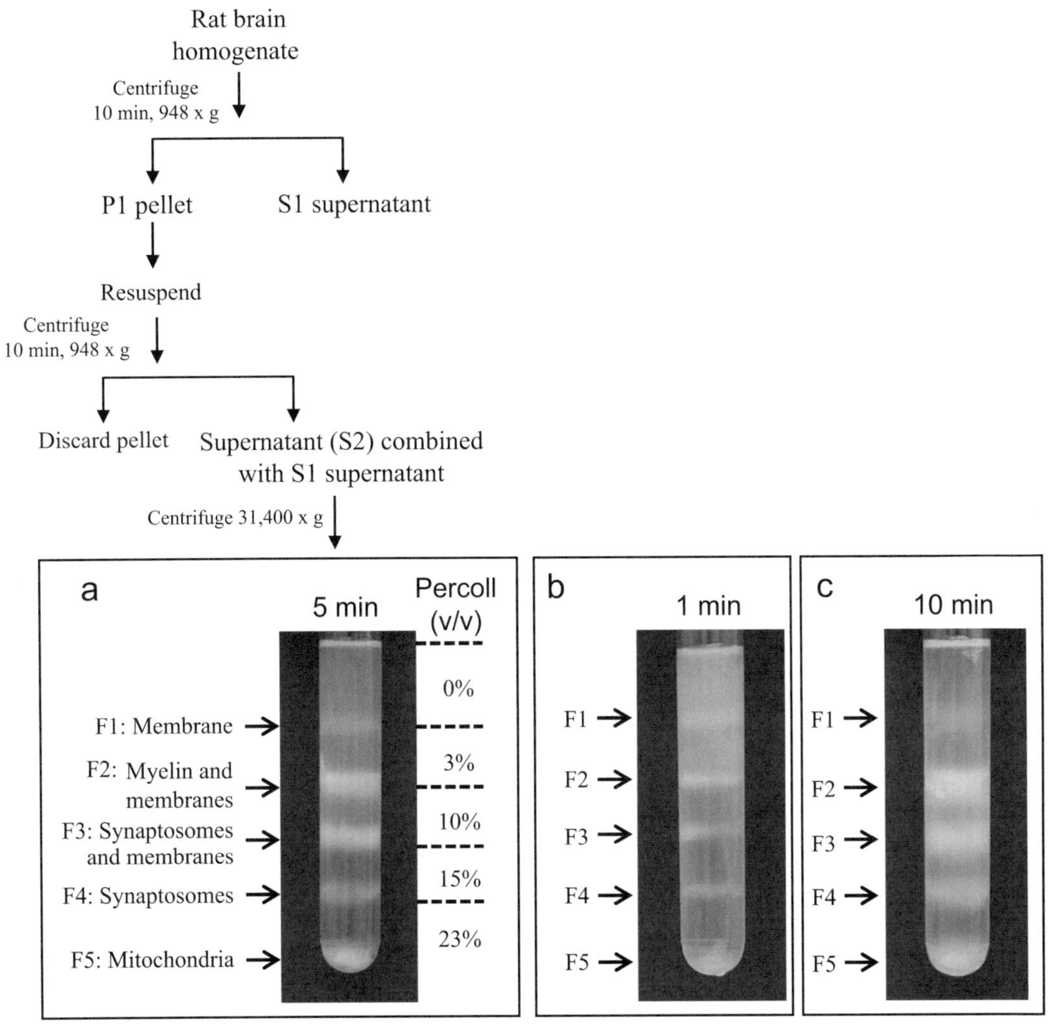

Fig. 3 The discontinuous Dunkley S1 Percoll gradient procedure. A rat brain homogenate is fractionated via centrifugation. The S1 and S2 fractions are applied to a discontinuous Percoll gradient for further fractionation and purification of synaptosomes. Five fractions F1–F5 are generated after a carefully timed centrifugation step. (**a**) 5-min centrifugation time at 31,400 × *g* is the optimal time to obtain uniform fractionated layers. Variations in centrifugation times at 1 min (**b**) or 10 min (**c**) show visible differences in the fractionated layers and thus purity of the isolated synaptosomes

JA25.50) for 30 min at 4°C with a slow deceleration brake centrifuge setting.

10. Remove the supernatant and combine the pellets into a single new centrifuge tube (*see* **Note 8**).
11. Resuspend the synaptosome pellet in 20 mL of ice-cold HBK without Ca^{2+} and centrifuge at 31,400 × *g* for 10 min at 4 °C for buffer exchange (*see* **Note 5**).

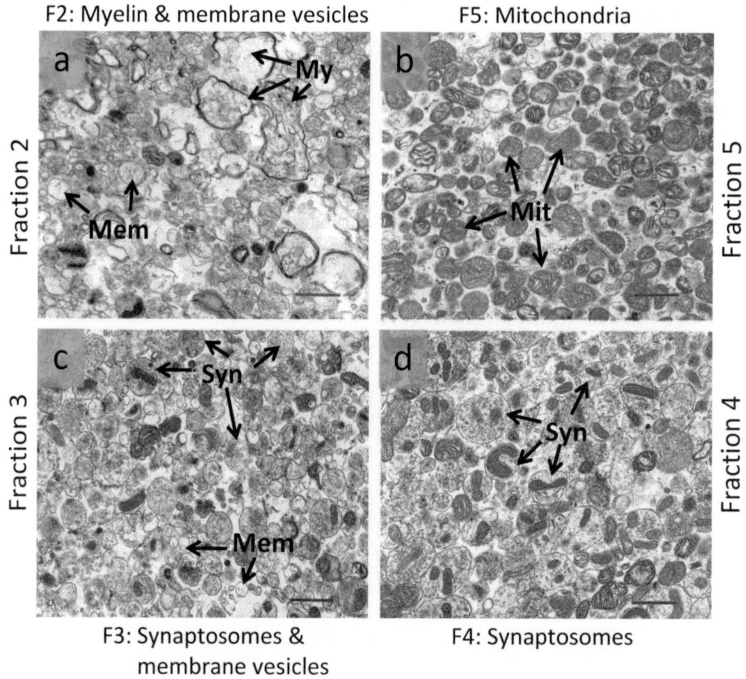

Fig. 4 Electron micrograph gallery of rat brain striatum synaptosomes collected from F2 to F5. (**a**) F2 contains myelin (My), membranes (Mem), membranous vesicles, and rare synaptosomes. (**b**) F5 contains predominantly extra-synaptic mitochondria (Mit). (**c**) F3 contains mainly sealed synaptosomes (Syn), numerous membrane (Mem), and membranous vesicles. (**d**) F4 contains predominantly large-diameter synaptosomes. Scale bar is 1 μm. (The figure has been adapted from Dunkley et. al. (2008), Nat Protoc, 3(11): 1718-28, Copyright © 2006 Nature Publishing Group)

12. A second buffer exchange step is optional by resuspending the pellets in 5–10 mL HBK without Ca^{2+} and recentrifuge in a 10 mL polycarbonate tube at 31,400 × *g* for 10 min at 4 °C.
13. For intact synaptosome studies the combined F3 and F4 pellets are finally resuspended in about 1.2 mL HBK (*see* **Note 9**). The samples are stored on ice until required for analysis.
14. In some cases, the synaptosomes are lysed to extract protein or further subcellular organelles. If this step is required, the combined F3 and F4 pellets are resuspended in about 600 μL of 5 mM Tris–HCl pH 7.4 (plus protease inhibitors) and further lysed in different buffers (*see* **Note 10**). Protein concentrations of the lysates can be checked using various protein quantitation methods. Lysates are stored at −80 °C until use.

4 Conclusions

Synaptosomes have become a widely used model system for investigating nerve terminal function. This is because synaptosomes are functionally similar and contain the same complement of subcellular organelles and proteins, cytoplasm, cytoskeleton, synaptic vesicle, and mitochondria as normally found in nerve terminals in situ [2, 4]. Since the initial discovery in the 1950s that synaptosomes can be an isolatable fraction(s), methods for making synaptosomes have evolved at each of the critical steps depending on its application. This chapter has focussed on the most common two preparations of synaptosomes that are widely used in most laboratories. The system has made an enormous contribution over 60 years as a model for the study of nerve endings and synaptic transmission. This is despite some technical limitations like short viability, unsuitability for transfections or knockdowns, difficulty in introducing new or mutant proteins, and their separation from the axon with which they normally interact. Despite this, researchers continue to develop new ways to adapt the model system to study aspects of neurotransmission and synaptic function in different mammalian species and neuronal disease states, using new technologies, including the latest cellular labeling techniques including fluorescent dyes and markers [72, 73], high-throughput drug screening strategies [59], and proteomic strategies [74, 75].

5 Notes

1. Rat species and strain are chosen for international consistency. Sprague-Dawley and Wistar rats are typically used. These are outbred strains of albino rats. The same procedures outlined in this chapter performed in rats can be performed in brains of any other mammalian species like mice. The volumes of buffers should be scaled to the amount of brain tissue. *One adult rat brain (8–16 weeks old) is about 1 g wet tissue weight.* Male animals are typically preferred because they have more consistent brain chemistry with no cyclic changes to their reproductive hormones. Often this does not matter for the process being measured. A few studies have reported sex differences in functional studies with synaptosomes that need to be considered in the context of experimental goals [76–78].
2. Many different procedures have been used to sacrifice the animals prior to the preparation of synaptosomes; for example, De Robertis used both anesthetics and decapitation [36]. We exclusively employ physical methods of euthanasia without sedation or tranquilization because chemical methods invalidate molecular studies of brain function by poisoning energy metabolism, ion channels, and/or synaptic transmission [79–81]. Alternative methods of sacrifice involving anesthesia or anoxia (e.g., CO_2) result in severe anoxia or carryover of the drug into the isolated tissue, particularly if the goal of the

study is to examine neurochemistry of the brain. Since all anesthetics are designed to stop synaptic transmission, which is the biological process we are investigating, no such sacrifice methods are possible. We immobilize and restrain the adult rats (larger size >150 g) in the plastic film tubes commercially called DecapiCone® (Braintree Scientific Inc., MA) and then euthanize by decapitation. Note that there are specific guidelines for each country that must be consulted and which are available on specific websites: for example the Guidelines on Euthanasia (2007) USA, from the American Veterinary Medical Association; or the Euthanasia of Animals Used for Scientific Purposes (ed. J. Reilly) 2nd edition (2001) (ISBN 0 9586821 4 3) from Australian and New Zealand Council for the Care of Animals in Research & Teaching.

3. The procedure described here is optimized for fresh mammalian whole-brain tissue which includes the cerebral cortex and cerebellum. Various brain regions, including cerebral cortex [53], hippocampus [17], striatum [82], cerebellum [83], and brain stem [37], have been used individually for isolation of synaptosome. Fresh tissues are strongly recommended, although frozen brain tissues have been used with specific modified conditions [84–86].
4. The brain must be rapidly removed, cooled to 0–4 °C by immersion in the pre-cooled homogenization buffer as soon as possible and all subsequent steps maintain the tissue at this temperature. Do not use saline-based buffers to wash the brain because the nerve terminals coacervate and will clump and will not fractionate after homogenization in 0.32 M sucrose. If using more than one animal, sacrifice and homogenize each separately and pool them together later if required. In other words, do not delay the time between animal sacrifice and the first homogenization step.
5. All HBK buffers must be saturated with O_2 before use by bubbling with carbogen 95% O_2 with 5% CO_2 for 1 h on ice using a sintered glass gas sparger. The presence or absence of calcium in the buffer at this point is determined only by its later role in experimental design. Omitting it at this stage provides flexibility in how much is used later in the biological experiments. Its omission may not be required.
6. The discontinuous Percoll gradients can be made the same day as the experiment or late on the day before the procedure and stored covered and undisturbed in a fridge overnight. Multiples of eight gradients will fill the JA25.50 rotor. *When making the discontinuous Percoll gradients with any of the three methods, extreme care and a slow and steady delivery of each gradient step must be taken to maintain good layers. The interface between gradient steps in the final tube should be seen as a clean sharp Schlieren line, visible for a short time when holding the tube up to the light. The line fades after*

approximately 20 min. Amongst the three methods, using the peristaltic pump is slower but provides more reliable gradient, using a manual pipet aid is fast but there is a greater chance of disrupting the gradient layers, whereas *the manual pipettor with the aid of a cork is the cheapest and fastest and has the advantage of less disturbance of the interface between layers.*

7. It is important to only start the timing of "5-min" centrifugation exactly when the rotor reaches the set centrifugation speed of 31,400 × *g*. The time at the maximum speed MUST be exactly 5 min. This is critical to the formation of the five major fractions on the discontinuous Percoll gradients, since this is not an equilibrium centrifugation (Fig. 3). At 5-min centrifugation time, the whitish layers of each fraction are evenly formed and separated (Fig. 3a). At shorter centrifugation time of 1 min (Fig. 3b), there is less material in each fraction evident, with less whitish material at each layer as the sample largely remains on the top of the 3% (v/v) Percoll gradient layer. This suggests that not enough material has been fractionated. Meanwhile, longer centrifugation time for 10 min (Fig. 3c) results in more material sedimenting in each fraction layer with the larger bands of whitish material at each layer and with a clear sample layer left on the top of 3% (v/v) gradient layer. This suggests that materials are merged between each of the fraction layers, potentially contaminating between fractions. Since centrifuges can vary in acceleration times, we suggest that trial runs are performed in 0.5-min steps around 5 min and the resultant separation visually compared with Fig. 3 to optimize conditions in your own laboratory. Similarly, use of a brake to slow the deceleration rate at the end of the cycle is critical to prevent mixing the layers when the centrifuge abruptly stops.
8. Be careful when the synaptosomes are collected and do not carry over any sedimented Percoll which occasionally forms a shiny layer under the synaptosomes (if the Percoll has clumped or has not been properly filtered).
9. The correct Ca^{2+} concentration in the HBK buffer needs to be selected for the appropriate experimental requirements. We use 0.1 mM Ca^{2+} for most experimental purposes, and in all experiments involving $^{32}P_i$ radioactive labeling. For endocytosis, exocytosis or other functional assays you may use 1.2 mM Ca^{2+} analogous to most physiological nerve cell conditions.
10. The composition of the lysis buffer is determined by the purpose of the experiments. For total synaptosome lysate we lyse with 5 mM Tris (pH 7.4), containing 150 mM NaCl, 1 mM EDTA, 1 mM EGTA, 1% Triton X-100, 1 mM phenylmethylsulfonyl fluoride (PMSF), 40 μg/mL leupeptin, complete protease — EDTA-free inhibitor tablets (Roche Life Sciences, Basel, Switzerland), and Phosphatase Inhibitor Cocktail (Merck-Millipore), for immunoprecipitation or pull-down experiments.

Acknowledgments

Peter Dunkley is thanked for unwavering support and assistance in these protocol designs. Lasse Bak, University of Copenhagen, Copenhagen, Denmark, is thanked for the idea of using cork to distribute the Percoll layers. Irma Villaflor, Chief Veterinary Officer and Head of BioResources at Children's Medical Research Institute in Sydney, Australia, is thanked for introducing us to the DecapiCone® and developing its application to avoid manual animal stunning. We wish to thank Emma Kettle and the Electron Microscope Facility, Westmead Hospital (Westmead, New South Wales, Australia), for technical assistance in obtaining the electron microscope images.

References

1. Whittaker VP, Michaelson IA, Kirkland RJ (1964) The separation of synaptic vesicles from nerve-ending particles ('synaptosomes'). Biochem J 90:293–303
2. Whittaker VP, Gray EG (1962) The synapse: biology and morphology. Br Med Bull 18:223–228
3. Gray EG, Whittaker VP (1962) Isolation of nerve endings from brain: an electron microscopic study of the cell fragments derived by homogenization and centrifugation. J Anat 96:79–88
4. Whittaker VP (1993) Thirty years of synaptosome research. J Neurocytol 22:735–742
5. van Spronsen M, Hoogenraad CC (2010) Synapse pathology in psychiatric and neurologic disease. Curr Neurol Neurosci Rep 10:207–214
6. Stephan KE, Baldeweg T, Friston KJ (2006) Synaptic plasticity and dysconnection in schizophrenia. Biol Psychiatry 59:929–939
7. Balu DT, Coyle JT (2011) Neuroplasticity signaling pathways linked to the pathophysiology of schizophrenia. Neurosci Biobehav Rev 35:848–870
8. Crabtree GW, Gogos JA (2014) Synaptic plasticity, neural circuits, and the emerging role of altered short-term information processing in schizophrenia. Front Synaptic Neurosci 6:28
9. Sudhof TC (2008) Neuroligins and neurexins link synaptic function to cognitive disease. Nature 455:903–911
10. Nelson SB, Valakh V (2015) Excitatory/inhibitory balance and circuit homeostasis in autism spectrum disorders. Neuron 87:684–698
11. Pfeiffer BE, Huber KM (2009) The state of synapses in fragile X syndrome. Neuroscientist 15:549–567
12. Soto D, Altafaj X, Sindreu C, Bayes A (2014) Glutamate receptor mutations in psychiatric and neurodevelopmental disorders. Commun Integr Biol 7:e27887
13. Selkoe DJ (2002) Alzheimer's disease is a synaptic failure. Science 298:789–791
14. Calo L, Wegrzynowicz M, Santivanez-Perez J, Grazia SM (2016) Synaptic failure and alpha-synuclein. Mov Disord 31:169–177
15. De Robertis E, Pelleggrino de Iraldi DE, Rodriguez GC, Gomez CJ (1961) On the isolation of nerve endings and synaptic vesicles. J Biophys Biochem Cytol 9:229–235
16. Dunkley PR, Jarvie PE, Heath JW, Kidd GJ, Rostas JA (1986) A rapid method for isolation of synaptosomes on Percoll gradients. Brain Res 372:115–129
17. Thorne B, Wonnacott S, Dunkley PR (1991) Isolation of hippocampal synaptosomes on Percoll gradients: cholinergic markers and ligand binding sites. J Neurochem 56:479–484
18. Hebb CO, Whittaker VP (1958) Intracellular distributions of acetylcholine and choline acetylase. J Physiol 142:187–196
19. Clementi F, Whittaker VP, Sheridan MN (1966) The yield of synaptosomes from the cerebral cortex of Guinea pigs estimated by a polystyrene bead "tagging" procedure. Z Zellforsch Mikrosk Anat 72:126–138
20. Dunkley PR, Jarvie PA, Robinson PJ (2008) A rapid percoll gradient procedure for preparation of synaptosomes. Nat Protoc 3:1718–1728
21. Sokolow S, Henkins KM, Williams IA, Vinters HV, Schmid I, Cole GM, Gylys KH (2012) Isolation of synaptic terminals from Alzheimer's disease cortex. Cytometry A 81:248–254
22. Quan A, McGeachie AB, Keating DJ, van Dam EM, Rusak J, Chau N, Malladi CS, Chen

C, McCluskey A, Cousin MA, Robinson PJ (2007) MiTMAB is a surface-active dynamin inhibitor that blocks endocytosis mediated by dynamin I or dynamin II. Mol Pharmacol 72:1425–1439

23. Tan TC, Valova VA, Malladi CS, Graham ME, Berven LA, Jupp OJ, Hansra G, McClure SJ, Sarcevic B, Boadle RA, Larsen MR, Cousin MA, Robinson PJ (2003) Cdk5 is essential for synaptic vesicle endocytosis. Nat Cell Biol 5:701–710
24. Polosa PL, Attardi G (1991) Distinctive pattern and translational control of mitochondrial protein synthesis in rat brain synaptic endings. J Biol Chem 266:10011–10017
25. Dunkley PR, Robinson PJ (1986) Depolarization-dependent protein phosphorylation in synaptosomes: mechanisms and significance. Prog Brain Res 69:273–293
26. Ashton AC, Ushkaryov YA (2005) Properties of synaptic vesicle pools in mature central nerve terminals. J Biol Chem 280:37278–37288
27. Morgan IG (1976) Synaptosomes and cell separation. Neuroscience 1:159–165
28. Erecinska M, Nelson D, Silver IA (1996) Metabolic and energetic properties of isolated nerve ending particles (synaptosomes). Biochim Biophys Acta 1277:13–34
29. Nicholls DG (2003) Bioenergetics and transmitter release in the isolated nerve terminal. Neurochem Res 28:1433–1441
30. Raiteri L, Raiteri M (2000) Synaptosomes still viable after 25 years of superfusion. Neurochem Res 25:1265–1274
31. Ghijsen WE, Leenders AG, Lopes Da Silva FH (2003) Regulation of vesicle traffic and neurotransmitter release in isolated nerve terminals. Neurochem Res 28:1443–1452
32. Verity MA, Brown WJ, Cheung M (1980) Isolation of ribosome containing synaptosome subpopulation with active in vitro protein synthesis. J Neurosci Res 5:143–153
33. Loguercio PP, Attardi G (1996) Mitochondrial protein synthesis in rat brain synaptosomes. Methods Enzymol 264:211–217
34. Eyman M, Cefaliello C, Ferrara E, De SR, Crispino M, Giuditta A (2007) Synaptosomal protein synthesis is selectively modulated by learning. Brain Res 1132:148–157
35. Gioio AE, Eyman M, Zhang H, Lavina ZS, Giuditta A, Kaplan BB (2001) Local synthesis of nuclear-encoded mitochondrial proteins in the presynaptic nerve terminal. J Neurosci Res 64:447–453
36. De Robertis E, Pellegrino de Iraldi A, Rodriquez de Lores Arniaz G, Salganicoff L (1962) Cholinergic and noncholinergic nerve endings in rat brain. I. Isolation and subcellular distribution of acetylcholine and acetylcholinesterase. J Neurochem 9:23–35
37. Tamir H, Rapport MM, Roizin L, Huang YL, Liu JC (1974) Preparation of synaptosomes and vesicles with sodium diatrizoate. J Neurochem 23:943–949
38. Abdel-Latif AA (1966) A simple method for isolation of nerve-ending particles from rat brain. Biochim Biophys Acta 121:403–406
39. Kurokawa M, Sakamoto T, Kato M (1965) A rapid isolation of nerve-ending particles from brain. Biochim Biophys Acta 94:307–309
40. Hayashi M, Raimondi A, O'Toole E, Paradise S, Collesi C, Cremona O, Ferguson SM, De Camilli P (2008) Cell- and stimulus-dependent heterogeneity of synaptic vesicle endocytic recycling mechanisms revealed by studies of dynamin 1-null neurons. Proc Natl Acad Sci U S A 105:2175–2180
41. Maurer UE, Sodeik B, Grunewald K (2008) Native 3D intermediates of membrane fusion in herpes simplex virus 1 entry. Proc Natl Acad Sci U S A 105:10559–10564
42. Fernandez-Busnadiego R, Zuber B, Maurer UE, Cyrklaff M, Baumeister W, Lucic V (2010) Quantitative analysis of the native presynaptic cytomatrix by cryoelectron tomography. J Cell Biol 188:145–156
43. Wilhelm BG, Mandad S, Truckenbrodt S, Krohnert K, Schafer C, Rammner B, Koo SJ, Classen GA, Krauss M, Haucke V, Urlaub H, Rizzoli SO (2014) Composition of isolated synaptic boutons reveals the amounts of vesicle trafficking proteins. Science 344:1023–1028
44. Whittaker VP (1968) The morphology of fractions of rat forebrain synaptosomes separated on continuous sucrose density gradients. Biochem J 106:412–417
45. Chan KY, Bunt AH (1978) An association between mitochondria and microtubules in synaptosomes and axon terminals of cerebral cortex. J Neurocytol 7:137–143
46. Gray EG, Burgoyne RD, Westrum LE, Cumming R, Barron J (1982) The enigma of microtubule coils in brain synaptosomes. Proc R Soc Lond B Biol Sci 216:385–396
47. Burgoyne RD, Cumming R (1983) Taxol stabilizes synaptosomal microtubules without inhibiting acetylcholine release. Brain Res 280:190–193
48. Kadota K, Kadota T (1973) Isolation of coated vesicles, plain synaptic vesicles, and flocculent material from a crude synaptosome fraction of Guinea pig whole brain. J Cell Biol 58:135–151
49. Leenders AG, Scholten G, de Lange RP, Lopes Da Silva FH, Ghijsen WE (2002) Sequential changes in synaptic vesicle pools

and endosome-like organelles during depolarization near the active zone of central nerve terminals. Neuroscience 109:195–206

50. Wenzel EM, Morton A, Ebert K, Welzel O, Kornhuber J, Cousin MA, Groemer TW (2012) Key physiological parameters dictate triggering of activity-dependent bulk endocytosis in hippocampal synapses. PLoS One 7:e38188

51. Gonatas NK, utilio-Gambetti L, Gambetti P, Shafer B (1971) Morphological and biochemical changes in rat synaptosome fractions during neonatal development. J Cell Biol 51: 484–498

52. Kanerva L, Hervonen A, Tissari AH (1974) Ultrastructure of synaptosomes from midterm human fetal brain. Acta Physiol Scand 92:286–288

53. Dunkley PR, Heath JW, Harrison SM, Jarvie PE, Glenfield PJ, Rostas JA (1988) A rapid Percoll gradient procedure for isolation of synaptosomes directly from an S1 fraction: homogeneity and morphology of subcellular fractions. Brain Res 441:59–71

54. Robinson PJ, Jarvie PA, Dunkley PR, Jarvie PE (1984) Depolarisation-dependent protein phosphorylation in rat cortical synaptosomes is inhibited by fluphenazine at a step after calcium entry. J Neurochem 43:659–667

55. Robinson PJ, Lovenberg W (1986) Dopamine and serotonin in two populations of synaptosomes isolated by percoll gradient centrifugation. Neurochem Int 9: 455–458

56. Michaelis ML, Jiang L, Michaelis EK (2017) Isolation of synaptosomes, synaptic plasma membranes, and synaptic junctional complexes. Methods Mol Biol 1538:107–119

57. Dodd PR, Hardy JA, Oakley AE, Edwardson JA, Perry EK, Delaunoy JP (1981) A rapid method for preparing synaptosomes: comparison, with alternative procedures. Brain Res 226:107–118

58. Robinson PJ, Gehlert DR, Sanna E, Hanbauer I (1989) Two fractions enriched for striatal synaptosomes isolated by percoll gradient centrifugation: synaptosome morphology, dopamine and serotonin receptor distribution, and adenylate cyclase activity. Neurochem Int 15:339–348

59. Daniel JA, Malladi CS, Kettle E, McCluskey A, Robinson PJ (2012) Analysis of synaptic vesicle endocytosis in synaptosomes by high content screening. Nat Protoc 7:1439–1455

60. Johansen L, Roberg B, Kvamme E (1987) Uptake and release for glutamine and glutamate in a crude synaptosomal fraction from rat brain. Neurochem Res 12:135–140

61. Robinson PJ, Sontag J-M, Liu JP, Fykse EM, Slaughter C, McMahon HT, Südhof TC (1993) Dynamin GTPase regulated by protein kinase C phosphorylation in nerve terminals. Nature 365:163–166

62. Anggono V, Smillie KJ, Graham ME, Valova VA, Cousin MA, Robinson PJ (2006) Syndapin I is the phosphorylation-regulated dynamin I partner in synaptic vesicle endocytosis. Nat Neurosci 9:752–760

63. Gylys KH, Bilousova T (2017) Flow cytometry analysis and quantitative characterization of tau in synaptosomes from Alzheimer's disease brains. Methods Mol Biol 1523:273–284

64. Xue J, Graham ME, Novelle AE, Sue N, Gray N, McNiven MA, Smillie KJ, Cousin MA, Robinson PJ (2011) Calcineurin selectively docks with the dynamin Ixb splice variant to regulate activity-dependent bulk endocytosis. J Biol Chem 286:30295–30303

65. Dunkley PR, Baker CM, Robinson PJ (1986) Depolarization-dependent protein phosphorylation in rat cortical synaptosomes: characterization of active protein kinases by phosphopeptide analysis of substrates. J Neurochem 46:1692–1703

66. Robinson PJ (1991) Dephosphin, a 96,000 Dalton substrate of protein kinase C in synaptosomal cytosol is phosphorylated in intact synaptosomes. FEBS Lett 282:388–392

67. Graham ME, Anggono V, Bache N, Larsen MR, Craft GE, Robinson PJ (2007) The in vivo phosphorylation sites of rat brain dynamin I. J Biol Chem 282:14695–14707

68. Nagy A, Delgado-Escueta AV (1984) Rapid preparation of synaptosomes from mammalian brain using nontoxic isoosmotic gradient material (Percoll). J Neurochem 43:1114–1123

69. Nichols RA, Chilcote TJ, Czernik AJ, Greengard P (1992) Synapsin I regulates glutamate release from rat brain synaptosomes. J Neurochem 58:783–785

70. Cousin MA, Robinson PJ (2000) Ca(2+) influx inhibits dynamin and arrests synaptic vesicle endocytosis at the active zone. J Neurosci 20:949–957

71. Cousin MA, Tan TC, Robinson PJ (2001) Protein phosphorylation is required for endocytosis in nerve terminals. Potential role for the dephosphins dynamin I and synaptojanin, but not AP180 or amphiphysin. J Neurochem 76:105–116

72. Budzinski KL, Sgro AE, Fujimoto BS, Gadd JC, Shuart NG, Gonen T, Bajjalieh SM, Chiu

DT (2011) Synaptosomes as a platform for loading nanoparticles into synaptic vesicles. ACS Chem Neurosci 2:236–241

73. Luquet E, Biesemann C, Munier A, Herzog E (2017) Purification of synaptosome populations using fluorescence-activated synaptosome sorting. Methods Mol Biol 1538:121–134
74. Mallei A, Failler M, Corna S, Racagni G, Mathe AA, Popoli M (2014) Synaptoproteomic analysis of a rat gene-environment model of depression reveals involvement of energy metabolism and cellular remodeling pathways. Int J Neuropsychopharmacol 18:1–21
75. Bosch PJ, Peng L, Kivell BM (2015) Proteomics analysis of dorsal striatum reveals changes in synaptosomal proteins following methamphetamine self-administration in rats. PLoS One 10:e0139829
76. Oztas B, Kocak H, Oner P, Kucuk M (2000) Sex-dependent changes in blood-brain barrier permeability and brain NA(+),K(+) ATPase activity in rats following acute water intoxication. J Neurosci Res 62:750–753
77. Mitrovic AD, Maddison JE, Johnston GA (1999) Influence of the oestrous cycle on L-glutamate and L-aspartate transport in rat brain synaptosomes. Neurochem Int 34:101–108
78. Miyasaka K, Kanai S, Ohta M, Funakoshi A (1997) Aging impairs release of central and peripheral cholecystokinin (CCK) in male but not in female rats. J Gerontol A Biol Sci Med Sci 52:M14–M18
79. Bodzenta A, Moniuszko-Jakoniuk J, Wisniewski K (1979) The effect of kallikrein on central effects of acetylcholine in rats receiving indomethacin and prostaglandin E1. Pol J Pharmacol Pharm 31:595–603
80. Wisniewski K, Bodzenta A (1975) Kinins and central effects of the acetylcholine. Acta Neurobiol Exp (Wars) 35:85–92
81. Braszko J, Koscielak M (1975) Effect of kinins on the central action of serotonin. Pol J Pharmacol Pharm 27:61–68
82. De Belleroche JA, Bradford HF (1975) The release of endogenous 3,4-dihydroxyphenethylamine from synaptosomes isolated from corpus striatum. Biochem Soc Trans 3:99–101
83. Clark M, Dar MS (1989) Release of endogenous glutamate from rat cerebellar synaptosomes: interactions with adenosine and ethanol. Life Sci 44:1625–1635
84. Whittaker VP, Greengard P (1971) The isolation of synaptosomes from the brain of a teleost fish, Centriopristes striatus. J Neurochem 18:173–176
85. Dodd PR, Watson WE, Morrison MM, Johnston GA, Bird ED, Cowburn RF, Hardy JA (1989) Uptake of gamma-aminobutyric acid and L-glutamic acid by synaptosomes from postmortem human cerebral cortex: multiple sites, sodium dependence and effect of tissue preparation. Brain Res 490:320–331
86. Dodd PR, Hardy JA, Baig EB, Kidd AM, Bird ED, Watson WE, Johnston GA (1986) Optimization of freezing, storage, and thawing conditions for the preparation of metabolically active synaptosomes from frozen rat and human brain. Neurochem Pathol 4:177–198
87. Hill TA, Odell LR, Edwards JK, Graham ME, McGeachie AB, Rusak J, Quan A, Abagyan R, Scott JL, Robinson PJ, McCluskey A (2005) Small molecule inhibitors of dynamin I GTPase activity: development of dimeric tyrphostins. J Med Chem 48:7781–7788

Part IV

Studying Synaptic Energy and Metabolism Using Synaptosomes

Chapter 7

Synaptosomal Bioenergetics and Glutamate Release

David G. Nicholls

Abstract

The isolated nerve terminal ("synaptosome") preparation combines the metabolic and bioenergetic functions of an intact neural cell with the complete machinery for the synthesis, exocytosis, and reuptake of neurotransmitters, and thus represents a "half-way house" between protein chemistry and electrophysiological approaches, combining the advantages, or disadvantages, of both approaches, depending on one's point of view. Transmitter release is an energetically demanding process, and techniques to monitor in situ mitochondrial function are reviewed. Two key techniques for studying exocytosis, the continuous fluorometric assay of released glutamate and the use of potassium channel inhibitors to induce spontaneous action potentials, are discussed, together with the application of these techniques to investigate the mode of action of presynaptic neurotoxins and G protein-coupled receptors.

Key words Synaptosomes, Mitochondria, Glutamate, 4-Aminopyridine, Calcium, Membrane potentials

1 Introduction: The Advantages and Limitations of Synaptosomal Preparations

The synaptosomal preparation is enriched in isolated, resealed, presynaptic nerve terminals, many of which retain their in vivo function. In addition to pathways for the synthesis, exocytosis, and reuptake of specific neurotransmitters, functional synaptosomes utilize glucose, via glycolysis and oxidative phosphorylation, to generate a high cytosolic ATP phosphorylation potential supplying, among other processes, plasma membrane Na^+/K^+- and Ca^{2+}-ATPases to maintain a polarized plasma membrane potential ($\Delta\psi_p$) and low cytosolic free calcium concentration ($[Ca^{2+}]_c$). Synaptosomes can thus be viewed as autonomous secretory "mini cells." Indeed, the initial synaptosomal investigations by our group were centered on the investigation of mitochondrial bioenergetics in an intact cellular environment, rather than any aspect of neurotransmitter release [1].

A major advantage of the preparation is the ability to prepare synaptosomes from animals of any age, free from the neonatal limitation inherent in primary cell culture, while their size and

Kathryn M. Murphy (ed.), *Synaptosomes*, Neuromethods, vol. 141,
https://doi.org/10.1007/978-1-4939-8739-9_7,

robustness allow them to be handled similarly to isolated mitochondria. On the other hand, the marked heterogeneity of the preparation, in terms of morphology, function, and neurotransmitter content, is problematical. For example, analysis at single-particle level of a mouse cerebrocortical Percoll gradient synaptosomal preparation labelled with tetramethylrhodamine methyl ester (TMRM) to monitor mitochondrial polarization, fura-4F (cytosolic free Ca^{2+}, $[Ca^{2+}]_c$), and Mitotracker Green (mitochondrial marker) revealed only about 66% of the fluorescent objects to possess polarized mitochondria and low $[Ca^{2+}]_c$, presumably reflecting functional, respiring, synaptosomes [2].

Transmitter heterogeneity depends on brain region. Classically, release of a specific transmitter was studied by prelabelling synaptosomes with radiolabelled precursor, but the assumption of full equilibration with the vesicular pool was not always met. A final problem is the choice of a plasma membrane depolarization protocol to initiate exocytosis that retains some of the key features of the in vivo axonal input. As I shall describe, the classical use of elevated KCl, or veratridine to prevent Na^+ channel inactivation, has serious shortcomings, notably in the clamping of $\Delta\psi_p$.

The focus of our research has been to develop and apply improved techniques to investigate synaptosomal bioenergetics, to depolarize synaptosomes, and to monitor the release of the major excitatory neurotransmitter, glutamate. Generic techniques, applied from other systems, are mentioned briefly, and the chapter is devoted to an overview of the novel methodologies we have developed and their application to investigate the control of presynaptic bioenergetic status and glutamate exocytosis. It should be appreciated that the majority of the techniques described were developed some 20 years ago, and that, while the principles remain valid, some details such as the source of reagents may have evolved. Obsolete techniques will only be mentioned in passing.

2 Synaptosomal Bioenergetics

2.1 Glycolysis

Synaptosomes have a highly aerobic metabolism, and favor oxidative phosphorylation over anaerobic glycolysis. However, when the latter source of ATP is inhibited by addition of protonophore [3] or chemical anoxia [4] the rate of glycolysis can increase more than tenfold in compensation. Glycolytic flux can be quantified by monitoring the exchange of tritium from D-[3-^{3}H]-glucose into H_2O at the level of the aldolase reaction.

Synaptosomes (4 mg protein/mL) are incubated in the presence of 5 mM glucose. D-[3-^{3}H]glucose is then added to give the final specific activity of about 0.4 Ci/mol. At defined times, 150 μL samples are mixed with 20 μL ice-cold perchloric acid to give a final concentration of 0.15 M. After centrifugation, ^3H_2O is determined by sublimation under vacuum and collection in a liquid nitrogen trap. For details see [3].

The Seahorse XF Extracellular Flux Analyzer detects extracellular acidification in addition to oxygen uptake. While the rate of acidification is sometimes taken to reflect the glycolytic production of lactate, it is important to correct for the effect of CO_2 production by metabolism. A protocol taking account of this correction, allowing glycolytic lactate production to be quantified, is contained in references [5, 6]. Further analysis of raw Seahorse data allows the absolute rates of glycolytic and oxidative ATP generation to be quantified [7].

2.2 Synaptosomal Respiration

As with more conventional "cells," analysis of synaptosomal respiration can provide a detailed picture of the bioenergetic status of the preparation [1, 8, 9]. Initial studies utilized conventional Clark-type oxygen electrode chambers [1, 8]. While this has yielded invaluable information, a major limitation is the amount of material required, bearing in mind that the in situ mitochondria account for only about 10% of the total synaptosomal protein [1]. Even with the use of micro oxygen electrode chambers [1] the number of experiments that can be performed is severely limited.

While the Seahorse Extracellular Flux Analyzer has revolutionized respiratory monitoring of attached cells, it is not suitable for cells or organelles in suspension. Synaptosomes may, however, be robustly attached to conventional Seahorse plates coated with polyethyleneimine (1:15,000 dilution from a 50% solution), by centrifugation [10]. In order to monitor respiration, it is important to optimize the yield of attached synaptosomes and to ensure that the attachment is sufficiently firm to avoid their displacement during the rather vigorous mixing that occurs prior to each determination in the respirometer. Synaptosomes aggregate when suspended in a medium with a significant ionic strength, and this facilitates attachment.

The synaptosomal band from a conventional Percoll gradient is diluted into a medium containing 20 mM HEPES, 10 mM D-glucose, 1.2 mM Na_2HPO_4, 1 mM $MgCl_2$, 5 mM $NaHCO_3$, 5 mM KCl, and 140 mM NaCl, pH 7.4 and centrifuged at 15,000 × g for 15 min at 4 °C to remove Percoll. The pellet is resuspended in the same medium. Synaptosomes (10 μg protein/well) are aliquoted into the wells of a 24-well polyethyleneimine-coated XF24 V7 cell culture plate, which is centrifuged at 3400 × g for 60 min at 4 °C in an A-4-62 rotor in an Eppendorf 5810R centrifuge. The supernatant is replaced with 700 μL of incubation medium (3.5 mM KCl, 120 mM NaCl, 1.3 mM $CaCl_2$, 0.4 mM KH_2PO_4, 1.2 mM Na_2SO_4, 2 mM $MgSO_4$, 15 mM D-glucose, 4 mg/mL bovine serum albumin, 37 °C). Plates are used immediately or stored on ice for not more than 3 h. The cell culture microplate is incubated and loaded into the Seahorse XF24 extracellular flux analyzer following the manufacturer's instructions.

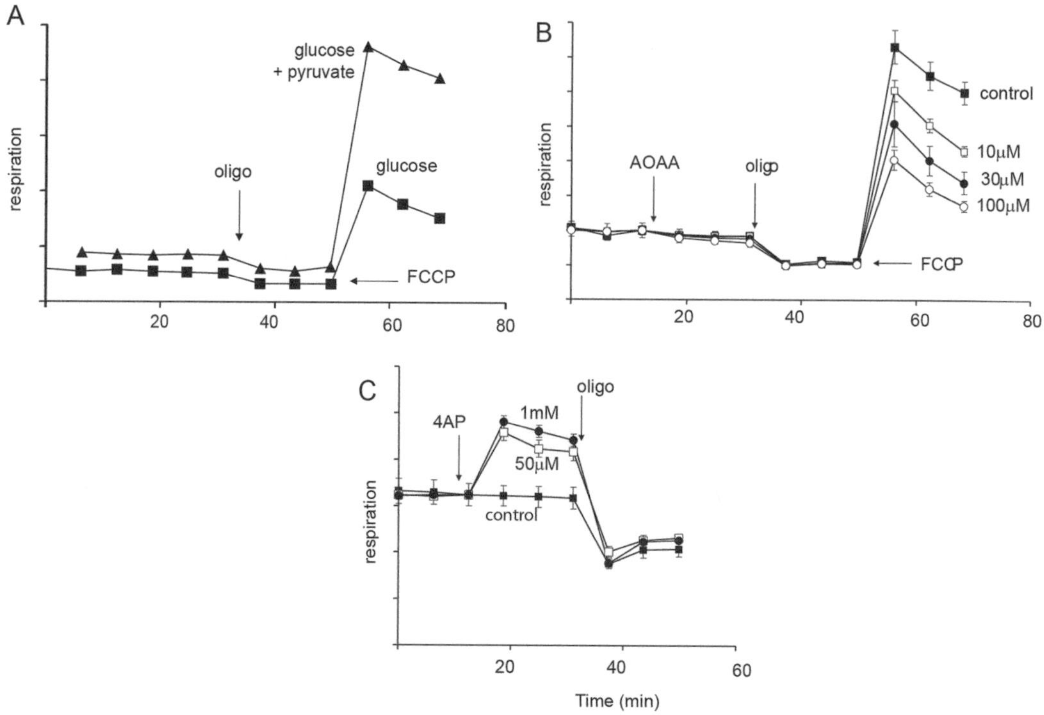

Fig. 1 Representative Seahorse respirometer experiments detailing approaches to synaptosomal function. (**a**) The classic cell respiratory control experiment (sequential additions of oligomycin to inhibit the ATP synthase followed by sufficient FCCP protonophore to just relieve respiratory control) with 10 mM glucose as substrate. The presence of exogenous pyruvate greatly enhances the maximal respiration, indicating that glycolytic pyruvate supply alone can be rate restricting. The higher oligomycin-resistant respiration, indicative of proton leak across the mitochondrial membrane, does not indicate that pyruvate is a weak "uncoupler," since parallel measurements indicate that $\Delta\psi_m$ hyperpolarizes, increasing the proton flux through the endogenous proton leak [3, 10]. (**b**) Aminooxyacetate (AOAA) is a competitive transaminase inhibitor. By inhibiting the malate/aspartate shuttle responsible for reoxidizing glycolytic NADH, AOAA diverts glycolytic pyruvate to lactate dehydrogenase in order to reoxidize NADH, effectively starving the mitochondrion of substrate [39]. However, this restriction is only apparent when downstream constraints are removed by the addition of protonophore. (**c**) The energetic demand of action potential firing. 4AP increases basal respiration, but not that following oligomycin, indicating that ATP demand (ATP-sensitive respiration) has increased. Data from [10]

The high sensitivity of the assay means that an extensive series of experiments can be performed on the synaptosomes obtained from a specific brain region of a mouse, greatly facilitating investigations when material is precious and limited, for example during transgenic manipulations [11, 12]. Examples of the approach are given in Fig. 1.

2.3 Plasma ($\Delta\psi_P$) and Mitochondrial ($\Delta\psi_M$) Membrane Potentials

The potential, $\Delta\psi_m$, across the inner membrane of the in situ mitochondria can be estimated from the accumulation of membrane-permeant, monovalent cations, assuming a Nernst equilibrium. However, in common with conventional cells, a complicating factor is that the probe is accumulated across both the plasma and mitochondrial membranes in response to their respective membrane potentials. Our initial attempt to monitor both potentials in parallel employed radioisotopes: $^{86}Rb^+$ as a potassium surrogate to monitor the potassium equilibrium potential (as an approximation to $\Delta\psi_p$), and [^{3}H]-methyltriphenylphosphonium, which is accumulated sequentially across both membranes in response to $\Delta\psi_p$ and $\Delta\psi_m$ [1]. Making plausible assumptions for the relative matrix volume and activity coefficients, it was estimated that the synaptosomal population maintained a $\Delta\psi_m$ in the region of 150 mV, while experimental manipulations predicted to affect either membrane potential produced the expected results [1]. Isotopic techniques have lost favor, and have been replaced by the use of ion-selective macro electrodes monitoring the decrease in external concentration of tetraphenylphosphonium (TPP^+) cations, or by optical monitoring of membrane-permeant fluorescent cations.

2.3.1 Tetraphenylphosphonium Selective Electrode

The principle of this technique is that an extracellular macro electrode, designed to monitor the concentration of TPP^+ in the incubation medium, will monitor the decrease in concentration as the cation is accumulated in the synaptosomes, sequentially across the plasma and inner mitochondrial membranes. The electrode is surprisingly sensitive, bearing in mind that the synaptosomal volume will be several hundredfold less than that of the surrounding incubation medium. In common with all such permanent cation approaches, the equilibrium accumulation will be equally responsive to changes in $\Delta\psi_p$ and $\Delta\psi_m$, and it is important to design experiments in such a way that one potential can be assumed with reasonable confidence to be relatively invariant during the course of the experiment (Fig. 2).

A TPP^+ electrode may be simply constructed in the laboratory. The following protocol was used in our group. A selectively permeable polyvinyl chloride membrane is formed by allowing a solution containing 0.34 mg of sodium tetraphenylboron, 16 mg of high-molecular-weight polyvinyl chloride, 57 μL of dioctyl phthalate, and tetrahydrofuran to a final volume of 500 μL, to evaporate on a glass plate constrained by a 2 cm diameter glass cylinder. The resulting membrane is glued to the end of a plastic tube equipped with a platinum electrode and filled with 10 mM tetraphenylphosphonium/10 mM NaCl [13]. The electrode is assembled into a thermostatted incubation chamber containing a bridge to a KCl reference electrode. A more detailed recipe for the construction of a TPP^+ electrode may be found in reference [14]. While the electrode was originally prepared for use with isolated mitochondria, it can, of course, be exploited for intact synaptosomes. Fig. 2 shows a proof of principle, where $\Delta\psi_p$ is depolarized by elevated KCl or by ouabain, and $\Delta\psi_m$ by FCCP or oligomycin plus rotenone.

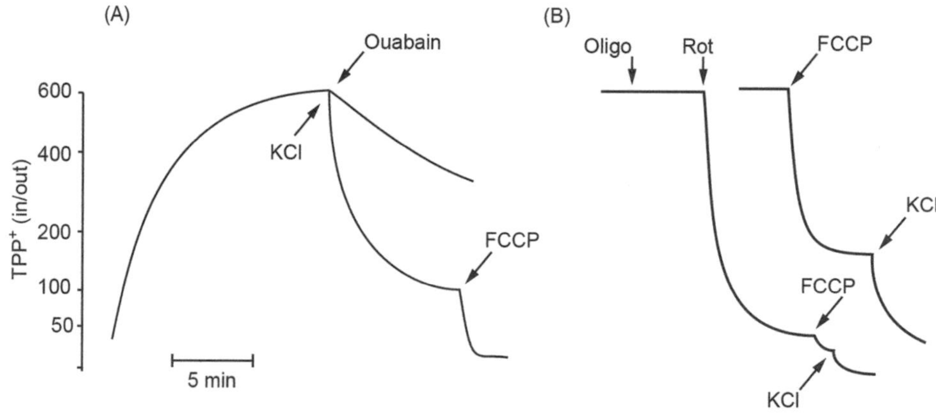

Fig. 2 Use of the TPP+ electrode to monitor changes in synaptosomal $\Delta\psi_p$ and $\Delta\psi_m$. (**a**) Time course of TPP+ accumulation by polarized synaptosomes, followed by $\Delta\psi_p$ depolarization by ouabain or 35 mM KCl, and $\Delta\psi_m$ depolarization by FCCP. (**b**) $\Delta\psi_m$ depolarization by rotenone plus oligomycin and/or FCCP, followed by KCl. Data from [13]

2.4 Fluorescent Membrane-Permeant Cations

Fluorescent membrane-permeant cations such as tetramethylrhodamine methyl ester (TMRM) are widely used in qualitative and semiquantitative estimates of $\Delta\psi_m$ in intact cells; for review see [15]. At low concentrations, <10 nM, the fluorescence is a function of both $\Delta\psi_p$ and $\Delta\psi_m$, meaning that, as in the case of the TPP+ electrode, there is a measure of ambiguity in the interpretation of fluorescence changes. With this proviso, time-dependent changes in $\Delta\psi_m$ have been monitored in fields of individual synaptosomes imaged in a fluorescence microscope after centrifugation onto coverslips [10]. In contrast to respirometry, where aggregation is advantageous to maximize yield, see above, this should be avoided for fluorescent imaging, by suspending synaptosomes in a very-low-ionic-strength medium (320 mM sucrose, 1 mM EDTA, 0.25 mM dithiothreitol, pH 7.4), plating onto Whatman 96-well glass-bottomed multi-well plates, and centrifuging, at 3400 × *g* for 1 h at 4 °C in an A-4-62 rotor in a Eppendorf 5810R centrifuge [10]. When using TMRM as a single indicator at low concentration, to avoid errors due to fluorescence quenching in the mitochondrial matrix, even the most carefully prepared synaptosomal preparations from defined brain areas display considerable heterogeneity in $\Delta\psi_m$ [10].

In many cells, the distribution of fluorescent lipophilic anions across the plasma membrane can be used to monitor $\Delta\psi_p$. We have used a proprietary oxonol from Molecular Devices, which we term PMPI ("plasma membrane potential indicator"), in a variety of cell culture systems [16, 17]. When used in parallel with TMRM, the contribution of $\Delta\psi_p$ to the latter signal can be corrected for, allowing $\Delta\psi_m$ to be monitored either semiquantitatively [16] or quantitatively [18]. However, for reasons that are unclear, the probe does not respond to potential changes in synaptosomes.

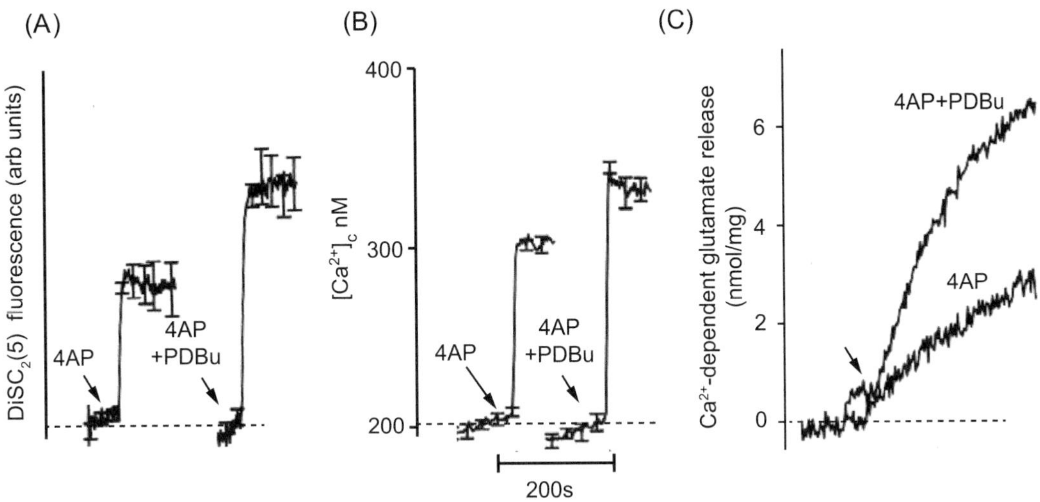

Fig. 3 Multi-parametric analysis of the phorbol ester potentiation of 4AP effects on synaptosomes. Experiments were performed in the absence and presence of 1 μM phorbol dibutyrate (PDBu). (**a**) $\Delta\psi_p$ monitored by $DiSC_2(5)$, upward deflection indicates depolarization. (**b**) $[Ca^{2+}]_c$ monitored with fura-2. (**c**) Net Ca^{2+}-dependent glutamate release. Data from [20]

The cationic probe $DiSC_2(5)$, at much higher concentrations than TMRM (5 μM instead of 10 nM), has been used to report predominantly $\Delta\psi_P$, due to concentration-dependent quenching of the emission signal in the cytosol by dye aggregation [19], and appears to reliably monitor plasma membrane depolarization (Fig. 3a) [20]. Note, however, that plasma membrane depolarization produces an *increase* in signal, due to dequenching in the cytosol, whereas TMRM at low nanomolar concentrations produces a decreased signal.

2.5 Adenine Nucleotides and Phosphocreatine

The maintenance of a polarized plasma membrane and low $[Ca^{2+}]_c$, transmitter uptake into synaptic vesicles, exocytosis itself, and plasma membrane transmitter transport are all ATP-dependent processes. It is therefore important to be able to monitor the terminal adenine nucleotide phosphorylation status. Our group utilized a luciferase-based assay to monitor ATP, ADP, and in some cases also AMP and phosphocreatine [4, 9].

Synaptosomal incubations are deproteinated with an equal volume of 1 M perchloric acid and 50 mM EDTA (sodium salt) at 0 °C and centrifuged. Supernatants are neutralized with 3 M KOH/1.5 M triethanolamine and assayed immediately. Extract (10 μL) is added to 143 μL of 100 mM Tris acetate (pH 7.75), 2 mM EDTA, 5 mM $MgCl_2$, 6 mM dithiothreitol, 2 mM KCl, and 57 μL of reconstituted luciferin reagent (one vial of LKB 1243-200 ATP monitoring reagent reconstituted with 10 mL of water). Bioluminescence generated by ATP is detected with an LKB Luminometer 1250 (LKB Wallac Ltd., Turku, Finland). After the reaction was complete, pyruvate kinase (1 U/mL) and 0.2 mM phosphoenolpyruvate are added to assay ADP. AMP is next assayed by addition of myokinase (50 U/mL) dialyzed against water. Finally, the assay is calibrated by addition of 0.1 μM ATP and 0.1 μM AMP as standards. Creatine phosphate is determined in an identical medium supplemented with 17 μM ADP and 10 μg/mL creatine kinase, using internal standards of creatine phosphate.

3 Depolarization Protocols

3.1 KCl and Veratridine

Early research on neurotransmitter release generally required the neurotransmitter pool to be prelabelled with the appropriate radioisotope, while the loss of physiological axonal stimulation meant that plasma membrane depolarization was induced, either by elevated external potassium or by veratridine to lock voltage-activated sodium channels in the open conformation. Each of these protocols has significant limitations. KCl depolarization relies on the dominant potassium conductance of the excitable plasma membrane, meaning that the plasma membrane potential, $\Delta\psi_p$, will approximate to the potassium diffusion potential. Increasing the external potassium concentration from 4 to 34 mM, for example, changes the diffusion potential from −92 to −34 mV, sufficient to activate a range of voltage-activated calcium channels and hence initiate exocytosis. However, $\Delta\psi_p$ is now clamped at this value, whereas the nerve terminal in situ undergoes complex potential oscillations driven by the axonal input, and modulated by presynaptic ion channels and receptors. In particular, KCl depolarization means that non-desensitizing ion channels will dominate at the expense of the multiplicity of channels that undergo rapid desensitization during this nonphysiological depolarization protocol.

Veratridine acts by inhibiting voltage-activated sodium channel desensitization. In the synaptosomal context, these channels appear to flicker occasionally, even under polarized conditions. Veratridine inhibits their desensitization, and thus rapidly leads to a condition in which the sodium channels are locked open. While this artifactual sodium conductance will depolarize the plasma membrane, $\Delta\psi_p$ is again clamped. In addition, the high sodium conductance results in a massive activation of the Na^+/K^+-ATPase, increasing the ATP demand on the mitochondria, as will be discussed later. Furthermore, the collapse of the sodium electrochemical potential across the plasma membrane results in the reversal of multiple sodium-coupled transport processes, notably the glutamate transporter, resulting in an enhanced non-exocytotic release from the cytosolic glutamate pool [21]. Finally, the raised cytosolic sodium concentration disturbs the balance between mitochondrial calcium uptake and release, raising the cytosolic free calcium concentration at which mitochondria become net calcium accumulators [22].

3.2 4-Aminopyridine

The nonselective K^+ channel inhibitor 4-aminopyridine (4AP) facilitates exocytosis at the neuromuscular junction and in the central nervous system [23]. The inhibitor was first used with synaptosomes by Tapia and Sitges to induce release of labelled GABA, glutamate, or acetylcholine from mouse synaptosomes [24]. The authors first speculated that 4AP functioned by enhancing calcium influx into the terminals [23], and later that 4AP facilitated the coupling between calcium entry and exocytosis [25]. Release of GABA and acetylcholine was not inhibited by tetrodotoxin. Agoston et al. [26] concluded that 4AP enhanced calcium entry into the synaptosomal preparation.

In 1985, in collaboration with the group of Oliver Dolly, we began investigating the effect of presynaptic neurotoxins on the bioenergetics of the guinea pig cerebrocortical synaptosomes [8]. Using our recently developed continuous fluorometric assay for glutamate release [27], see below, we found that dendrotoxin and 4AP each induced the calcium-dependent release of glutamate from guinea pig cerebrocortical synaptosomes [28]. In contrast to the earlier findings discussed above, we found that the 4AP-induced elevation in $[Ca^{2+}]_c$ and glutamate exocytosis were inhibited by 1 μM tetrodotoxin [29]. This was of crucial importance, since the only way in which the rapidly desensitizing tetrodotoxin-sensitive voltage-activated sodium channels could have such a dramatic affect would be if they were being continuously and repetitively activated and contributing to the depolarization. This in turn led us to conclude that 4AP, as well as dendrotoxin, was destabilizing $\Delta\psi_p$ and initiating spontaneous action potential-like firing of the preparation. If the plasma membrane potential was being depolarized transiently, on a millisecond timescale with intervening quiescent periods, then the time average depolarization of the entire population would be expected to be greatly reduced. Thus the tetraphenylphosphonium accumulation reported only a 5 mV $\Delta\psi_p$ depolarization, as well as no perceptible change in the potassium equilibrium potential monitored by $^{86}Rb^+$ [28]. Interestingly, tetrodotoxin also had a slight inhibitory effect on the spontaneous release of glutamate in low potassium medium, suggesting that the plasma membrane potential of individual synaptosomes undergoes occasional spontaneous flickering. This would explain why veratridine, which acts on open sodium channels to prevent desensitization, is able to act as an effective depolarizing agent for the preparation.

The significance of the spontaneous action potential firing in the presence of 4AP is crucial. The previous artifactual, clamped, depolarization protocols involving elevated potassium or veratridine resulted in a clamped, invariant depolarization in which only non-desensitizing ion channels were able to play a role. In contrast, the spontaneous action potentials recruited the physiological diversity of desensitizing potassium, sodium, and calcium voltage-activated channels, as well as revealed a wealth of presynaptic G protein-coupled receptor-mediated modulation of these channels, and hence glutamate exocytosis [30]; see Fig. 3. Experimental protocols to investigate receptor-mediated control of glutamate exocytosis will be reviewed later.

4 Cytosolic Free Ca^{2+} ($[Ca^{2+}]_c$)

The synaptosomal cytosolic free calcium concentration, $[Ca^{2+}]_c$, can be determined by conventional techniques, for example by loading with fura-2-acetoxymethyl ester (fura-2-AM). Typically, synaptosomes are resuspended in incubation medium and preincubated with 5 μM fura-2-AM for 30 min at 37 °C and resuspended in fresh medium prior to transfer to stirred thermostatic cuvettes in the fluorimeter. Inclusion of albumin (1 mg/mL) during loading helps to keep the hydrophobic fura-2-AM in solution and improves uptake. Figure 3b shows the enhanced Ca^{2+} elevation produced by phorbol ester potentiation of 4AP-mediated plasma membrane potential depolarization.

4.1 The Role of Localized and the Bulk Calcium Elevation in the Control of Exocytosis

There is persuasive evidence that the fast release of small synaptic vesicles is triggered by a low-affinity, localized, interaction between calcium entry through voltage-activated calcium channels and the exocytotic machinery clustered at active zones, while, in contrast, large dense-core vesicles are released at ectopic sites outside these active zones. Guinea pig hippocampal synaptosomes contain amino acid, catecholamine, and neuropeptide transmitters, and in order to test this concept we took the approach that this high, localized, calcium concentration will be present at active zones during KCl- or 4AP-mediated depolarization, but not following the careful addition of the $Ca^{2+}/2H^{+}$ exchange ionophore ionomycin, which mediates uniform calcium entry across the plasma membrane [31]. Calcium indicators such as fura-2 report the bulk cytosolic free calcium concentration, and thus do not detect high, localized calcium. Addition of KCl, 4AP, or ionomycin to produce the same bulk calcium elevation should help to establish whether a given neurotransmitter was released at active zones or ectopically. The latter mechanism should result in comparable bulk cytosolic Ca^{2+} sensitivity for transmitter release by all three protocols, whereas ionomycin would be an ineffective trigger of small synaptic vesicle exocytosis. Figure 4 compares the net calcium-dependent release of glutamate and the representative neuropeptide cholesystokinin-8. The neuropeptide is released by equal effectiveness by ionomycin and the bulk calcium elevation induced by KCl or 4AP, whereas the fast neurotransmitter glutamate requires localized calcium. A word of caution, however, is that excess ionomycin induces a dissipative Ca^{2+} cycling across the inner mitochondrial membrane that “uncouples” and can lead to bioenergetic failure [16].

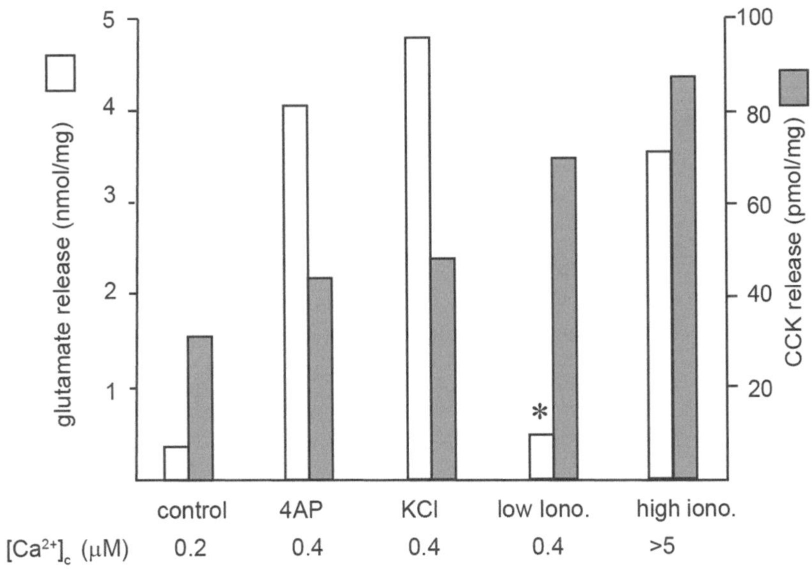

Fig. 4 Localized coupling of Ca^{2+} to glutamate but not cholecystokinin (CCK) exocytosis. KCl, 4AP, and ionomycin were added to give equal elevations in $[Ca^{2+}]_c$ to 400 nM. Ca^{2+}-dependent release of glutamate and CCK was monitored over 400 s. KCl and 4AP, which depolarize and activate presynaptic release-coupled Ca^{2+} channels, effectively release glutamate, while ionomycin, which creates a uniform increase in cytosolic Ca^{2+}, is ineffective (*) unless $[Ca^{2+}]_c$ is raised to very high levels. Data from [31]

5 Glutamate Release

5.1 Continuous Fluorometric Assay of Glutamate Release

Glutamate is the major excitatory neurotransmitter in the mammalian central nervous system. It follows that the ability to monitor glutamate release in real time from synaptosomal preparations facilitates investigation of the pathways controlling exocytosis. In 1986 we introduced a continuous fluorometric assay utilizing the enzymatic oxidation of glutamate to 2-oxoglutarate by glutamate dehydrogenase added to the synaptosomal suspension [27]. With slight modifications [32] the assay is still current, e.g., [33].

Glutamate dehydrogenase can utilize both NAD^+ and $NADP^+$, the latter being less susceptible to reduction by side reactions. The equilibrium of the reaction is somewhat unfavorable in this direction, and does not go to completion. Since the reaction evolves ammonia, ammonium sulfate suspensions of the enzyme must be avoided; in addition, we found it advantageous to dialyze the enzyme to remove traces of contaminating glutamate. The assay can be performed in a conventional fluorimeter equipped with temperature control and the ability to stir the contents of the cuvette. With fluorimeters possessing a 90° light path, we have found that a conventional spectrophotometer cuvette can be used, with a frosted side towards the excitation source, resulting in diffusion of the light beam within the sample and decreased noise.

Typically, synaptosomal pellets (0.5–1.0 mg protein) are resuspended in 1 mL of incubation medium. The maintenance of physiological ion gradients is, of course, dependent on the metabolism and the maintenance of cytosolic ATP levels. Since these are compromised in the anoxic synaptosomal pellet, it is recommended that the synaptosomal suspension is preincubated in the cuvette for at least 5 min before starting the assay. Since any adenosine leaking from the preparation may inhibit exocytosis via the inhibitory adenosine A1 receptor (see below), it is advisable to include adenosine deaminase (1 U/mg protein) in the incubation. 1 mM NADP and 50 U of glutamate dehydrogenase are then added. When the emission signal has stabilized, the experiment can be started by the addition of appropriate agents.

As was reviewed above, a variety of methods can be used to induce glutamate release and/or exocytosis, although action potential firing in the presence of 4AP approximates most closely to the physiological. Depolarization protocols, as well as factors compromising the bioenergetics of the preparation, can also initiate non-exocytotic glutamate release. If exocytosis is defined as the calcium-dependent component of release, then non-exocytotic release is estimated in a parallel experiment in which calcium is excluded from the incubation medium. The difference between the two release time courses can therefore be used to calculate exocytosis (Fig. 5). It is sufficient to lower the extra synaptosomal free calcium concentration to about 0.2 μM to inhibit exocytosis. Excessive depletion, for example by the inclusion of EGTA, raises the ionic permeability of the synaptosomal plasma membrane [34].

At the end of the experiment, a standard addition of glutamate, typically 4–5 nmol, is made. Since the reaction does not go to completion, this serves as a calibration. If the primary interest of the experiment is to monitor the extent, rather than the rate, of glutamate release, this is sufficient. However, even with this relatively high amount of enzyme, released glutamate is not assayed instantaneously, with the result that there is a significant lag in the fluorescence signal. The time course with which the final glutamate standard is assayed can be used to quantify the enzyme kinetics in the previous assay, allowing the lag to be corrected for and eliminated. The detailed protocol for this correction is contained in reference [35]. The most noticeable effect of the kinetic correction is to reveal a biphasic release of glutamate, for example following KCl depolarization. A typical experiment with guinea pig cerebrocortical synaptosomes revealed a rapid exocytosis of about 1 nmol glutamate/mg protein in response to 30 mM KCl, followed by a slow exocytotic phase with a $t_{1/2}$ of 50 s accounting for a further 4 nmol/mg. These synaptosomes possess a total glutamate content of about 20 nmol/mg, suggesting that some 5% of the total glutamate content is contained within a fast releasable vesicular pool, 20% in a slow pool, with the remainder comprising a non-vesicular

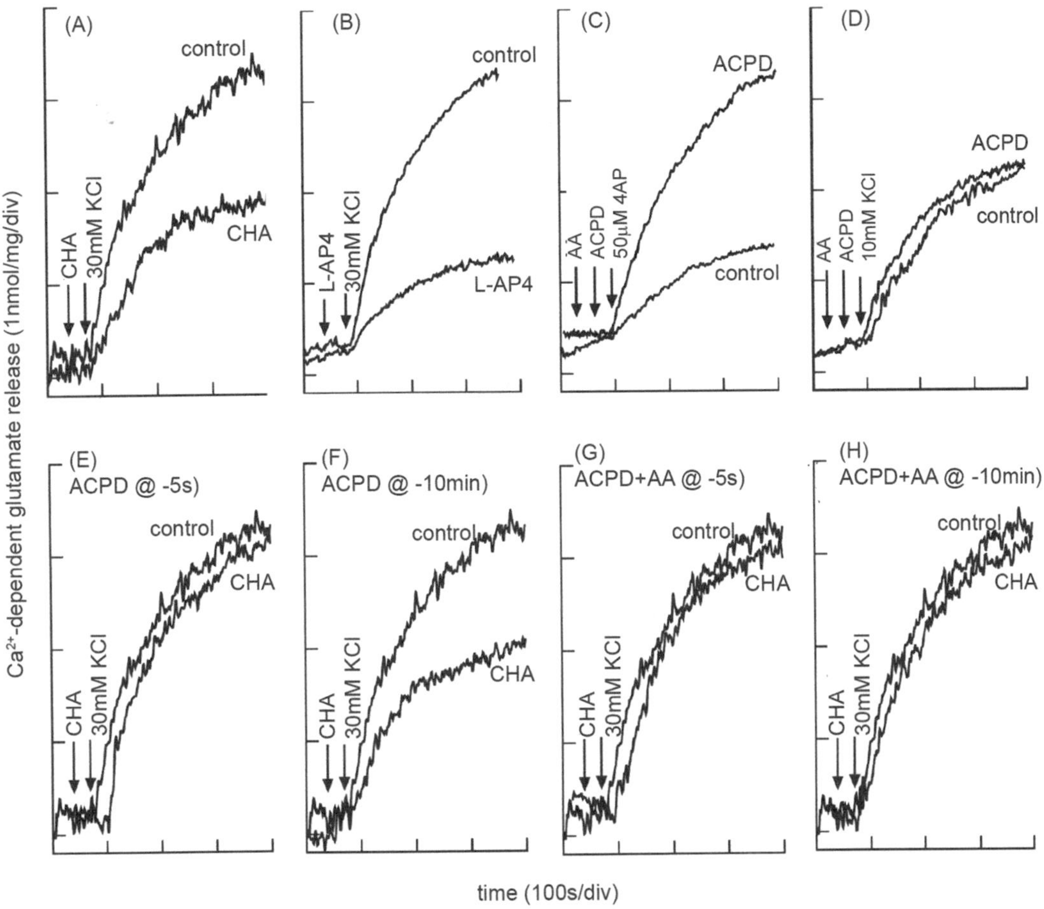

Fig. 5 Experimental design to detect regulation of glutamate by inhibitory receptors coupled to Ca^{2+} channel activation, by facilitatory receptors coupled to action potential potentiation and by transient versus sustained decoupling. The adenosine A1 agonist CHA (**a**) and the inhibitory mGluR agonist L-AP4 (**b**) each reduces Ca^{2+}-dependent glutamate release evoked by 30 mM KCl. Since KCl creates a clamped depolarization, action potential modulation is not involved. Note that L-AP4 is only effective in synaptosomes from young mice [47]. The facilitatory mGluR agonist (1S, 3R)-ACPD, in combination with arachidonate (AA), acts via PKC to strongly enhance exocytosis evoked by low (50 μM) 4AP (**c**), but is without effect on suboptimal KCl-evoked release (**d**), indicating facilitation by action potential enhancement. An additional effect of (1S, 3R)-ACPD is to decouple the CHA-mediated inhibition shown in A. In the absence of AA the decoupling is transient, being seen when the mGluR agonist is added 5 s before CHA (**e**), but not when added 10 min prior to CHA (**f**). In the presence of AA decoupling persists for at least 10 min (**g**, **h**). Parallel monitoring of $\Delta\psi_p$ and $[Ca^{2+}]_c$ was used to substantiate interpretation. Data based on [20, 29, 49]

metabolic pool of the amino acid. Glutamate dehydrogenase is allosterically activated by leucine, or the non-metabolizable analog, BCH (2-amino 2-norbornane carboxylic acid), but I'm not aware of any release assays in which the latter compound has been added to enhance the kinetics of the assay.

6 Applications

The techniques, detailed above, have been applied to a wide range of questions. I have tried to emphasize the importance of multiparametric approaches. For example, with receptor modulation of glutamate release, parallel determination of KCl- and 4AP-mediated release, $[Ca^{2+}]_c$ elevation, and $\Delta\psi_p$ depolarization, together with bioenergetic determinations of respiration, $\Delta\psi_m$, and ATP/ADP ratios, can provide a robust foundation for proposed mechanisms. Since a comprehensive review is not suitable for this chapter, I shall restrict discussion to generic aspects of the methodological approaches while giving key references to the literature.

6.1 Bioenergetic Manipulations

The combination of rotenone and iodoacetate to inhibit both oxidative and glycolytic ATP generation collapses the ATP/ADP ratio with a $t_{1/2}$ of about 15 s [35]. Calcium-dependent glutamate release is inhibited almost immediately, despite $[Ca^{2+}]_c$ rising due to the failure of the Ca^{2+}-ATPase [9, 35].

Conversely, cytosolic glutamate release is calcium independent and facilitated by energy depletion, the resulting failure of the Na^+/K^+-ATPase, and decline of the sodium gradient across the plasma membrane causing reversal of the plasma membrane glutamate transporter. Using these criteria, the release of glutamate induced by kainate [32, 36], α-latrotoxin [37], and aspartate [32] together with a significant proportion of the release induced by veratridine [21] originates from the cytosol. In contrast, that evoked by KCl when Ca^{2+} is replaced by Ba^{2+} is exocytotic [38].

Mitochondrial dysfunction is implicated in a wide range of neurodegenerative disorders. The synaptosomal preparation provides a means of investigating the bioenergetics of presynaptic mitochondria in a quasi-physiological environment, while the Seahorse respirometer reduces the amount of material required sufficiently to allow a detailed analysis to be performed on microgram quantities of synaptosomal protein from defined brain areas of wild-type and transgenic mice [10]. As with other cells, carefully designed respiratory experiments can yield detailed information on energy supply, utilization, and capacity. It is essential to design the experiment in such a way that the component of interest exerts some control over the overall respiration rate. This is illustrated in Fig. 1b investigating the effects of aminooxyacetic acid, AOAA, a transaminase inhibitor, which inhibits the malate aspartate shuttle facilitating the mitochondrial oxidation of glycolytic NADH, and thus diverts pyruvate from the mitochondrion to lactate dehydrogenase to reoxidize NADH, starving the mitochondrion of substrate and inducing a functional “hypoglycemia” [39]. Basal respiration is controlled predominantly by the inherent mitochondrial proton leak and mitochondrial ATP demand, and shuttle activity is sufficiently in excess that its partial inhibition by AOAA has no effect on respiration. Only when downstream rate restriction is removed by the addition of the protonophore FCCP does the malate aspartate shuttle become rate restricting, revealing the effect of the inhibitor [10].

The energetic demands of action potential firing and neurotransmitter release can be visualized by the addition of 4AP (Fig. 1c). The increase in basal respiration is entirely sensitive to the ATP synthase inhibitor oligomycin, illustrating that 4AP enhances ATP demand, rather than increasing the (oligomycin insensitive) mitochondrial proton leak. While standard practice is to provide glucose as synaptosomal substrate, glycolytic pyruvate is insufficient to satisfy the demands of the mitochondrial electron transport chain in the presence of protonophore. The further addition of exogenous pyruvate can double the maximal respiration [10], and it is advisable to include both substrates routinely (Fig. 1a).

While the respirometer provides detailed information, it only measures rates and not driving forces. For example, comparison of respiratory rates with glucose and glucose plus pyruvate reveals that the oligomycin-insensitive respiratory rate is increased (Fig. 1a). It thus appears that pyruvate is acting as a weak "uncoupler." However, parallel determinations of the mitochondrial membrane potential show that the additional substrate produces a significant hyperpolarization [3, 10], such that the increased leak can be ascribed to an increased driving force, rather than change in the proton conductance of the inner membrane.

Armed with this technology, it has been possible to test for presynaptic mitochondrial dysfunction in a variety of murine disease models. As proof of principal, deletion of the superoxide dismutase 2 (SOD2) gene, which generates severe oxidative damage, greatly restricted the maximum respiratory capacity of the in situ mitochondria [11]. Negative findings, while less exciting, can be equally important in testing well-established hypotheses. Thus Choi et al. [12] examined synaptosomes from a variety of Alzheimer's related transgenic mice, without finding significant bioenergetic defects, suggesting that hypotheses relating to presynaptic mitochondrial dysfunction might require reevaluation.

6.2 Presynaptic Neurotoxins

Bioenergetic, ion transport, and glutamate release assays can be combined to investigate the mechanisms of a range of presynaptic neurotoxins. The bioenergetic component is important, since mitochondrial dysfunction and consequent energy depletion both inhibit exocytosis and facilitate Ca^{2+}-independent glutamate release [9]. Thus, while β-bungarotoxin and dendrotoxin each facilitates glutamate release, the phospholipase activity of the former results in plasma membrane damage, mitochondrial dysfunction, and massive release of cytosolic glutamate pools [8, 28, 40, 41].

Similarly, α-latrotoxin depolarizes the synaptosomal plasma membrane [13] and induces massive $[Ca^{2+}]_c$ elevation and glutamate release associated with a collapse of the ATP/ADP ratio, indicative of extensive damage [37]. In contrast, the Clostridial toxins botulinum neurotoxin types A and B and tetanus toxin each inhibits glutamate exocytosis, but with no effect on key bioenergetic parameters or Ca^{2+} entry [42, 43].

6.3 Presynaptic Modulation of Glutamate Release

The development of the continuous fluorometric glutamate release assay, together with the alternate strategies to trigger transmitter release, elevated KCl, or 4AP, opened up a new field of research in which the presynaptic mechanisms controlling exocytotic and nonexocytotic glutamate release can be investigated without interference from effects at the presynaptic membrane or cell body. While the results obtained tend to confirm electrophysiological investigations, the greater simplicity of the synaptosomal preparation greatly facilitates biochemical investigations of the underlying mechanisms. Both inhibitory and facilitatory G protein-coupled receptors are expressed in cerebrocortical synaptosomes, and the careful use of the two modes of depolarization has allowed the elaboration of a remarkably detailed set of interlocking regulatory mechanisms (reviewed in [30]); see Fig. 6. There is an extensive literature on presynaptic inhibitory and facilitatory receptor mechanisms in synaptosomes, and I shall illustrate the generic methodology applicable to both classes.

6.3.1 Inhibitory Receptors Acting on Release-Coupled Calcium Channels

The first step is to establish whether the receptor is influencing action potential firing, for example by influencing 4AP-insensitive desensitizing potassium channels, or is restricting calcium entry via voltage-activated calcium channels. Calcium-dependent glutamate release is quantified using both elevated KCl and 4AP to trigger release. If both release modes produce comparable inhibitions of exocytosis, then modulation of channel-related action potential firing can be excluded. A calcium channel locus should also result in a decreased $[Ca^{2+}]_c$ elevation in response to either trigger. Adenosine A1 receptor-mediated inhibition of synaptosomal glutamate release fits this category [44], Fig. 5a.

Adenosine deaminase (1 U/mg protein) should be present in the incubation medium present to degrade contaminating adenosine. CHA (N^6-cyclohexyladenosine) is a representative adenosine A1 receptor agonist. The agonist can be added 1–2 min prior to the depolarizing trigger. In the case of cerebrocortical synaptosomes, the agonist decreases the rate and extent of Ca^{2+}-dependent glutamate release (Fig. 5a), and decreases the plateau $[Ca^{2+}]_c$ elevation, evoked by both 4AP and KCl [44]. To control off-target effects, it is important to establish that the effect of the agonist can

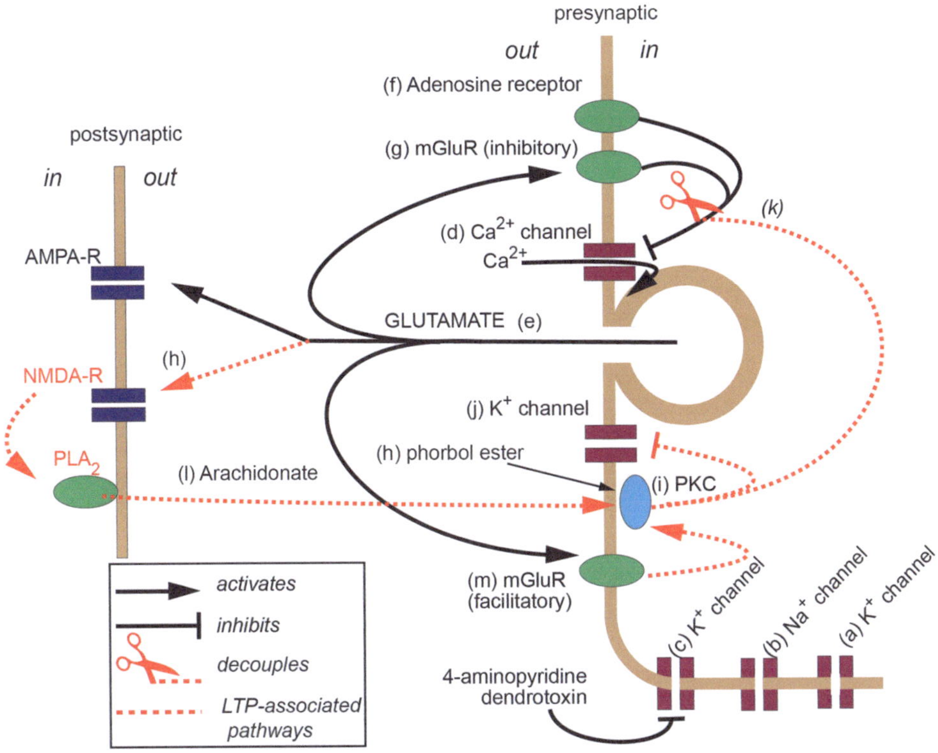

Fig. 6 Pathways of synaptosomal glutamate release regulation. A high-conductance constitutive K^+ channel (**a**) maintains the resting synaptosomal $\Delta\psi_p$ close to the K^+ diffusion potential, allowing elevated KCl to depolarize the membrane. Artifactual activation of the voltage-activated Na^+ channel (**b**) by veratridine also depolarizes. Inhibition of a distinct K^+ channel (**c**) by dendrotoxin or 4AP destabilizes $\Delta\psi_p$ and initiates spontaneous action potentials. Each of these depolarization protocols activates release-coupled Ca^{2+} channels (**d**) inducing glutamate exocytosis (**e**). A presynaptic adenosine A_1 receptor (**f**) can be activated by cyclohexyladenosine (CHA), reducing exocytosis via Ca^{2+} channel inhibition. An inhibitory metabotropic glutamate autoreceptor (**g**) activated by L-AP4 has a similar action. Phorbol esters (**h**) transiently activate protein kinase C (**i**), which inhibits a 4AP-insensitive K^+ channel (**j**), enhancing glutamate exocytosis, and transiently "decouples" (**k**) the inhibitory adenosine and mGluR receptor coupling to the Ca^{2+} channel. During long-term potentiation in vivo, it has been hypothesized that the NMDA receptor activates postsynaptic phospholipase A_2, leading to the generation of arachidonic acid (**l**), and that this may act as a retrograde messenger. Arachidonate potentiates and prolongs the phorbol ester activation of PKC, and its downstream effects. A facilitatory mGluR (**m**) autoreceptor synergizes with arachidonate to activate PKC

be negated by an appropriate antagonist. If it is accepted that fluorescence of the cationic probe $DiSC_2(5)$ at high concentrations is a valid qualitative indicator of $\Delta\psi_p$ in the preparation, then a further check is to establish that the agonist has no significant effect on the plasma membrane depolarization evoked by 4AP. Other inhibitory receptors acting on synaptosomal release-coupled calcium channels include WIN55,212-2-sensitive cannabinoid receptors [45], GABA-B receptors [46], and an inhibitory metabotropic glutamate receptor, MGluR7, activated by L-AP4 and apparent in synaptosomes from young rats [47], Fig. 5b.

6.3.2 Protein Kinase C and Facilitatory Receptors

Protein kinase C activation by the phorbol ester,4β-phorbol dibutyrate (4βPDBu) enhances 4AP-induced glutamate exocytosis from cerebrocortical synaptosomes and raises the plateau elevation in $[Ca^{2+}]_c$ [20, 48]; see Fig. 1. Conversely, the PKC inhibitor Ro-31-8220 inhibits 4AP-evoked release [48]. In contrast, the phorbol ester is without effect on the release of glutamate evoked by KCl, even though protein kinase C is still able to phosphorylate target proteins such as MARCKS (myristoylated alanine-rich C kinase substrate) [48]. This indicates that protein kinase is modulating 4AP-induced spontaneous action potentials, and this is supported by the enhanced 4AP-mediated plasma membrane potential depolarization in the presence of $DiSC_2(5)$.

The mGluR agonist, (1S,3R)-ACPD, is active at the inhibitory mGluR7 [47]. However, in the additional presence of 2 μM arachidonic acid, inhibition is changed to facilitation for 4AP (Fig. 5c) but not KCl-mediated (Fig. 5d) release [49]. Arachidonate acts by prolonging PKC activation [50, 51].

An additional effect of PKC activation by phorbol ester [44] or (1S,3R)-ACPD is to transiently reverse, or decouple, the adenosine-mediated inhibition of exocytosis. Decoupling is transient in the absence of AA (Fig. 5e, f), but persists in the presence of AA (Fig. 5g, h). Note that decoupling is independent of the ester's action potential potentiation, since it can be observed with KCl-evoked glutamate release.

7 Conclusions

The development and application of multiple techniques to monitor synaptosomal bioenergetics and neurotransmitter glutamate release, together with the exploitation of K^+ channel inhibitors to induce spontaneous action potential firing, greatly expanded the depth with which the underlying mechanisms could be studied. While this chapter is of necessity a largely historical review, the techniques and principles remain current.

Acknowledgments

The techniques reviewed here were aided by numerous colleagues, including, but not limited to, Karl Åkerman, Anne Barrie, Sung Choi, David Budd, Eleanor Coffey, Michael Cousin, Risto Kauppinen, Harvey McMahon, Jennifer Pocock, Michela Rugolo, Jose Sanchez-Prieto, Talvinder Sihra, Gareth Tibbs, and Mattheis Verhage.

References

1. Scott ID, Nicholls DG (1980) Energy transduction in intact synaptosomes: influence of plasma- membrane depolarization on the respiration and membrane potential of internal mitochondria determined in situ. Biochem J 186:21–33
2. Choi SW, Gerencser AA, Lee D et al (2011) Intrinsic bioenergetic properties and stress-sensitivity of dopaminergic synaptosomes. J Neurosci 31:4524–4534
3. Kauppinen RA, Nicholls DG (1986) Synaptosomal bioenergetics: the role of glycolysis, pyruvate oxidation and responses to hypoglycaemia. Eur J Biochem 158:159–165
4. Kauppinen RA, Nicholls DG (1986) Failure to maintain glycolysis in anoxic nerve terminals. J Neurochem 47:1864–1869
5. Mookerjee SA, Goncalves RL, Gerencser AA et al (2015) The contributions of respiration and glycolysis to extracellular acid production. Biochim Biophys Acta 1847:171–181
6. Mookerjee SA, Nicholls DG, Brand MD (2016) Determining maximum glycolytic capacity using extracellular flux measurements. PLoS One 11:e0152016
7. Mookerjee SA, Gerencser AA, Nicholls DG et al (2017) Quantifying rates, pathways and flexibility of intracellular ATP production and consumption using extracellular flux measurements. J Biol Chem 292:7189–7207
8. Nicholls DG, Snelling RM, Dolly JO (1985) Bioenergetic actions of β-bungarotoxin, dendrotoxin and bee venom phospholipase A2 on Guinea-pig synaptosomes. Biochem J 229:653–662
9. Kauppinen RA, McMahon HT, Nicholls DG (1988) Ca^{2+}-dependent and Ca^{2+}-independent glutamate release, energy status and cytosolic free Ca^{2+} concentration in isolated nerve terminals following metabolic inhibition: possible relevance to hypoglycaemia and anoxia. Neuroscience 27:175–182
10. Choi SW, Gerencser AA, Nicholls DG (2009) Bioenergetic analysis of isolated cerebrocortical nerve terminals on a microgram scale: spare respiratory capacity and stochastic mitochondrial failure. J Neurochem 109:1179–1191
11. Flynn JM, Choi SW, Day NU et al (2011) Impaired spare respiratory capacity in cortical synaptosomes from Sod2 null mice. Free Radic Biol Med 50:866–873
12. Choi SW, Gerencser AA, Ng R et al (2012) No consistent mitochondrial bioenergetic defects in presynaptic nerve terminals isolated from mouse models of Alzheimer's disease. J Neurosci 32:16775–16784
13. Nicholls DG, Rugolo M, Scott IG et al (1982) Latrotoxin of black widow spider venom depolarizes the plasma membrane, induces massive Ca influx and stimulates transmitter release in Guinea-pig synaptosomes. Proc Natl Acad Sci U S A 79:7924–7927
14. Moreno AJ, Santos DL, Magalhaes-Novais S et al (2015) Measuring mitochondrial membrane potential with a tetraphenylphosphonium-selective electrode. Curr Protoc Toxicol 65:25–16
15. Nicholls DG (2012) Fluorescence measurement of mitochondrial membrane potential changes in cultured cells. Methods Mol Biol 810:119–133
16. Nicholls DG (2006) Simultaneous monitoring of ionophore- and inhibitor-mediated plasma and mitochondrial membrane potential changes in cultured neurons. J Biol Chem 281:14864–14874
17. Gerencser AA, Mulder H, Nicholls DG (2015) Calcium modulation of exocytosis-linked plasma membrane potential oscillations in INS-1 832/13 cells. Biochem J 471:111–122
18. Gerencser AA, Chinopoulos C, Birket MJ et al (2012) Quantitative measurement of mitochondrial membrane potential in cultured cells: calcium-induced de- and hyperpolarization of neuronal mitochondria. Am J Physiol Cell Physiol 590:2845–2871
19. Bunting JR, Phan TV, Kamali E et al (1989) Fluorescent cationic probes of mitochondria. Metrics and mechanism of interaction. Biophys J 56:979–993
20. Barrie AP, Nicholls DG, Sánchez-Prieto J et al (1991) An ion channel locus for the protein kinase C potentiation of transmitter glutamate release from Guinea pig cerebrocortical synaptosomes. J Neurochem 57:1398–1404
21. Sánchez-Prieto J, Sihra TS, Nicholls DG (1987) Characterization of the exocytotic release of glutamate from Guinea-pig cerebral cortical synaptosomes. J Neurochem 49:58–64
22. Nicholls DG, Scott ID (1980) The regulation of brain mitochondrial calcium-ion transport: the role of ATP in the discrimination between kinetic and membrane-potential-dependent Ca efflux mechanisms. Biochem J 186:833–839
23. Thesleff S (1980) Aminopyridines and synaptic transmission. Neuroscience 5:1413–1419
24. Tapia R, Sitges M (1982) Effect of 4-aminopyridine on transmitter release in synaptosomes. Brain Res 250:291–299

25. Tapia R, Sitges M, Morales E (1985) Mechanism of the calcium-dependent stimulation of transmitter release by 4-aminopyridine in synaptosomes. Brain Res 361:373–382
26. Agoston D, Hargittai P, Nagy A (1983) Effects of a 4-aminopyridine in calcium movements and changes of membrane potential in pinched-off nerve terminals from rat cerebral cortex. J Neurochem 41:745–751
27. Nicholls DG, Sihra TS (1986) Synaptosomes possess an exocytotic pool of glutamate. Nature 321:772–773
28. Tibbs GR, Dolly JO, Nicholls DG (1989) Dendrotoxin, 4-aminopyridine, and beta-bungarotoxin act at common loci but by two distinct mechanisms to induce Ca^{2+}- dependent release of glutamate from Guinea-pig cerebrocortical synaptosomes. J Neurochem 52:201–206
29. Tibbs GR, Barrie AP, Van-Mieghem F, McMahon HT, Nicholls DG (1989) Repetitive action potentials in isolated nerve terminals in the presence of 4-aminopyridine: effects on cytosolic free Ca^{2+} and glutamate release. J Neurochem 53:1693–1699
30. Sanchez-Prieto J, Budd DC, Herrero I, Vazquez E, Nicholls DG (1996) Presynaptic receptors and the control of glutamate exocytosis. Trends Neurosci 19:235–239
31. Verhage M, McMahon HT, Ghijsen WEJM et al (1991) Differential release of amino acids, neuropeptides and catecholamines from nerve terminals. Neuron 6:517–524
32. McMahon HT, Barrie AP, Lowe M et al (1989) Glutamate release from Guinea-pig synaptosomes: stimulation by reuptake-induced depolarization. J Neurochem 53:71–79
33. Ferrero JJ, Ramirez-Franco J, Martin R et al (2016) Cross-talk between metabotropic glutamate receptor 7 and beta adrenergic receptor signaling at cerebrocortical nerve terminals. Neuropharmacology 101:412–425
34. Kauppinen RA, Sihra TS, Nicholls DG (1986) Divalent cation modulation of the ionic permeability of the synaptosomal plasma membrane. Biochim Biophys Acta 860:178–184
35. McMahon HT, Nicholls DG (1991) Transmitter glutamate release from isolated nerve terminals: evidence for biphasic release and triggering by localized Ca^{2+}. J Neurochem 56:86–94
36. Pocock JM, Murphie HM, Nicholls DG (1988) Kainic acid inhibits the synaptosomal plasma membrane glutamate carrier and allows glutamate leakage from the cytoplasm but does not affect glutamate exocytosis. J Neurochem 50:745–751
37. McMahon HT, Rosenthal L, Meldolesi J et al (1990) α-Latrotoxin releases both vesicular and cytoplasmic glutamate from isolated nerve terminals. J Neurochem 55:2039–2047
38. McMahon HT, Nicholls DG (1993) Barium-evoked glutamate release from Guinea-pig cerebrocortical synaptosomes. J Neurochem 61:110–115
39. Kauppinen RA, Sihra TS, Nicholls DG (1987) Aminoxyacetic acid inhibits the malate-aspartate shuttle in isolated nerve terminals and prevents the mitochondria from utilizing glycolytic substrates. Biochim Biophys Acta 930:173–178
40. Rugolo M, Dolly JO, Nicholls DG (1986) The mechanism of action of B-bungarotoxin at the presynaptic membrane. Biochem J 223:519–523
41. Tibbs GR, Nicholls DG, Dolly JO (1989) Dendrotoxin and charybdotoxin increase the cytosolic concentration of free Ca2+ in cerebrocortical synaptosomes an effect not shared by apamin. FEBS Lett 255:159–162
42. Sánchez-Prieto J, Sihra TS, Evans D et al (1987) Botulinum toxin A blocks glutamate exocytosis from Guinea-pig cerebral cortical synaptosomes. Eur J Biochem 165:675–681
43. McMahon HT, Foran P, Dolly JO et al (1992) Tetanus toxin and botulinum toxins type A and B inhibit glutamate, gamma-aminobutyric acid, aspartate, and met-enkephalin release from synaptosomes. Clues to the locus of action. J Biol Chem 267:21338–21343
44. Barrie AP, Nicholls DG (1993) Adenosine A1 receptor inhibition of glutamate exocytosis and protein kinase C-mediated decoupling. J Neurochem 60:1081–1086
45. Godino MC, Torres M, Sanchez-Prieto J (2005) Inhibition of N- and P/Q-type Ca^{2+} channels by cannabinoid receptors in single cerebrocortical nerve terminals. FEBS Lett 579:768–772
46. Martin R, Ladera C, Bartolome-Martin D, Torres M et al (2008) The inhibition of release by mGlu7 receptors is independent of the Ca^{2+} channel type but associated to GABAB and adenosine A1 receptors. Neuropharmacology 55:464–473
47. Vazquez E, Budd DC, Herrero I et al (1995) Co-existence and interaction between facilitatory and inhibitory metabotropic glutamate receptors and the inhibitory adenosine A1 receptor in cerebrocortical nerve terminals. Neuropharmacology 34:919–927
48. Coffey ET, Sihra TS, Nicholls DG (1993) Protein kinase C and the regulation of gluta-

mate exocytosis from cerebrocortical synaptosomes. J Biol Chem 268:21060–21065

49. Herrero I, Miras-Portugal MT, Sanchez-Prieto J (1992) Positive feedback of glutamate exocytosis by metabotropic presynaptic receptor stimulation. Nature 360:163–166

50. Budd DC, Nicholls DG (1998) Arachidonic acid potentiates the duration of the metabotropic, protein kinase C-mediated suppression of the inhibitory adenosine A1 receptor pathway in glutamatergic nerve terminals. Neurosci Lett 244:133–136

51. Coffey ET, Herrero I, Sihra TS et al (1994) Glutamate exocytosis and MARKS phosphorylation are enhanced by a metabotropic glutamate receptor coupled to a protein kinase C synergistically activated by diacylglycerol and arachidonic acid. J Neurochem 63:1303–1310

Chapter 8

Synaptosome Bioenergetics and Calcium Handling: Aging Response

Silvia Lores-Arnaiz, Georgina Rodríguez de Lores Arnaiz, Analía G. Karadayian, and Juanita Bustamante

Abstract

Synaptic function and role of mitochondria inside nerve terminals can be studied by the isolation of an enriched fraction of synaptosomes, which consist in nerve-ending particles that are formed during homogenization of brain tissue. Different procedures have been described for the isolation of an enriched synaptosomal fraction, most of them based on the use of gradients.

Neuronal function seems to be critically dependent on the energy provided by mitochondrial respiration. The determination of bioenergetic parameters such as mitochondrial membrane potential, respiratory rates, ATP content, and mitochondrial Ca^{2+} uptake in synaptosomal preparations can provide useful information for the understanding of the importance of mitochondrial function in neurotransmission.

Synaptic nerve terminals are constantly exposed to complex Ca^{2+} fluxes. At the presynaptic terminal, the recovery from calcium oscillations critically depends on the proper mitochondrial function to generate ATP and modulate Ca^{2+} homeostasis together with an efficient endoplasmic reticulum function.

The differential characteristics of synaptic and non-synaptic mitochondria in terms of bioenergetics and free radical production as well as the response to aging are discussed.

Key words Synaptosome, Bioenergetics, Calcium uptake, Oxygen free radicals, Nitric oxide, Aging

Abbreviations

DAF-2	4,5-Diaminofluorescein
DAF-2-DA	4,5-Diaminofluorescein diacetate
EDTA	Ethylenediaminetetraacetic acid
ER	Endoplasmic reticulum
FCCP	Carbonyl cyanide-4-(trifluoromethoxy)phenylhydrazone
GSNO	*S*-nitrosoglutathione
H_2O_2	Hydrogen peroxide
L-NNA	N_ω-nitro-L-arginine
MPT	Mitochondrial permeability transition

Kathryn M. Murphy (ed.), *Synaptosomes*, Neuromethods, vol. 141,
https://doi.org/10.1007/978-1-4939-8739-9_8, © Springer Science+Business Media, LLC, part of Springer Nature 2018

NAO	10-*N*-nonylacridine orange
NMDA	*N*-methyl-D-aspartate
NMDAR2B	NMDA receptor NR2B subunit
nNOS	Neuronal nitric oxide synthase
NO	Nitric oxide
NOS	Nitric oxide synthase
PSD-95	Postsynaptic density protein 95 kDa
RNS	Reactive nitrogen species
ROS	Reactive oxygen species
SYTs	Synaptotagmins
TMRE	Tetramethylrhodamine, ethyl ester
VDAC	Voltage-dependent anion channel

1 Introduction: The Synaptosome

The term "synaptosome" was proposed by Whittaker et al. to name nerve-ending particles that are formed during homogenization of brain tissue. These particles preserve the main features of the presynaptic nerve terminals, including mitochondria, synaptic vesicles, and cytoplasm as well as the postsynaptic density where the postsynaptic membranes frequently remain attached. Synaptosomes contain functional ion channels, receptors, enzymes, and proteins involved in release, uptake, and storage of neurotransmitters [1].

The synaptosomes were known to behave as miniature cells [2], inside which mitochondria provide ATP to the cytoplasm and plasma membrane, preserving the metabolism, plasma membrane excitability, and all the key machinery for the exo- and endocytosis of neurotransmitters [3]. Therefore, synaptic function and role of mitochondria inside nerve terminals can be studied by the isolation of an enriched fraction of these structures.

The typical characteristic of the nerve terminals is that they are subjected to high energy demands and extensive calcium fluxes. Neuronal function is critically dependent on the energy provided by mitochondrial respiration. At the pre-synapsis energy is required for vesicle trafficking and neurotransmitter exocytosis and release from synaptic vesicles [4]; at the post-synapsis where Ca^{2+} and Na^{+} fluxes activate ATPases, a high amount of energy production is also required, mainly for reversing ion influxes underlying action potential signaling [5]. Therefore, an adequate function of the mitochondrial electron transport chain and the maintenance of calcium homeostasis are essential to accomplish these important neuronal functions.

2 Preparation of Synaptosomes

The first original evidences for the presence of nerve endings in brain crude mitochondrial fractions were reported by De Robertis et al., 1961, based on electron microscope studies which showed the images of numerous intact endings filled with synaptic vesicles, some of these endings containing small mitochondria within [6]. This observation offered an explanation for the finding of acetylcholine in a fraction which sediments together with mitochondria [7, 8].

Since then, several attempts were made to isolate nerve endings from rat brain [9] or from guinea pig forebrain [10], both based on the use of discontinuous sucrose density gradients. Tissues were first homogenized in 0.32 M sucrose, and then a fraction containing mitochondria, synaptosomes, small myelin fragments, and some microsomes (the so-called P2 fraction) was obtained by centrifugation at 10,000 × g for 20 min, and further fractionation was carried out by means of a discontinuous density gradient consisting of equal volumes of 0.8 and 1.2 M sucrose, centrifuged at 100,000 × g for 60 min [1].

According to De Robertis et al. five subfractions were obtained: (A) myelin, (B) membranes and fragmented cholinergic endings, (C) cholinergic nerve endings, (D) non-cholinergic nerve endings, and (E) free mitochondria [9]. While fractions B and C were rich in acetylcholine and acetylcholinesterase, fraction D resulted to be poor in these components. Succinate dehydrogenase activity was absent in A and B, very low in C, and high in D, reaching its highest value in E [9]. Also, a methodology to prepare a membrane fraction enriched in synaptic plasma membrane from an osmotically shocked crude mitochondrial fraction was originally developed by Rodríguez de Lores Arnaiz et al., 1967 [11].

Procedures involving sucrose-density-gradient centrifugation present the disadvantage that metabolic integrity of intrasynaptosomal mitochondria can be compromised by the effect of hyperosmotic sucrose concentration in the gradient [12].

Many attempts have been carried out to isolate synaptosomal fractions; some of them require many steps that at the end of the process can cause decline in respiratory function and viability, mechanical damage, and reduced function of the synaptosomes. A fractionation method for the isolation of more intact and homogeneous nerve endings from brain based on the use of discontinuous density gradients of Ficoll was originally described, resulting in a yield of 1.5–2.0 mg/g brain wet weight [13].

The methodology proposed by Lai and Clark for the preparation of synaptosomal fractions consisted of an initial Ficoll gradient separation with 7.5 and 13% steps in buffer 0.32 M sucrose, 1 mM EDTA (potassium salt), 10 mM Tris–HCI (pH 7.4), followed by purification with another Ficoll gradient containing 3 and 6% Ficoll medium (Ficoll, 0.24 M mannitol, 60 mM sucrose, 50 μM EDTA, and 10 mM Tris–HCI, pH 7.4), which allowed also the isolation of other fractions, e.g., non-synaptic mitochondria, microsomal fraction, and myelin [12].

In line with this methodology, and based on a previous Ficoll gradient separation [14], synaptosome isolation from mouse cerebral cortex was carried out. The tissue was homogenized in a medium consisting of 0.23 M mannitol, 0.07 M sucrose, 5 mM HEPES, and 1 mM EDTA, pH 7.4. The procedure implied the preparation of a crude mitochondrial fraction by differential centrifugation (600 × g for 10 min to discard nuclei and cell debris followed by centrifugation of the supernatant at 8000 × g for 10 min). The resulting pellet was resuspended and layered on a Ficoll gradient containing steps of 13, 8, and 3%. After centrifugation of the gradients (11,500 × g for 30 min), two fractions were collected: a pellet at the bottom of the tube corresponds to a fraction of heavy mitochondria, which are mainly non-synaptic, and the fraction lying at 8% contains synaptosomes [15]. In this synaptosome preparation, mitochondrial protein content was estimated to be between 4 and 6.5 mg/g tissue [16].

Procedures incorporating density gradient centrifugation usually produce highly enriched fractions of mitochondria but typically result in yields of 10% or less of the original organelles [17]. On the basis of MAO activity measurements mitochondrial yield is roughly 9.3% for synaptosomal fractions versus data in cerebral cortex homogenates [16].

Another gradient procedure commonly used for synaptosome preparations is based on the use of Percoll. For instance, Dunkley et al. reported a rapid procedure in which, after brain tissue homogenization and low-speed centrifugation, the supernatant fraction is collected and directly applied to a discontinuous Percoll gradient, comprising layers of 3, 10, 15, and 23% (v/v) Percoll [18]. From this procedure, five major subcellular fractions are separated, mainly consisting of (1) small unidentified membranous material; (2) myelin, membranes, and membrane vesicles; (3) synaptosomes and membrane vesicles; (4) synaptosomes; and (5) extrasynaptosomal mitochondria. This procedure yields 2.5–4 mg of synaptosomal protein per gram rat brain and presents advantages over other procedures in terms of producing relatively homogeneous synaptosomes, minimizing the presence of synaptic and glial plasma membranes and extrasynaptosomal mitochondria [18].

3 Characterization of Synaptosomal Fractions

Regarding previous characterization of the synaptosomal preparations, it was observed that fractions containing membranes and cholinergic nerve endings were rich in acetylcholinesterase [9]. Specific proteins present in synaptosomal membranes were also detected by Western blot analyses to determine specificity of the synaptosome preparation procedure. These individual synaptic proteins are N-methyl-D-aspartate receptor 2B subunit

(NMDAR2B), PSD95, GluR2/3/4, α-amino-3-hydroxy-5-methyl-4-isoxazolepropionic acid (AMPA) receptor, and synaptophysin [19].

Synaptosomes are considered a primarily presynaptic preparation with a diameter of 0.5–1 μm, containing one or more small mitochondria, and present an area of postsynaptic membrane adhering to it [20]. Therefore, both presynaptic and postsynaptic protein elements may be observed at electron microscope images [21, 22].

A size-based gating analysis by flow cytometry provides a simple and cost-effective method to monitor different fluorescence responses in synaptic regions. From the total number of particles, 66% were positive for a synaptic marker, SNAP-25 [23]. This docking protein on the presynaptic membrane of synapses is found only in neuronal terminals [24]. Also, levels of a postsynaptic marker, PSD-95, which is part of a scaffolding protein complex that anchors the NMDA receptor within synapses, can also be found in synaptosomal preparations [25, 26]. Cytometric quantification of the postsynaptic component in fresh synaptosomes indicates that a majority (76%) of the large particles are PSD-95 positive [27].

According to electron microscopy studies, the free mitochondria are usually smaller than 0.5 μm [22] and may be excluded from further examination by drawing an analysis gate based on size standards [23].

Flow cytometric assays allow the detection of intrasynaptosomal mitochondria through the measurement of specific markers for cardiolipin and evaluation of mitochondrial membrane potential in intact nerve endings. This procedure is based on a data analysis gated on forward scatter in a selected size range and allows the exclusion of most of the non-synaptosomal elements identifying the intact subpopulation of interest. The quantitative characterization of the crude synaptosomal fraction by flow cytometry showed that synaptosomes are among the largest particles in the crude synaptosomal preparation, with a size standard calibration around 6 μm in diameter [23]. Assays of brain cortex synaptosomal fractions labeled with NAO and TMRE and analyzed by contour plot of forward-angle scatter versus side-angle scatter indicate that the highest level of viable mitochondria (NAO- and TMRE-positive particles) is inside the largest synaptosomal population [15].

4 Mitochondria Heterogeneity

Mitochondria supply ATP to various biological activities essential for neuronal growth, survival, and function. Due to unique morphological features, neurons face exceptional challenges to maintain ATP and Ca^{2+} homeostasis. Neurons require specialized mechanisms distributing mitochondria to distal areas where energy and Ca^{2+} buffering are in high demand, such as synapses and axonal branches [28]. These distal compartments also undergo development- and activity-dependent remodeling, thereby altering mitochondrial trafficking and distribution. Mitochondria move bidirectionally, pause briefly, and move again, frequently changing their direction. In mature neurons, only one-third of axonal mitochondria are motile [29–32]. Stationary mitochondria serve as local energy sources and for intracellular Ca^{2+} buffering. The balance between motile and stationary mitochondria responds quickly to changes in axonal and synaptic physiology [33]. Like many other neuronal organelles, mitochondria are thought to arise mainly in the neuronal cell body, but they not only need to be transported to the axon terminal but also are retrieved back to the soma when they become dysfunctional, to be degraded by mitophagy [32]. A strong coordination of mitochondrial mobility and synaptic activity is crucial for neuronal development and synaptic plasticity [32]. Considering the relevant role of mitochondria in energy production, calcium homeostasis, and cell death signaling pathways, defects in axonal transport are of special interest in relation to neuropathies and degenerative diseases of the nervous system.

Important differences in synaptic and non-synaptic mitochondria have been described. For instance, synaptic mitochondria are smaller than mitochondria present in neuronal bodies, and also seem to respond differently to different stimulus. Mitochondria isolated from rat cortical synaptosomes are more prone to undergo mitochondrial permeability transition (MPT) in response to Ca^{2+} insult as compared to non-synaptic mitochondria [34]. Differences in mitochondrial Ca^{2+} buffering capacity between synaptic mitochondria and non-synaptic mitochondria could be attributed to a higher cyclophilin D content in synaptic mitochondria than in non-synaptic mitochondria [35]. Also, reactive oxygen species production and mitochondrial membrane potential dissipation induced by 100 and 200 μM Ca^{2+} overload appeared to be higher in synaptic mitochondria than in non-synaptic mitochondria [36].

5 Bioenergetics

Synaptosomes are accessible to many of the techniques employed for the evaluation of bioenergetic parameters including mitochondrial membrane potential, respiratory rates, and ATP generation and storing, as well as mitochondrial Ca^{2+} uptake [3].

When synaptosomes are disrupted by digitonin treatment, centrifuged, and resuspended, a preparation of synaptic mitochondria can be obtained. The permeabilization procedure should be carefully performed, in order to optimize time and concentration of digitonin effects. When synaptosomes are treated for 5 min with 0.01% digitonin, centrifuged, and resuspended to obtain a synaptic mitochondrial fraction, the respiratory control measured in the presence of malate-glutamate is 29% higher than that measured in intact synaptosomes, suggesting that malate, glutamate, and ADP could be more rapidly incorporated to the mitochondria when synaptosomes are previously permeabilized [16]. However, if studies require structurally intact mitochondria, whole synaptosomal fractions must be used to avoid the risk of partial disruption of the outer mitochondrial membrane due to permeabilization procedures.

Respiratory activity can be evaluated in synaptosomes in a reaction medium consisting of 120 mM NaCl, 3.5 mM KCl, 1.3 mM $CaCl_2$, 0.4 mM KH_2PO_4, 15 mM D-glucose, 10 mM pyruvate, 0.4% (w/v) fatty acid-free bovine serum albumin, and 10 mM HEPES, pH 7.4 [37].

Respiration driving proton leak can be determined after addition of 4 μg/mL oligomycin, an inhibitor of ATP synthase, while maximal respiration is achieved after addition of 4 μM FCCP. Respiration driving ATP synthesis can be calculated as the basal respiration minus respiration driving proton leak [37].

A low respiratory rate in the presence of oligomycin indicates that the intrasynaptosomal mitochondria maintain a sufficiently high proton motive force to restrict electron transport. A higher respiratory rate in the presence of oligomycin indicates altered proton leakiness in vivo [38].

Spare respiratory capacity is defined as a percentage ratio between maximum and basal respiration and is a critical factor in maintaining a reserve of ATP-generating capacity [39].

Brain cortex synaptosomes isolated by Ficoll gradient procedures exhibited a basal respiration of 69.9 ± 6.5 ng at O/min mg protein, while respiration driving ATP synthesis and maximal respiration rates were 12.7 ± 0.7 and 82.9 ± 7.1 ng- at O/min mg protein, respectively, thus allowing the estimation of a spare respiratory capacity of 119.1 ± 4.9% [15].

The presence of enriched levels of ATP has been described in presynaptic nerve terminals [40]. Surprisingly, this observation was not correlated with an abundance of presynaptic mitochondria. In fact, the possible explanation for the smaller number of presynaptic mitochondria detected by confocal imaging could be due to their smaller size as compared with the extra-synaptic mitochondria, being more difficult to be detected by immunofluorescence [40]. It is likely that a high level of ATP required in the synaptic environment should be produced, stored, and packed in the vesicle lumen in order to rapidly accomplish neurotransmission, suggesting that the possible low mitochondrial content at the presynaptic terminal may play a very active role which should be more profoundly studied.

Mitochondrial membrane potential ($\Delta\psi_m$) can also be determined in brain cortex synaptosomes by flow cytometry using potentiometric probes such as DiOC6 [16] or tetramethylrhodamine ethyl ester (TMRE) [15]. Both probes at nM concentration have been used for direct measurement of transmembrane potential in cells and isolated mitochondria from different sources [41]. The mitochondrial uncoupler FCCP used as positive control for mitochondrial depolarization produces a significant decrease in $\Delta\psi_m$ when assays are carried out in synaptosomal fractions [15, 16].

In addition, intrasynaptosomal $\Delta\Psi m$ may be determined by the fluorescent dye rhodamine-123, demonstrating that incubation of rat brain synaptosomes with 100 μM glutamate leads to a decrease of $\Delta\Psi m$ [42].

Changes in mitochondrial and plasma membrane potentials have also been monitored after glucose deprivation of rat brain synaptosomes using the fluorescent dyes DiSC3 [5] and JC-1, showing that glucose removal from incubation medium induces a shift of membrane potentials leading to mitochondrial polarization [43].

6 Synaptosomal Calcium Handling

In the central nervous system, Ca^{2+} is a ubiquitous cell messenger that participates in the regulation of neurotransmitter release and nerve excitability, and is closely associated with cellular differentiation and migration, synaptic plasticity, neurite growth, and neuronal apoptosis [44–46].

Under normal physiological conditions, there is a spatiotemporal combination pattern in intracellular calcium signaling transduction in neuronal cells: the transduction starts from a local cytosolic [Ca^{2+}] elevation and is then transmitted within the cells in a recurrent and transient pattern. This periodical appearance of calcium peaks is called calcium oscillation. Studies have shown that one calcium oscillation consists of a series of steps: (1) the extracellular Ca^{2+} enters the cytosol through the simultaneous activation or the synergistic effect of ligand-gated (such as NMDA receptor) and voltage-sensitive calcium channels; (2) the Ca^{2+} that enters the cell can produce a calcium-releasing effect at the endoplasmic reticulum (ER) through the activation of inositol 1,4,5-triphosphate receptors; (3) a portion of the released Ca^{2+} is uptaken by the mitochondria; the mitochondrial Ca^{2+} can return to the cytosol through the Na^+-Ca^{2+} exchanger and induce the release of calcium from the ER until the mitochondrial calcium is depleted; (4) after the messenger effects are completed, the increased cytosolic Ca^{2+} will be rapidly exported outside the cells by Ca^{2+}-ATP pumps and Na^+-Ca^{2+} exchangers localized in the cell membrane or rapidly removed by Ca^{2+}-ATPase in the ER; and therefore (5) the cytosolic Ca^{2+} levels can return to the baseline [47–50].

The significance of interactions between ER and mitochondria at nerve terminals necessary to modulate synaptic and dendritic $[Ca^{2+}]_I$ transients is still obscure. Both organelles are transported along microtubules by the molecular motors of the kinesin family [51]. The presence of ER is observed as isolated vesicles in dendrites that moved and slowed down in the vicinity of active synapses where they often established long-lasting contacts with mitochondria. Both presynaptic organelles respond with changes in functional variables (lumenal $[Ca^{2+}]$ and mitochondrial potential) in concert with synaptic activity. Exocytosis induced by local membrane depolarizations shows a clear dependence on the presence of functional ER vesicles or mitochondria and the efficacy of the Ca^{2+} exchange between the organelles [52]. Pharmacological exclusion of ER or mitochondria from Ca^{2+} buffering modifies synaptic potentials [53–56].

Synaptic nerve terminals should recover rapidly from calcium elevation-induced excitation by restoring the transmembrane ion gradients, replenishing energy substrates, and removing reactive oxygen species (ROS). This recovery depends critically on the proper mitochondrial function to generate ATP and buffer Ca^{2+} transients together with an efficient ER function [52, 57, 58].

The exposure of synaptosomal mitochondria to high Ca^{2+} concentrations can affect mitochondrial membrane potential, and therefore the ability of mitochondria to handle nerve terminal calcium homeostasis [15, 16]. It is well known that calcium exerts a regulatory role on mitochondrial function involving both calcium entry into the mitochondria and activation of dehydrogenases (DH) and ATP synthase (the so-called mitochondrial calcium uniporter MCU-DH pathway). Mitochondrial calcium uniporter is activated at high calcium concentrations in the micromolar range.

Mitochondrial calcium handling can be evaluated by using flow cytometry techniques through the use of the acetoxymethyl ester derivative of the green fluorescent indicator FLUO-4 (FLUO-4AM). Evaluation of calcium uptake was carried out in isolated hippocampal mitochondria after loading the fractions with 50 nM FLUO-4AM and incubating them with different calcium concentrations, in a reaction medium containing MSH buffer (0.23 M mannitol, 0.07 M sucrose, 5 mM HEPES), supplemented with 5 mM malate, 5 mM glutamate, and 1 mM phosphate. The ability of isolated brain hippocampal mitochondria to accumulate calcium was markedly affected by different conditions. Hippocampal mitochondria isolated from ketamine (a noncompetitive NMDA receptor antagonist)-injected animals exposed to different external $[Ca^{2+}]$ from 5 to 100 μM failed to accumulate calcium, as compared with mitochondria from saline-injected animals, in which a higher and sustained Ca^{2+} fluorescent signal indicates an efficient mitochondrial ability to accumulate and retain Ca^{2+} up to the highest Ca^{2+} concentration (100 μM). Also, the ability to acquire and retain Ca^{2+} from the extracellular medium is markedly decreased in mitochondria from untreated animals exposed to 50 mM H_2O_2 even at the lowest calcium concentration (5 μM) [59].

Moreover, mitochondrial exposure to high calcium concentrations results in a persistent elevation of NADH within the matrix. When not used as a substrate for ATP generation, NADH transmits to the cytoplasm to influence protein acetylation and gene expression. Mitochondrial Ca^{2+}-driven bioenergetics participates as a regulator of inflammatory events. A fundamental mechanism by which mitochondria regulate cytosolic NAD^+/NADH metabolism and initiate signaling, epigenetic regulation, and metabolic adaptation has been suggested. The physiological importance of the mitochondrial Ca^{2+} handling has been highlighted by research of the first human disease associated with dysfunctional mitochondrial calcium uptake harboring loss-of-function M1CU1 mutations [60].

It has been suggested that mitochondrial Ca^{2+} uptake contributing to respiratory stimulation is important in highly energy-demanding tissues, especially when exposed to high workloads. The importance of mitochondrial respiratory function in maintaining neuronal ATP homeostasis has been extensively studied [61]. However, the role of mitochondria in Ca^{2+} modulation at the presynaptic compartment is not totally understood.

It is well known that learning and memory are thought to rely on changes in the strength of the connections between nerve cells. Fast synaptic communication in the brain requires synchronous vesicle fusion that is evoked by action potential-induced Ca^{2+} influx. Synaptic transmission is initiated when an action potential invades a nerve terminal, opening Ca^{2+} channels, which gate a highly localized, transient increase in intracellular Ca^{2+} at the active zone. Ca^{2+} triggers synaptic vesicle exocytosis, thereby releasing the neurotransmitters contained in the vesicles and initiating synaptic transmission. This fundamental mechanism was discovered in pioneering work on the neuromuscular junction by Katz and Miledi [62].

Different studies using synaptosomal membranes have allowed the understanding of neurotransmitter release and the molecular mechanism(s) underlying presynaptic regulation by different molecules. The amazing speed and precision of Ca^{2+}-triggered neurotransmitter release in a few hundred microseconds [63] on the same timescale as the gating of an ion channel is mediated, at least in part, by the properties of the Ca^{2+} sensor synaptotagmin (SYT) and its cofactor complexin, as described in Fig. 1. SYTs function as synaptic Ca^{2+} sensors for neurotransmitter release and require complexin as a cofactor in Ca^{2+}-triggered exocytosis. These proteins attach the presynaptic membrane by a transmembrane region [64].

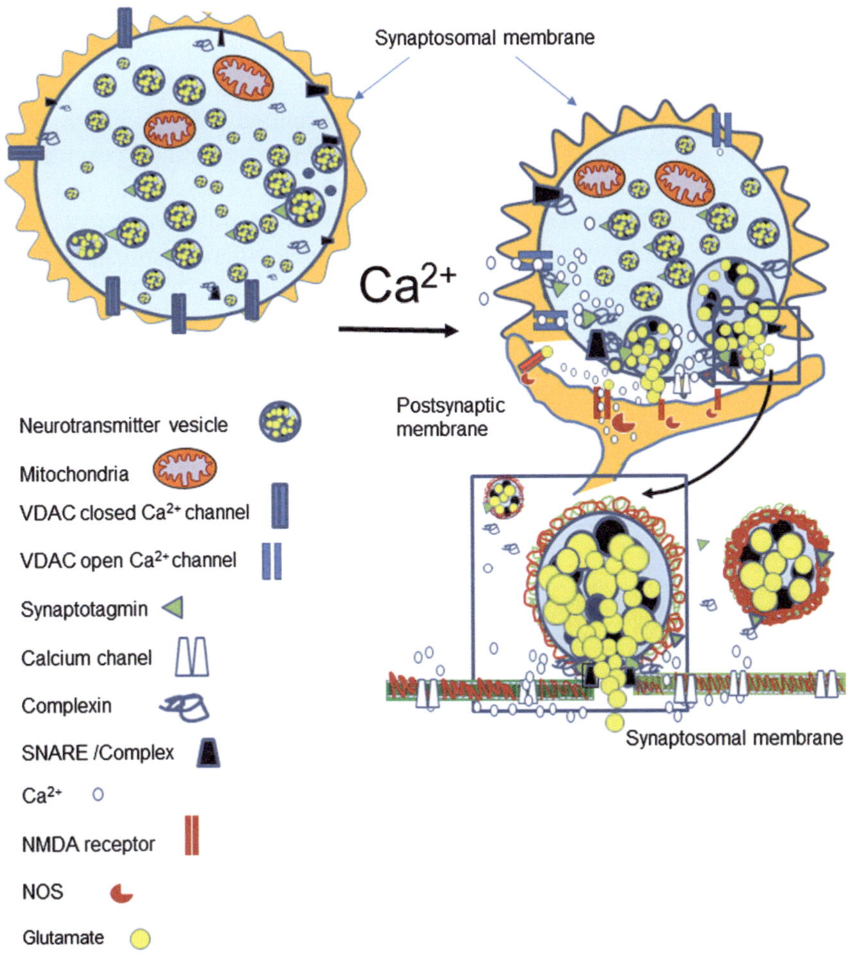

Fig. 1 Schematic diagram of synaptosomal neurotransmitter release. On the left top a synaptosome with a cytosolic content: vesicles, mitochondria, calcium channels, synaptotagmin, complexin, SNARE complex. On the right top, a synaptosome with vesicles approaching the synaptosomal membrane attached to a postsynaptic membrane portion in order to start exocytosis after calcium exposure. In the postsynaptic membrane calcium influx mediated by glutamate NMDA receptor activation induces NOS activation producing NO. On the right bottom the amplification of the vesicles and synaptosomal membrane fusion describes the synaptosomal membrane and the Ca^{2+}-dependent fusion machinery, which is composed by the calcium sensor synaptotagmin, interacting with complexin in order to activate and clamp the SNARE complex to the vesicle, and finally triggers fusion-pore opening. The objects in the scheme are not in scale

Over the last two decades, a universal mechanism by which Ca^{2+} binding to SYTs triggers fusion of synaptic vesicles to presynaptic membranes has emerged. This process involves SYTs and complexin, which act simultaneously on SNARE complexes (complexin, synaptobrevin, and SNAP-25) and phospholipid membranes [65]. This mechanism ensures the millisecond control of neurotransmitter release from presynaptic terminals on arrival of an action potential, but goes far beyond the synapse. This starts a series of events that ultimately lead to new receptors being inserted into the membrane of this second cell, which strengthens the connection between the two cells. Ca^{2+} transients have been detected in hippocampal neurons in the absence of action potentials with physiological levels of extracellular Mg^{2+}. These $[Ca^{2+}]$ transients are the key to understanding the signaling events that include Ca^{2+} release from internal sources occurring at rest, resulting in the regulation of protein translation and gene transcription associated with synaptic plasticity [66]. ER vesicles together with mitochondria make them optimally suited to dynamically adjust a local handling of $[Ca^{2+}]_i$ in neurons at such strategic "hot spots" as synapses. Modulation of their traffic in the regions of intense neuronal activity [52] may increase the probability for mitochondria and ER vesicles to meet and act in a choreographed interplay, shaping the Ca^{2+} signals in neural processes.

7 Oxygen Free Radicals

Superoxide levels can be measured in intact brain cortex synaptosomes by flow cytometry using the fluorescent dye Mitosox, a membrane-permeant and rapidly targeted mitochondrial superoxide indicator. Also, mitochondrial hydrogen peroxide (H_2O_2) production has been measured in freshly isolated synaptosomes by a fluorometric method based on the use of scopoletin-horseradish peroxidase in the presence of mitochondrial respiratory substrates malate and glutamate, in different experimental models, including aging [16]. The activity of monoamine oxidase (MAO), an enzyme present in mitochondrial membranes, is also capable of generating H_2O_2 as a by-product of dopamine catabolism, thus contributing to the total H_2O_2 production.

Increased rates of H_2O_2 production have been observed in synaptosomes from 18-month-old animals compared to those from 6-month-old rats [67]. In accordance, evidences of increased H_2O_2 production are found in synaptosomes from cerebral cortex of 14-month-old mice, as compared with young animals [16].

ROS production from cortical and hippocampal synaptosomes has been associated with the triggering of the exocytotoxic process by the activation of presynaptic glutamate receptors, which appeared to be, at least partially, mediated by the efflux of Ca^{2+} ions from internal stores [68].

Agonists of glutamate receptors NMDA and kainate may stimulate ROS production in rat brain cortical synaptosomes, and specific antagonists of NMDA and kainate/AMPA receptors, dizocilpine hydrogen maleate (MK-801) and 6-cyano-7-nitroquinoxaline-2,3-dione (CNQX), significantly inhibit the ROS increase [69]. A high contribution of NADPH oxidase to this process has been suggested, with spontaneous as well as agonist-evoked ROS production being effectively attenuated by the potent inhibitor of

NADPH oxidase diphenyleneiodonium (DPI) [69]. In fact, though mitochondrial rates of oxygen consumption in forebrain freshly isolated synaptosomes were substantially higher than NADPH oxidase-dependent oxygen consumption, the latter was associated with greater H_2O_2 production, suggesting that NADPH oxidase rather than mitochondrial respiration is the major source of synaptic H_2O_2 generation [70].

In agreement, Tarasenko et al. reported that the process of spontaneous synaptosomal ROS generation is markedly higher in the presence of 10 mM glucose than in the presence of 4 mM pyruvate. As glucose metabolism is coupled with regeneration of NADPH required for NADPH oxidase activity, this finding suggested an important contribution of NADPH oxidase to total synaptosomal ROS production [69].

An increased ROS production, dependent on endogenous dopamine metabolism, has also been detected by flow cytometry analysis in methamphetamine-treated rat striatal synaptosomes [71].

Sipos et al. investigated the quantitative relationship between inhibition of the respiratory chain complexes and ROS formation in isolated nerve terminals, addressing to what extent Complex I, III, and IV, respectively, should be inhibited to cause ROS generation. Threshold inhibition of Complexes III and IV for generating ROS formation (around 70%) was much greater than the one of Complex I (16%) as measured by decrease in aconitase activity [72].

Interestingly, ROS generation of in situ mitochondria in nerve terminals in response to inhibition of Complex I or Complex III seems to be independent of $\Delta\psi$, as neither of the uncouplers FCCP (250 nM) or DNP (50 μM) had a significant effect on the rate of H_2O_2 production induced by rotenone or antimycin [73].

8 Nitric Oxide

Studies on subcellular localization of nNOS have shown that this enzyme is predominately attached to membranes of the endoplasmic reticulum, and in presynaptic regions [74] as well as on the external membrane of mitochondria, in pre- and postsynaptic profiles and also diffusely distributed in the cytosol [75].

Western blot assays allow disclosing a protein of 157 kDa able to react with anti-nNOS (amino terminal) antibody in mouse brain synaptosomes [76].

NOS activity can be detected in mouse brain cortex synaptosomal fractions being 0.9 ± 0.1 nmol/min mg protein. Changes in synaptosomal NOS activity have been reported in several experimental models. For instance, incubation of brain synaptosomes with 10, 25, and 50 μM deprenyl inhibited NOS activity in a concentration-dependent manner [76].

Also, synaptosomal NO production was 44% decreased in 14-month-old mice, as compared with young animals [16]. Nitric oxide is capable of reversibly inhibiting oxygen consumption of brain synaptosomes by competing with oxygen at cytochrome oxidase [77].

Studies on the effects of a NO donor on mitochondria of rat cortical synaptosomes have shown that NO and $ONOO^-$ are responsible for the decrease of O_2 consumption and ATP production and for mitochondrial membrane depolarization, observed in synaptosomal mitochondria [26].

When glutamate binds to NMDA receptor, calcium entry to the cytosol is induced, promoting the activation of neuronal NOS. Nitric oxide levels can be detected in synaptosomal fractions by a cytometric method based on the use of fluorescent probe DAF-2-DA [78]. This probe is deacetylated by intracellular esterases to DAF-2, which reacts with NO to yield the highly fluorescent triazolofluorescein (DAF-2T; λexc = 495 nm, λem = 515 nm). Figure 2 shows a typical experiment in which it can be observed that NO levels were significantly increased after incubation of brain cortex synaptosomes with 2.5 mM glutamate. Assays are carried out as follows: synaptosome fractions (25 μg protein/mL) were incubated at 37 °C for 30 min in MSH buffer (0.23 M mannitol, 0.07 M sucrose, 5 mM HEPES) supplemented with 5 mM malate, 2 mM phosphate, 1 mM $MgCl_2$, and 1.3 mM $CaCl_2$, in the presence of 10 μM DAF-2-DA. As described in Fig. 2a, dot blot of FSC versus SSC indicates the evaluated synaptosomal population. Figure 2b shows a typical histogram for control sample. Different in vitro treatments were assayed in control samples in order to evaluate NO metabolism (Fig. 2c–e). As expected, DAF-2T relative fluorescence intensity was 37% lower than controls after preincubation with the NOS inhibitor L-NNA ($p < 0.05$, Fig. 2c, f). The addition of the NO donor, GSNO, induces a 2-fold increment in DAF-2T fluorescence ($p < 0.01$, Fig. 2d, f).

The incubation with 2.5 mM glutamate also produced a 74% enhancement of NO production, as observed by an increase in DAF-2 fluorescence intensity (Fig. 2e, f).

Fig. 2 (continued) fluorescence taking control as 100%. (**a**) Typical dot blot of FSC-H versus SSC-H indicating a gated synaptosome population (R1) for control. (**b**) Typical histogram for control sample. Autofluorescence (sample without probe) is presented in the inset. (**c–e**) Typical histograms of gated events versus relative DAF-2T fluorescence intensity (FL-1) of synaptosomal fractions after incubation with 1 mM L-NNA, 5 mM GSNO, and 2.5 mM glutamate, respectively. Each histogram represents a typical experiment, which was performed in triplicate. (**f**) Bars graph quantification of the amount of DAF-2T relative fluorescence. Fluorescence events were quantified as the number of events which drop under a common marker M1 (fixed at the median value of the control histogram) taking control fluorescence events as 100%. Bars represent the mean ± SEM. Student's ***t***-test *$p < 0.05$, **$p < 0.001$ compared with control sample

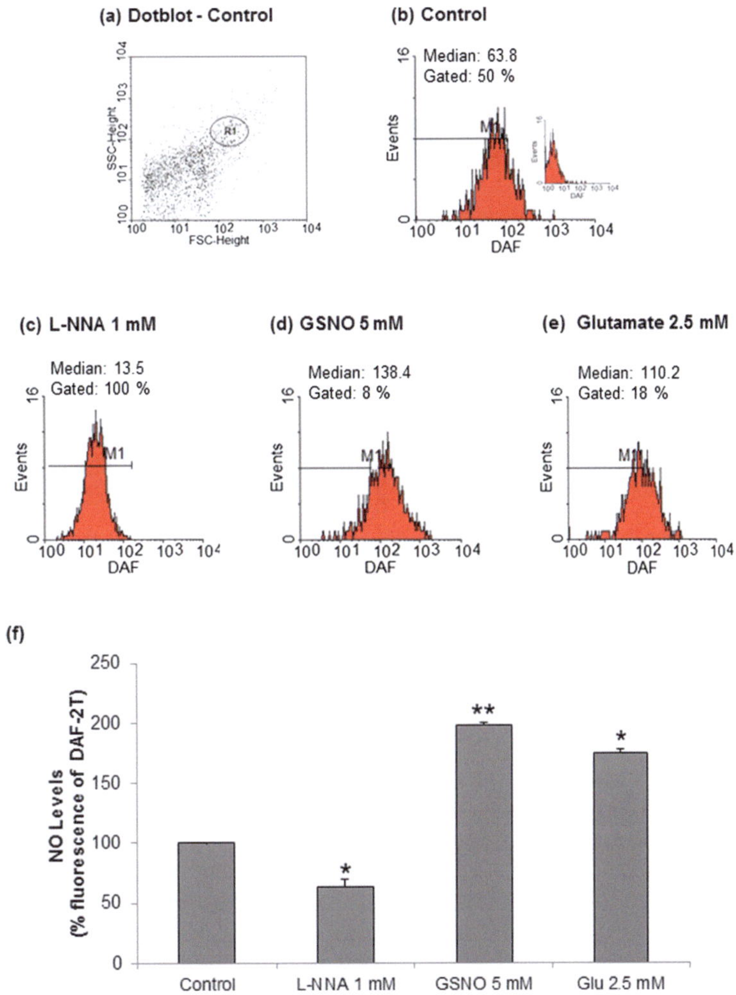

Fig. 2 Nitric oxide levels by DAF-2T fluorescence in mice brain cortex synaptosomes. Brain cortex synaptosomal fractions were loaded with the probe DAF-2-DA and direct measurements of NO levels were obtained by flow cytometry. Assays were carried out as follows: synaptosome fractions (25 μg protein/mL) were incubated at 37 °C for 30 min in MSH buffer supplemented with 5 mM malate, 2 mM phosphate, 1 mM $MgCl_2$, and 1.3 mM $CaCl_2$, in the presence of 10 μM DAF-2-DA (λexc = 495 nm, λem = 515 nm). Samples were protected from light until acquired by the FACSCalibur flow cytometer (Becton-Dickinson, USA). Autofluorescence was evaluated in samples without the probe. Negative and positive controls were conducted using 1 mM L-NNA and 5 mM GSNO, respectively. DAF-2T fluorescence was analyzed using the median value of the distribution of fluorescence events from each treatment. A common marker (M1) was fixed on control median value representing 50% of the fluorescent events. Histogram differences in DAF-2T fluorescence were quantified as the number of events which drop under the median value of the relative fluorescence distribution corresponding to M1. A lower number of DAF-2T fluorescence events under M1 would reflect increased nitric oxide levels. Quantification of results was shown as bar graphs in which data were expressed as the percentage of DAF-2T

9 Synaptosomes and the Response to Aging

The aging process involves increased ROS production together with alterations in Ca^{2+} homeostasis and mitochondrial dysfunction. Synaptic mitochondria from aged animals seem to be partially dysfunctional, probably leading to ATP deficiency, alterations of calcium homeostasis, and decreased NO levels, while non-synaptic mitochondria would be more resistant to age-related dysfunction and oxidative damage partially due to its location at the cell soma as compared with those present at the nerve terminals where high calcium movements occur [16].

Decreases in the activity of mitochondrial respiratory chain complexes have been implicated in aging and neurodegenerative disorders. A decreased NADH oxidation was observed in brain cortex non-synaptic mitochondria from old rats, while no major changes in NADH oxidation rate were found in synaptic mitochondria from brain cortex [79].

The aging-associated energy decline has been attributed to somatic mutations of mitochondrial DNA induced by oxygen radicals, with Complex I being mostly affected and strongly rate limiting for electron transfer [80]. Decreased enzyme activities of NADH dehydrogenase and cytochrome oxidase in whole-brain mitochondria have been observed during aging [81].

Differences in the response of synaptosomal mitochondria to functional impairment associated to the aging process have been characterized, as compared with non-synaptic mitochondria. Mitochondrial respiratory dysfunction is observed in brain cortex synaptosomes in 14-month-old animals, with decreased activity of mitochondrial respiratory Complexes II–III and IV as compared with young animals [16].

An impaired function of these electron transport chain components had been previously reported in brain synaptosomes from 18-month-old animals, compared to those from 6-month-old rats [67]. Davey et al. investigated the extent to which a complex activity could be inhibited before causing impairment of mitochondrial energy metabolism and examined the relative contribution of individual mitochondrial respiratory chain complexes to the control of NAD-linked substrate oxidative phosphorylation in synaptic mitochondria [82]. Complex I, III, and IV activities can be reduced by approximately 25, 80, and 70%, respectively, in synaptic mitochondria, before major changes in rates of oxygen consumption and ATP synthesis are observed. The reduced threshold value for Complex I, compared to the other mitochondrial complexes, suggests that, in mitochondria of synaptic origin, Complex I activity has a major control of oxidative phosphorylation, such that when a threshold of 25% inhibition is exceeded, energy metabolism is severely impaired, resulting in a reduced synthesis of ATP [82].

Interestingly, protein expression of cytochrome oxidase subunits I, II/III, and IV exhibited an age-related increase from 2 months up to 24 months old in rat cerebral cortex [83]. Also, evidence of increased expression of mitochondrial encoded genes in Complexes I, III, IV, and V of the respiratory chain in 12- and 18-month-old C57BL6 mice compared to 2-month-old mice suggests a compensatory mechanism that allows the production of proteins involved in the electron transport chain [84]. In agreement, activity of mitochondrial respiratory Complexes I–III was found increased in brain cortex synaptosomes from 14-month-old animals as compared with young mice [16]. Regarding complexes II–III and IV, enzymatic activities were increased by 54% and 63%, respectively, in synaptosomes from 17-month-old animals, as compared with young mice [15].

Impairment in oxygen consumption rates has been observed in synaptosomes from 17-month-old mice, as compared with young animals, exhibiting a decreased basal respiration and a tendency to decrease in ATP generation (oligomycin-sensitive mitochondrial respiration) [15].

A decrease in proton leak (oligomycin-insensitive mitochondrial respiration) can be associated with UCP activity. UCP-2, a protein known to be responsible for a mild uncoupling of oxidative phosphorylation, has been shown to be increased in aging in different tissues [85]. In fact, UCP-2 protein expression is significantly higher in synaptosomal membranes from 17-month-old mice, as compared with young animals. UCP-2 upregulation seems to provide a regulatory physiological mechanism by which a mild electron leakage with a slight decrease in energy production could allow the maintenance of low oxygen free radical generation [15].

Even though a tendency to decrease is found in respiration driving ATP synthesis and in maximal respiration in aged mice, spare respiratory capacity was not modified by aging. Therefore, it can be assumed that the ability of the synaptosome to adjust oxygen consumption in response to intracellular ATP demands was preserved during aging [15]. In agreement, Stauch et al. have shown that while differences can be observed in coupling and electron flow during aging, mitochondrial capacity for substrate oxidation, ATP turnover, and proton leak was not found to be significantly different between ages in the synaptic mitochondria [86].

Differences in $\Delta\psi$ during aging have been previously reported in different tissues. For instance, Kokoszka et al. reported a 10 mV decrease in mitochondrial membrane potential in mouse liver mitochondria from old mice relative to young and middle-aged animals [87]. On the contrary, Choi et al. have reported that synaptosomes from aged mice maintained membrane potential as effectively as those from young mice [37].

According to our results, mitochondrial membrane potential was preserved or slightly diminished in brain cortex synaptosomes from 14- or 17-month-old mice, respectively, as compared with young animals. However, a decrease in $\Delta\psi$ was found in 14- and 17-month-old mice, after incubation of brain cortex synaptosomes with 100 μM calcium [15, 16].

Acknowledgments

This research was supported by grants from Consejo Nacional de Investigaciones Científicas y Técnicas (CONICET, PIP 112-20110100271) and Universidad de Buenos Aires (UBA, 20020130100255).

References

1. Whittaker VP, Michaelson IA, Kirkland RJ (1964) The separation of synaptic vesicles from nerve-ending particles ('synaptosomes'). Biochem J 90:293–303
2. Rodríguez de Lores Arnaiz G, De Robertis E (1972) Properties of the isolated nerve endings. In: Bronner F, Kleinzeller A (eds) Current topics in membranes and transport, vol III. Academic Press, New York, pp 237–272
3. Nicholls DG (2010) Stochastic aspects of transmitter release and bioenergetic dysfunction in isolated nerve terminals. Biochem Soc Trans 38:457–459
4. Ly CV, Verstreken P (2006) Mitochondria at the synapse. Neuroscientist 12:291–299
5. Attwell D, Laughlin SB (2001) An energy budget for signaling in the grey matter of the brain. J Cereb Blood Flow Metab 21:1133–1145
6. De Robertis E, Pellegrino de Iraldi A, Rodríguez de Lores Arnaiz G et al (1961) On the isolation of nerve endings and synaptic vesicles. J Biophys Biochem Cytol 9:229–235
7. Whittaker VP (1959) The isolation and characterization of acetylcholine-containing particles from brain. Biochem J 72:694–706
8. Bellamy D (1959) The distribution of bound acetylcholine and choline acetylase in rat and pigeon brain. Biochem J 72:165–168
9. De Robertis E, Pellegrino de Iraldi A, Rodríguez de LoresArnaiz G et al (1962) Cholinergic and non-cholinergic nerve endings in rat brain. I. Isolation and subcellular distribution of acetylcholine and acetylcholinesterase. J Neurochem 9:23–35
10. Gray EG, Whittaker VP (1962) The isolation of nerve endings from brain: an electron-microscopic study of cell fragments derived by homogenization and centrifugation. J Anat 96:79–88
11. Rodríguez de Lores Arnaiz G, Alberici M, De Robertis E (1967) Ultrastructural and enzymic studies of cholinergic and noncholinergic synaptic membranes isolated from brain cortex. J Neurochem 14:215–225
12. Lai JC, Clark JB (1976) Preparation and properties of mitochondria derived from synaptosomes. Biochem J 154:423–432
13. Cotman CW, Matthews DA (1971) Synaptic plasma membranes from rat brain synaptosomes: isolation and partial characterization. Biochim Biophys Acta 249:380–394
14. Rodríguez de Lores Arnaiz G, Girardi E (1977) The increase in respiratory capacity of brain subcellular fractions after the administration of the convulsant 3-mercaptopropionic acid. Life Sci 21:637–646
15. Lores-Arnaiz S, Lombardi P, Karadayian AG et al (2016) Brain cortex mitochondrial bioenergetics in synaptosomes and non-synaptic mitochondria during aging. Neurochem Res 41:353–363
16. Lores-Arnaiz S, Bustamante J (2011) Age-related alterations in mitochondrial physiological parameters and nitric oxide production in synaptic and non-synaptic brain cortex mitochondria. Neuroscience 188:117–124
17. Anderson MF, Sims NR (2000) Improved recovery of highly enriched mitochondrial fractions from small brain tissue samples. Brain Res Protocol 5:95–101
18. Dunkley PR, Jarvie PE, Robinson PJ (2008) A rapid Percoll gradient procedure for prepara-

tion of synaptosomes. Nat Protoc 3:1718–1728
19. Tenreiro P, Rebelo S, Martins F et al (2017) Comparison of simple sucrose and percoll based methodologies for synaptosome enrichment. Anal Biochem 517:1–8
20. Nicholls DG (1993) The glutamatergic nerve terminal. Eur J Biochem 212:613–631
21. Dodd PR, Hardy JA, Oakley AE et al (1981) A rapid method for preparing synaptosomes: comparison, with alternative procedures. Brain Res 226:107–118
22. Dunkley PR, Jarvie PE, Heath JW et al (1986) A rapid method for isolation of synaptosomes on Percoll gradients. Brain Res 372:115–129
23. Gylys KH, Fein JA, Cole GM (2000) Quantitative characterization of crude synaptosomal fraction (P-2) components by flow cytometry. J Neurosci Res 61:186–192
24. Sudhof TC (1995) The synaptic vesicle cycle: a cascade of protein-proteininteractions. Nature 375:645–653
25. Kornau HC, Schenker LT, Kennedy MB et al (1995) Domain interaction between NMDA receptor subunits and the postsynaptic density protein PSD-95. Science 269:1737–1740
26. Blanco Garcia J, Aldinucci C, Maiorca SM et al (2009) Physiopathological effects of the NO donor 3-Morpholinosydnonimine on rat cortical synaptosomes. Neurochem Res 34:931–941
27. Gylys KH, Fein JA, Yang F et al (2004) Enrichment of presynaptic and postsynaptic markers by size-based gating analysis of synaptosome preparations from rat and human cortex. Cytometry A 60:90–96
28. Hollenbeck PJ, Saxton WM (2005) The axonal transport of mitochondria. J Cell Sci 118:5411–5419
29. Guo X, Macleod GT, Wellington A et al (2005) GTPase dMiro is required for axonal transport of mitochondria to Drosophila synapses. Neuron 47:379–393
30. Ma H, Cai Q, Lu W et al (2009) KIF5B motor adaptor syntabulin maintains synaptic transmission in sympathetic neurons. J Neurosci 29:13019–13029
31. Stowers RS, Megeath LJ, Górska-Andrzejak J et al (2002) Axonal transport of mitochondria to synapses depends on Milton, a novel Drosophila protein. Neuron 36:1063–1077
32. Lin MY, Sheng ZH (2015) Regulation of mitochondrial transport in neurons. Exp Cell Res 334:35–44
33. Sheng ZH (2014) Mitochondrial trafficking and anchoring in neurons: new insight and implications. J Cell Biol 204:1087–1098
34. Brown MR, Sullivan PG, Geddes JW (2006) Synaptic mitochondria are more susceptible to Ca^{2+} overload than non-synaptic mitochondria. J Biol Chem 281:11658–11668
35. Naga KK, Sullivan PG, Geddes JW (2007) High cyclophilin D content of synaptic mitochondria results in increased vulnerability to permeability transition. J Neurosci 27:7469–7475
36. Yarana C, Sanit J, Chattipakorn N et al (2012) Synaptic and non synaptic mitochondria demonstrate a different degree of calcium-induced mitochondrial dysfunction. Life Sci 90:808–814
37. Choi SW, Gerencser AA, Lee DW et al (2011) Intrinsic bioenergetic properties and stress sensitivity of dopaminergic synaptosomes. J Neurosci 31:4524–4534
38. Brand MD, Nicholls DG (2011) Assessing mitochondrial dysfunction in cells. Biochem J 435:297–312
39. Choi SW, Gerencser AA, Nicholls DG (2009) Bioenergetic analysis of isolated cerebrocortical nerve terminals on a microgram scale: spare respiratory capacity and stochastic mitochondrial failure. J Neurochem 109:1179–1191
40. Chavan V, Willis J, Walker SK et al (2015) Central presynaptic terminals are enriched in ATP but the majority lack mitochondria. PLoS One 10(4):e0125185. https://doi.org/10.1371/journal.pone.0125185
41. Bustamante J, Di Libero E, Fernandez-Cobo M et al (2004) Kinetic analysis of thapsigargin-induced thymocyte apoptosis. Free Radic Bio Med 37:1490–1498
42. Alekseenko AV, Lemeshchenko VV, Pekun TG et al (2012) Glutamate-induced free radical formation in rat brain synaptosomes is not dependent on intrasynaptosomal mitochondria membrane potential. Neurosci Lett 513:238–242
43. Hrynevich SV, Pekun TG, Waseem TV et al (2015) Influence of glucose deprivation on membrane potentials of plasma membranes, mitochondria and synaptic vesicles in rat brain synaptosomes. Neurochem Res 40:1188–1196
44. Berridge MJ (1997) Elementary and global aspects of calcium signalling. J Physiol 499:291–306
45. Bootman MD, Collins TJ, Peppiatt CM et al (2001) Calcium signalling-an overview. Semin Cell Dev Biol 12:3–10
46. Bucurenciu I, Bischofberger J, Jonas P (2010) A small number of open Ca2+ channels trigger transmitter release at a central GABAergic synapse. Nat Neurosci 13:19–21

47. Bacci A, Verderio C, Pravettoni E et al (1999) Synaptic and intrinsic mechanisms shape synchronous oscillations in hippocampal neurons in culture. Eur J Neurosci 11:389–397
48. Berridge MJ (1998) Neuronal calcium signaling. Neuron 21:13–26
49. Berridge MJ, Lipp P, Bootman MD (2000) The versatility and universality of calcium signalling. Nat Rev Mol Cell Biol 1:11–21
50. Ishii K, Hirose K, Iino M (2006) Ca2+ shuttling between endoplasmic reticulum and mitochondria underlying Ca2+ oscillations. EMBO Rep 7:390–396
51. Bannai H, Inonie H, Nakayama T et al (2004) Kinesin dependent, rapid, bi-directional transport of ER sub-compartment in dendrites of hippocampal neurons. J Cell Sci 117:163–175
52. Mironov SL, Symonchuk N (2006) ER vesicles and mitochondria move and communicate at synapses. J Cell Sci 119:4926–4934
53. Markram H, Helm PJ, Sakmann B (1995) Dendritic calcium transients evoked by single back-propagating action potentials in rat neocortical pyramidal neurons. J Physiol 485:1–20
54. Billups B, Forsythe ID (2002) Presynaptic mitochondrial calcium sequestration influences transmission at mammalian central synapses. J Neurosci 22:5840–5847
55. Medler K, Gleason EL (2002) Mitochondrial Ca(2+) buffering regulates synaptic transmission between retinal amacrine cells. J Neurophysiol 87:1426–1439
56. Belair EL, Vallee J, Robitaille R (2005) Long-term in vivo modulation of synaptic efficacy at the neuromuscular junction of Rana pipiens frogs. J Physiol 569:163–178
57. Mattson MP (2008) Glutamate and neurotrophic factors in neuronal plasticity and disease. Ann N Y Acad Sci 1144:97–112
58. Nicholls DG (2008) Oxidative stress and energy crises in neuronal dysfunction. Ann N Y Acad Sci 1147:53–60 o
59. Bustamante J, Czerniczyniec A, Lores-Arnaiz S (2016) Ketamine effect on intracellular and mitochondrial calcium mobilization. Biocell 40:11–14
60. Logan CV, Szabadkai G, Sharpe JA et al (2014) Loss-of-function mutations in MICU1 cause a brain and muscle disorder linked to primary alterations in mitochondrial calcium signaling. Nat Genet 46:188–193
61. Rangaraju V, Calloway N, Ryan TA (2014) Activity-driven local ATP synthesis is required for synaptic function. Cell 156:825–835
62. Katz B, Miledi R (1967) Ionic requirements of synaptic transmitter release. Nature 215:651
63. Sabatini BL, Regehr WG (1996) Timing of neurotransmission at fast synapses in the mammalian brain. Nature 384:170–172
64. Perin MS, Fried VA, Mignery GA et al (1990) Phospholipid binding by a synaptic vesicle protein homologous to the regulatory region of protein kinase C. Nature 345:260–263
65. Südhof TC (2012) Calcium control of neurotransmitter release. Cold Spring Harb Perspect Biol 4:a011353
66. Kavalali ET (2015) The mechanisms and functions of spontaneous neurotransmitter release. Nat Rev Neurosci 16(1):5–16
67. Kilbride SM, Telford JE, Davey GP (2008) Age-related changes in H2O2 production and bioenergetics in rat brain synaptosomes. Biochem Biophys Acta 1777:783–788
68. Tarasenko A, Krupko O, Himmelreich N (2011) Presynaptic kainate and NMDA receptors are implicated in the modulation of GABA release from cortical and hippocampal nerve terminals. Neurochem Int 59:81–89
69. Tarasenko A, Krupko O, Himmelreich N (2012) Reactive oxygen species induced by presynaptic glutamate receptor activation is involved in [3H] GABA release from rat brain cortical nerve terminals. Neurochem Int 61:1044–1051
70. Abdel-Rahman EA, Mahmoud AM, Aaliya A et al (2016) Resolving contributions of oxygen-consuming and ROS-generating enzymes at the 2016 synapse. Oxidative Med Cell Longev 2016:1. https://doi.org/10.1155/2016/1089364
71. Pubill D, Chipana C, Camins A et al (2005) Free radical production induced by methamphetamine in rat striatal synaptosomes. Toxicol Appl Pharmacol 204:57–68
72. Sipos I, Tretter L, Adam-Vizi V (2003) Quantitative relationship between inhibition of respiratory complexes and formation of reactive oxygen species in isolated nerve terminals. J Neurochem 84:112–118
73. Sipos I, Tretter L, Adam-Vizi V (2003) The production of reactive oxygen species in intact isolated nerve terminals is independent of the mitochondrial membrane potential. Neurochem Res 28:1575–1581
74. Faber-Zuschratter H, Seidenbecher T, Reymann K et al (1996) Ultrastructural distribution of NADPH-diaphorase in the normal hippocampus and after long-term potentiation. J Neural Transm 103:807–817
75. Batista CM, de Paula KC, Cavalcante LA et al (2001) Subcellular localization of neuronal nitric oxide synthase in the superficial gray layer of the rat superior colliculus. Neurosci Res 41:67–70

76. Czerniczyniec A, Bustamante J, Lores-Arnaiz S (2006) Modulation of brain mitochondrial function by deprenyl. Neurochem Int 48:235–241
77. Brown GC, Cooper CE (1994) Nanomolar concentrations of nitric oxide reversibly inhibit synaptosomal respiration by competing with oxygen at cytochrome oxidase. FEBS Lett 356:295–298
78. López-Figueroa MO, Caamaño C, Morano MI et al (2000) Direct evidence of nitric oxide presence within mitochondria. Biochem Biophys Res Commun 272:129–133
79. Genova ML, Bovina C, Marchetti M et al (1997) Decrease of rotenone inhibition is a sensitive parameter of complex I damage in brain non-synaptic mitochondria of aged rats. FEBS Lett 410:467–469
80. Lenaz G, D'Aurelio M, Merlo Pich M et al (2000) Mitochondrial bioenergetics in aging. Biochem Biophys Acta 1459:397–404
81. Navarro A, Boveris A (2007) Brain mitochondrial dysfunction in aging: conditions that improve survival, neurological performance and mitochondrial function. Front Biosci 12:1154–1163
82. Davey GP, Peuchen S, Clark JB (1998) Energy thresholds in brain mitochondria. Potentialinvolvement in neurodegeneration. J Biol Chem 273:12753–12757
83. Nicoletti VG, Tendi EA, Lalicata C et al (1995) Changes of mitochondrial cytochrome c oxidase and FoF1 ATP synthase subunits in rat cerebral cortex during aging. Neurochem Res 20:1465–1470
84. Manczak M, Jung Y, Park BS et al (2005) Time course of mitochondrial gene expressions in mice brains: implications for mitochondrial dysfunction, oxidative damage, and cytochrome c in aging. J Neurochem 92:494–504
85. Barazzoni R, Nair KS (2001) Changes in uncoupling protein-2 and -3 expression in aging rat skeletal muscle, liver, and heart. Am J Physiol Endocrinol Metab 280:E413–E419
86. Stauch KL, Purnell PR, Fox HS (2014) Aging synaptic mitochondria exhibit dynamic proteomic changes while maintaining bioenergetic function. Aging (Albany NY) 6:320–334
87. Kokoszka JE, Coskun P, Esposito LA et al (2001) Increased mitochondrial oxidative stress in the Sod2 (+/−) mouse results in the age-related decline of mitochondrial function culminating in increased apoptosis. Proc Natl Acad Sci USA 98:2278–2283

Chapter 9

Isolation of Highly Purified and Functionally Active Synaptic Mitochondria

Sergei V. Baranov, Nicolas K. Khattar, Svitlana Yablonska, Oxana V. Baranova, Eric S. Kretz, Timothy M. Larkin, Diane L. Carlisle, and Robert M. Friedlander

Abstract

Isolation of synaptic mitochondria is an important step in assessing synaptic bioenergetics. Purity and functional activity of mitochondria are important to characterize the mitochondrial proteome and determine the contribution of intrinsic factors to mitochondrial activity. We have developed and validated an isolation protocol to obtain highly purified synaptosomal mitochondria from brain tissue. The method combines Percoll step gradient centrifugation to obtain the synaptosomal fraction and nitrogen cavitation to release neuronal mitochondria and using anti-mitochondrial outer membrane protein antibodies conjugated to magnetic nanobeads to extract the mitochondria. The final products of isolation are non-synaptosomal mitochondria, which are a mixture of mitochondria isolated from different brain cells (i.e., neurons, astrocytes, oligodendrocytes, microglia) and synaptic mitochondria, which are of neuronal origin. The procedure produces mitochondria that (1) have minimal cytoplasmic contamination and (2) are functionally active based on measurements of mitochondrial matrix protein import; the process depends on the mitochondrial integrity and presence of high membrane potential. The procedure requires approximately 4 h for the isolation of synaptosomal mitochondria and can be used to isolate mitochondria from other species, including rat, monkey, and human brain tissue as well. This method allows researchers to study disease implications on highly enriched synaptosomal mitochondria without the confounding effect of cellular and organelle contaminants.

Key words Mitochondria, Isolation, Synaptosomes, Neurons

1 Introduction: Functional Mitochondria in Brain Synaptosomes

Mitochondria play an important role in neuronal homeostasis by providing energy and participating in numerous signaling pathways [1–5]. Pure mitochondria isolations are essential to appropriately assess mitochondrial function in the brain. Given the heterogeneity of brain tissue, isolation of highly purified functional neuronal mitochondria can be very challenging. To provide mitochondria predominantly of the neuronal origin, mitochondria are extracted

Kathryn M. Murphy (ed.), *Synaptosomes*, Neuromethods, vol. 141,
https://doi.org/10.1007/978-1-4939-8739-9_9,

from synaptosomes. Synaptosomes are vesicles formed from neuronal terminals during mechanical homogenization of brain tissue and encapsulate neuronal mitochondria. It has been shown that synaptosomal mitochondria respond differently to environmental insults such as Ca^{2+} overload as compared with non-synaptosomal mitochondria [6]. Our group has also reported that mitochondrial protein import is impaired exclusively in synaptosomal mitochondria of Huntington's disease mouse models [2].

1.1 Methods to Study Brain Mitochondria In Vitro

The most robust protocols to isolate synaptosomal mitochondria include Percoll (or Ficoll) discontinuous gradient centrifugation followed by synaptosomal disruption with either digitonin or nitrogen cavitation pressurization [7, 8]. Neither method, however, allows for the elimination of contaminants, including synaptic vesicles, lysosomes, and peroxisomes, which previously proved extremely challenging. Whole synaptosomes are also still found in the final mitochondrial fraction using these methods [8]. To mitigate this obstacle, we used specific anti-mitochondrial outer membrane protein TOM22 antibodies to purify synaptosomal mitochondria without loss of function. These antibodies are conjugated with magnetic nanobeads for use in magnetic activated cell sorting (MACS). Using a magnetic field, the nanobeads are able to selectively retain mitochondria as they bind to the TOM22 antibodies. Any non-mitochondrial fraction of the sample will therefore be washed out. The exclusive use of the MACS technique was shown to be sufficient to obtain brain mitochondria of different sources (neuronal body, glia). However, these mitochondria are not purely of neuronal origin, as it requires disruption of the synaptosomes before application of the antibodies.

Our method enables us to obtain highly purified and functional neuronal (synaptic) mitochondria. We have successfully applied the method to characterize the mitochondrial status of human cortex and mouse brain. The protocol focuses on obtaining pure synaptosomal/neuronal mitochondria but it also yields purified non-synaptosomal mitochondria that can further be used for functional or structural analyses.

2 Materials

2.1 Animal Preparation

To obtain synaptosomal mitochondria, we surgically extracted fresh brain tissue from mice. We have used a brain tissue from several strains, including C57/Bl6 and B6CBA mice, transgenic, and wild type, with no differences in results among mouse strains.

2.2 Buffer Preparation

Isolation of functional mitochondria requires the use of buffers that are optimized to prevent loss of mitochondrial function during isolation. A stock solution of 200 mM ethylene-bis (oxyethylenenitrilo)tetraacetic acid (EGTA, pH 7.4) is needed. Dissolve 7.6 g of EGTA in 80 mL of bi-distilled water, adjust pH to 7.4 with KOH, and bring the solution to 100 mL with deionized water. We use two isolation buffers, IB-1 and IB-2. IB-1 contains 225 mM sucrose (Sigma, cat. no. S0389), 75 mM mannitol (Sigma, cat. no. M9546), 1 mM EGTA (EGTA; Sigma, cat. no. E4378), and 5 mM HEPES (Sigma, cat. no. H3375) adjusted to pH 7.4 with Tris base (Sigma, cat. no. T1503). IB-2 is similar to IB-1 but without the EGTA. IB-2 contains 225 mM sucrose, 75 mM mannitol, and 5 mM HEPES adjusted to pH 7.4 with Tris base with the same chemical source as for IB-1. All solutions are stored at 4 °C.

2.3 Tissue Disruption and Differential Centrifugation

To disrupt the brain tissue, we used a motor-driven Teflon Potter-Elvehjem homogenizer (Sigma, cat. no. Z403903). For initial centrifugation we used 16 mL polycarbonate round-bottom centrifuge tubes in a standard benchtop refrigerated centrifuge held at 4 °C. To make the Percoll for differential centrifugation, we prepare three Percoll solutions, 40%, 24%, and 15%. The 40% Percoll solution is made with 80 mL of Percoll (Sigma, cat. no. 17-0891-01) dissolved in 120 mL of isolation buffer I, pH adjusted to 7.4 with HCl. The 24% Percoll solution is 30 mL of 40% Percoll with the total volume brought to 50 mL with isolation buffer I. The 15% Percoll solution is 17.5 mL of 40% Percoll with the total volume brought to 50 mL with isolation buffer I. Prepare the Percoll gradient by creating in 50 mL round-bottom transparent centrifuge tubes (Nalgene, cat. no. 3138-0050) a discontinuous layer of 1.5 mL of 40% Percoll at the bottom of the tube with 4 mL of 24% Percoll layered on top of it. Differential centrifugation in the Percoll gradient requires a refrigerated high-speed centrifuge with fixed-angle rotor (Thermo, Sorvall RC-6, F21-8x50y rotor) held at 4 °C.

2.4 Disruption of Synaptosomes and Isolation of Synaptosomal Mitochondria

Synaptosomes are disrupted to obtain synaptosomal mitochondria using a cell disruption vessel (Parr Instrument Company, cat. no. 4639) that is capable of maintaining nitrogen gas at 1100 psi for 15 min at 4 °C. After disruption of the synaptosomes, mitochondria are isolated using magnetic sorting. We used the mitochondria isolation kit for mouse tissue, which includes anti-TOM22 microbeads, 10× separation buffer, MACS separation LS Columns, and magnetic Quadro MACS Separator (Miltenyi Biotec, cat. no. 130-097-039).

3 Methods

This procedure will yield non-synaptosomal and synaptosomal (neuronal) mitochondria. In general, the workflow of the isolation is to obtain fresh brain tissue, disrupt the tissue, and separate non-

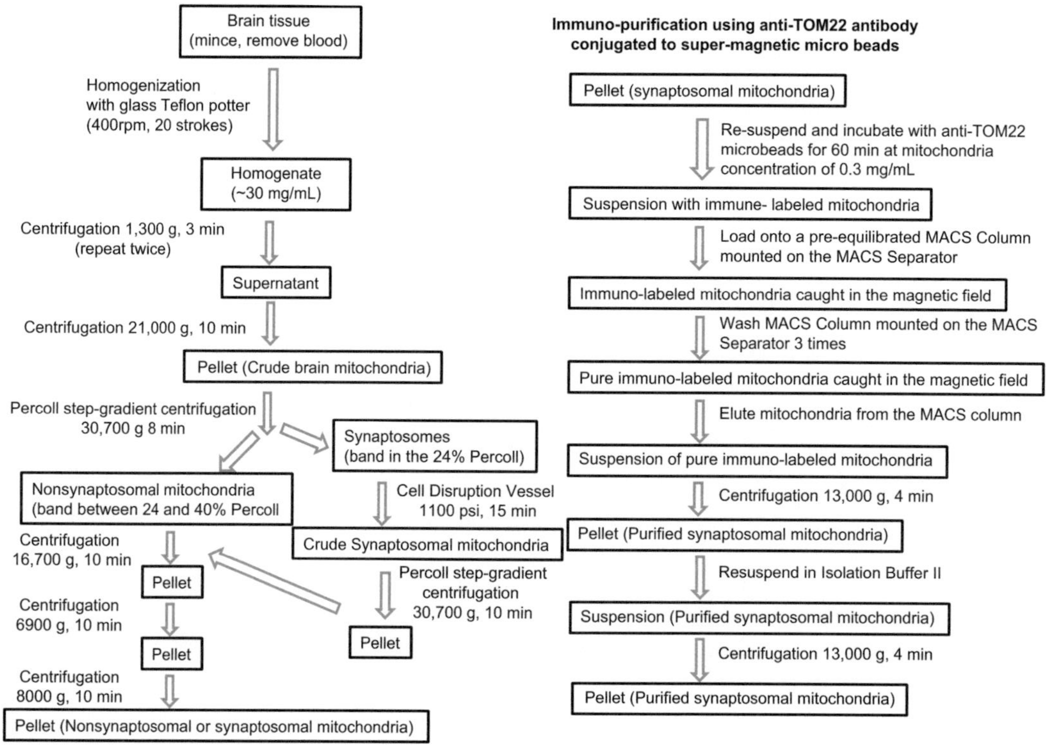

Fig. 1 Flowchart of the mitochondrial isolation technique

synaptosomal mitochondria and synaptosomes. Synaptosomes are then disrupted and magnetic sorting is used to obtain synaptosomal mitochondria. The general workflow is shown in Fig. 1.

3.1 Isolation of Non-synaptosomal Mitochondria and Synaptosomes from Mouse Forebrain

In order to have enough material to work with in the Percoll gradient and subsequent steps, we start with pooling fresh brain tissue from five mice. Mitochondria were isolated from fresh brain tissue. On a regular basis, we use five mouse brains to get enough synaptosomal mitochondria to perform protein analysis and functional characterization. We extract mouse forebrains, put them immediately to 10–15 mL of ice-cold IB1, and place the container on ice during transport back to the workbench. Tissues are rinsed using ice-cold IB1. After rinsing, we transfer the washed tissue into a beaker with fresh ice-cold IB-1. Using scissors we mince the tissue into small pieces to get rid of the blood trapped inside the specimen. This is a critical step because trapped blood in the specimen may compromise mitochondrial activity.

Tissues are then transferred to the glass potter for homogenization. We homogenize the tissue using 20 slow up and down strokes with a Teflon pestle operated at 400 rpm. To preserve mitochondria, we keep the volume of added IB-1 at five to ten times the estimated volume of the tissue sample. Slow movement of the pestle during homogenization is essential to avoid disruption of the mitochondria by inducing too low pressure in homogenate during the upward movement. It is preferable to perform all the steps to the isolation at 4 °C.

We then transfer the homogenate to a 16 mL centrifuge tube and centrifuge at 1300 × *g* for 3 min at 4 °C. Non-synaptosomal mitochondria from broken cells are in the supernatant along with other cellular components, and so the supernatant is carefully decanted into an empty polycarbonate centrifuge tube and placed on ice. The remaining pellet contains additional intact cells, and in order to disrupt the cell membrane the pellet is resuspended in 5–10 mL of IB-1 and centrifuged again at 1300 × *g* for 3 min at 4 °C. The supernatant is added to the previous supernatant and in order to pellet the non-synaptosomal mitochondria and the synaptosomes the pooled supernatant is centrifuged at 21,700 × *g* for 10 min at 4 °C. It is important to make sure that no loose pellet is transferred for hi-speed centrifugation. Blood and other impurities often found in the pellet may compromise mitochondrial activity.

Non-synaptosomal mitochondria and synaptosomes are isolated from other cellular components using a Percoll gradient. Thus, during the centrifugation step, we prepared the Percoll step gradient by creating a discontinuous layer of 1.5 mL of 40% Percoll at the bottom of the tube with 4 mL of 24% Percoll layered on top of it.

The non-synaptosomal mitochondria and the synaptosomes are in the pellet due to the high-speed centrifugation. We discarded the supernatant and resuspended the pellet in 15% Percoll. We then gently transfer 3.5 mL of resuspended pellet in the 15% Percoll on the top of 24% Percoll to create a third discontinuous layer. To infuse the resuspended pellet into the Percoll gradient, we centrifuged the suspension at 30,700 × *g* for 8 min at 4 °C using slow acceleration/deceleration regimes. Due to its density, the synaptosomal fraction (SN) will accumulate between the 15 and 24% layers and it is collected for further processing.

The non-synaptosomal mitochondria are on top of the 40% Percoll density and are collected and can be used in additional experiments.

It is critical that we do not pick up any material lying on the bottom of the tube; it is most likely debris and/or blood that can compromise mitochondrial activity. We adjust the volume of the NS fraction to 10–12 mL with IB1, then centrifuge at 16,700 × *g* for 10 min at 4 °C to remove Percoll, carefully aspirate the supernatant, and resuspend the loose pellet in 10 mL of IB1. Mitochondria are pelleted by centrifugation at 6900 × *g* for 10 min at 4 °C. To wash the pellet, decant the supernatant, resuspend the pellet in 1 mL of

IB2, and centrifuge at 8000 × *g* for 10 min at 4 °C. The non-synaptosomal mitochondria are ready for functional assays.

In order to obtain synaptosomal mitochondria after collecting synaptosomes from the Percoll gradient, we need to disrupt the synaptosomes without disrupting the mitochondria contained within them. For this step, we dilute the SN fraction at a 1:1 ratio with IB1 and incubate in the cell disruption vessel at 1100 psi for 15 min at 4 °C using a magnetic stirrer. Careful regulation of the pressure is important, since increasing the pressure above 1100 psi would result in decreased amount/activity of the synaptosomal mitochondria. Now that synaptosomes are disrupted, we collect the SN fraction from the cell disruption vessel and centrifuge at 16,700 × *g* for 10 min at 4 °C to remove Percoll. To wash the pellet to remove any remaining Percoll, we carefully aspirate the supernatant, resuspend the loose pellet in 10 mL of IB1, centrifuge at 6900 × *g* for 10 min at 4 °C, decant the supernatant, resuspend the pellet in 1 mL of IB2, and centrifuge at 8000 × *g* for 10 min at 4 °C. We then decant the supernatant and resuspend the pellet into minimal volume of IB2. The SN fraction now contains mitochondria, but may also contain other organelles that were trapped in synaptosomes such as endoplasmic reticulum. Synaptosomal mitochondria are ready to be further purified with the MACS isolation kit.

3.2 Purification of Synaptosomal Mitochondrial Fraction Using MACS Mitochondria Isolation Kit

The obtained mixture contains the contents of synaptosomes; however, it is not highly enriched mitochondria. To separate the mitochondria from other synaptosome contents, we use a magnetic bead-based mitochondrial purification system. To use our synaptosomal contents with the system, we resuspend the disrupted SN fraction in 10 mL freshly diluted 1× separation buffer and incubate it on the nutator shaker (Clay Adams, cat. no. 421105) at a speed of 12 rpm in the presence of 75 μL anti-TOM22 microbeads for 60 min at mitochondrium concentration of 0.3 mg/mL at 4 °C. We mount the LC column on the magnetic field of MACS separator and pre-equilibrate the column by washing it once with 3 mL of 1× separation buffer and apply 3.3 mL of mitochondrial suspension into the reservoir of LS column. Wait until the liquid is filtered out (approximately 5 min) and repeat this step twice. We find that it is possible to increase the yield of mitochondria by repeating this step with flow-through solution. The column is washed three times with 1× separation buffer. To elute the mitochondria, add 1 mL of IB2 into the column reservoir and place a 1.5 mL tube under the column nozzle. Remove the column from the separator and flush out the mitochondria by firmly pushing the liquid through the column using the plunger (provided with each LC column). To pellet the mitochondria, we centrifuge the eluted solution at 13,000 × *g* for 4 min at 4 °C. To minimize the loss of mitochondria, we repeat the flushing of the column with IB2 and subsequent centrifugation twice or until the liquid flushed from the column becomes clear.

We find that the separation buffer can affect subsequent mitochondrial studies; thus, it is critical that we wash the collected fraction to remove the separation buffer. To do so, we resuspend the mitochondria in IB2 by centrifuging at 13,000 × *g* for 4 min at 4 °C. We find that synaptosomal mitochondria are best maintained at high concentration in the suspension; thus, after centrifugation, we resuspend the mitochondria in IB2 using no more than 50 μL of IB2 buffer. To use synaptosomal mitochondria in functional assays, we transfer the mitochondrial suspension into a 1 mL test tube and store on ice until used for functional tests, no longer than 4 h after isolation. If mitochondria are needed for other assay types, such as protein characterization or other molecular biological assays, it may be stored at −80 °C or, preferably, in liquid nitrogen until used.

This method is specifically focused to provide highly enriched synaptic mitochondria that exhibit good functional activity. In order to make sure that the yielded mitochondria fulfill these criteria, both the synaptosomal and non-synaptosomal mitochondria are tested for purity and functionality.

3.3 Assessment of Mitochondrial Purity

Purity of isolated mitochondria is tested using immunoblotting against known markers for mitochondrial and cytosolic fractions. Mitochondria at 15 μg of protein obtained from the different isolation methods (non-synaptosomal and synaptosomal) were resolved by 4–12% gradient SDS-PAGE and transferred onto polyvinylidene fluoride-FL (Millipore, cat. no. IPFL-00010) membranes using a wet transfer apparatus (Bio-Rad; Hercules, CA) at 40 mA overnight at 4 °C. Nonspecific bindings were blocked with Odyssey® Blocking Buffer (Li-Cor, cat. no. 927–40,000) at room temperature for 1 h. The PVDF-FL membrane was then incubated with primary antibodies (1:1000 dilution in the Odyssey® Blocking Buffer) at 4 °C overnight. The primary antibodies used included total OXPHOS Rodent WB antibody Cocktail (Abcam, cat. no. ab110413) which is a mixture of five antibodies against different components of the oxidative phosphorylation chain: Respiratory Complex I, subunit NDUFB8 20 kDa (ab110242); Respiratory Complex II, subunit SDHB 30 kDa (ab14714); Respiratory Complex III-Core protein 2, subunit UQCRC2 50 kDa (ab14745); Respiratory Complex IV, COX I (ab14705); and ATP-synthase (Complex V), alpha subunit ATP5A (ab14748). Other antibodies used were anti-α-tubulin (Sigma, cat. no. T5168), anti-synaptophysin (Sigma, cat. no. S5768), and anti-calreticulin (Abcam, cat. no. ab2907). After washing with phosphate-buffered saline, the membranes were further incubated with goat anti-mouse or anti-rabbit infrared-labeled IRDye® secondary antibodies (Li-Cor) (1:30,000 dilution in Odyssey® Blocking Buffer) at room temperature for 1 h. The protein bands were visualized on

Odyssey CLx imager (Li-Cor) and quantitated using the Image Studio Light software. Figure 2 shows that Percoll-isolated synaptosomal mitochondria retain substantial amounts of α-tubulin, a cytosolic marker, and calreticulin, which is a marker of endoplasmic reticulum contamination. We found that additional purification with the MACS apparatus decreased both markers indicating an increase of the purity.

3.4 Assessment of Mitochondrial Functionality

Functionality of isolated mitochondria was assessed by measuring mitochondrial protein import because import of mitochondrial matrix proteins depends on the mitochondrial membrane potential (which indicates the structural integrity of the mitochondria and the activity of the mitochondrial respiratory chain) and functional activity of the mitochondrial enzymes involved in the protein uptake and processing. The processing of the pre-proteins includes cleavage reaction done by certain mitochondrial proteases; therefore the presented protocol does not contain any inhibitors of proteases. A pre-ornithine transcarbamoylase (pOTC) cDNA in pGEM-3Zf(+)-pOTC plasmid was transcribed and translated in vitro using the TNT® T7 Coupled Transcription/Translation system (Promega, cat. no. L1170) in the presence of L-[^{35}S]-methionine (Perkin Elmer, cat. no. NEG709A500UC). Following translation, [^{35}S]-methionine-labeled pOTC was incubated with isolated mitochondria at 25 °C for different time points as shown, and mitochondria containing imported OTC were collected by centrifugation (9000 × *g* for 10 min at 4 °C) and separated using SDS-PAGE. The radioactively labeled proteins are visualized by exposing dried gel to a phospho screen (Bio-Rad) and scanned on PMITM Personal Molecular Imager system (Bio-Rad). Cleaved mature OTC (mOTC), which represents the completion of import into the mitochondrial matrix, and uncleaved OTC (pOTC) were quantified using the provided software (Quantity One, version 4.6.9, Bio-Rad). Cleaved mOTC (35.2 kDa) has a smaller molecular weight than pOTC (39.9 kDa) and therefore migrates faster on the gel. This allows the quantification of the efficiency of the mitochondrial import machinery based on the ratio of mOTC to pOTC. Since the isolation of synaptic mitochondria includes cavitation process we tested how does it affect functional activity of the isolated mitochondria. Therefore, NS mitochondria were also subjected to the incubation in the cell disruption vessel at 1100 psi for 15 min before the experiment. Figure 2 represents the activity of protein import tested in isolated mitochondria. The lower bands represent cleaved mOTC and show conclusively that the isolated mitochondria are able to import proteins into the matrix. Non-synaptosomal mitochondria have been used as a reference for import activity [2]. Non-synaptosomal mitochondria subjected to the cell disruption vessel showed a decreased import activity. Synaptosomal mitochondria isolated using the traditional Percoll

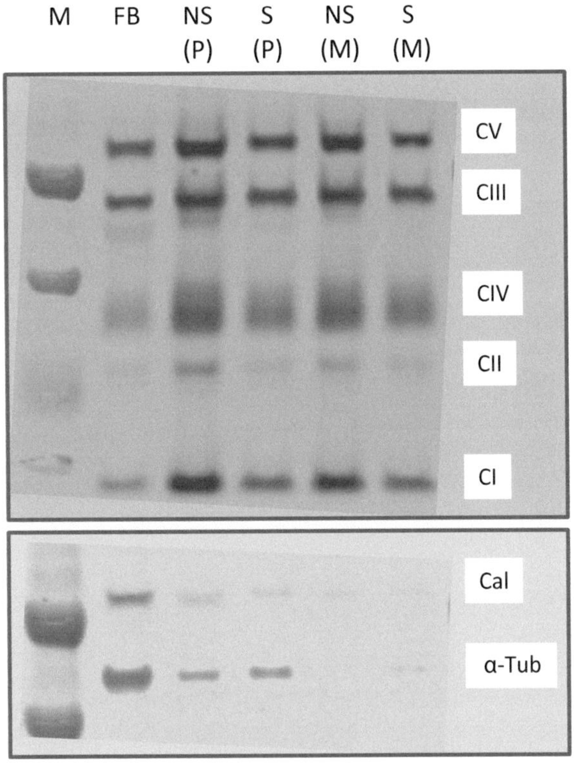

Fig. 2 Assessment of purity using our mitochondrial isolation method. As mentioned in the text, immunoblotting was done using 5 μg of protein to compare the purity of the mitochondrial samples obtained using both the traditional Percoll method and our new method. FB is the sample from forebrain homogenate, which was used as a control to the mitochondrial enrichment methods. NS (P) and S (P) represent samples from Non-synaptosomal and synaptosomal mitochondria, respectively, isolated using the traditional Percoll method. NS (M) and S (M) represent samples from non-synaptosomal and synaptosomal mitochondria, respectively, isolated using our new MACS-based method. The samples were resolved by 4–12% gradient SDS-PAGE and transferred onto PVDF-FL membranes using a wet transfer apparatus at 40 mA overnight at 4 °C. The membranes were then incubated for 1 h with Odyssey® blocking buffer to prevent nonspecific binding. The PVDF-FL membrane was then incubated with primary antibodies overnight at 4 °C. After washing with phosphate-buffered saline (PBS), the membranes were further incubated with IRDye® Goat secondary antibodies at 1:30,000 dilution in Odyssey® blocking buffer at room temperature for 1 h. Following further washing with PBS, immunoreactive protein bands were visualized on Odyssey CLx imager. The NS (P) samples show a much smaller contamination level compared to S (P) samples. Using our MACS-based method, α-tubulin (α-Tub) and calreticulin (Cal) are decreased in the S (M) samples indicating improved mitochondrial purity

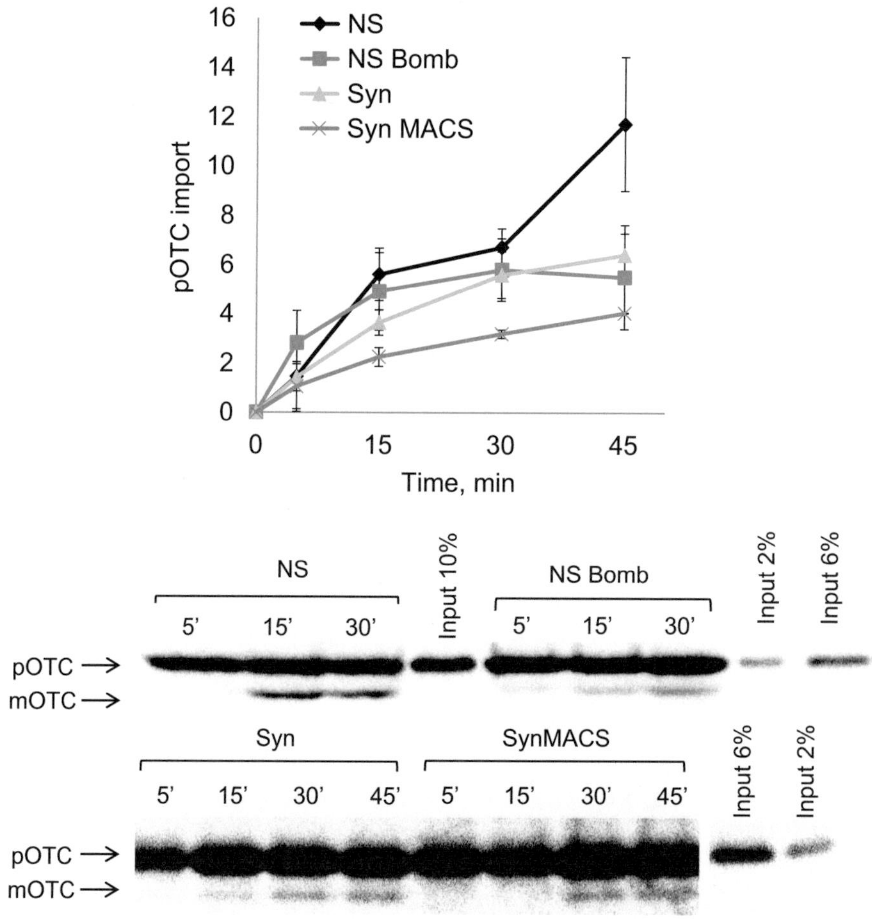

Fig. 3 Assessment of the mitochondrial function in vitro. Isolated mitochondria were subjected to pOTC import assays. Representative results of pOTC import reaction using non-synaptosomal mitochondria (NS (P), non-synaptosomal mitochondria additionally subjected to incubation at 1100 psi for 15 min (NS Bomb), synaptosomal mitochondria isolated the traditional Percoll method (Syn (P), and synaptosomal mitochondria isolated using our method (Syn MACS). Representative gel images used for quantification are shown. Import reaction times are indicated. The error bars reported on this graph represent the standard error of the mean (SEM) associated with the dataset

method or our new method show decreased import activity compared to non-synaptosomal mitochondria. Based on the decreased activity of non-synaptosomal mitochondria subjected to the cavitation, we attributed a decreased activity to the exposure of mitochondria to 1100 psi pressure in the cell disruption vessel. On the other hand, synaptosomal mitochondria isolated using both aforementioned methods showed similar import activity (45-min time point, Student's t-test, $p = 0.2$) (Fig. 3). Given that the functionality of the mitochondria is intact, any further functional or structural studies to examine differences between neuronal and non-neuronal mitochondria can be conducted.

4 Conclusion

Using this method, we demonstrate the ability to isolate highly enriched and purified functional synaptosomal mitochondria. The use of our method will allow investigators to probe both structural and functional properties of the neuronal mitochondria with no interactions/contaminations from other cellular contaminants.

Acknowledgments

This work was supported by US National Institutes of Health grants R01 NS039324 and NS077748 (to R.M.F.), the DSF Charitable Foundation (R.M.F.), the Pittsburgh Brain Institute (to R.M.F.), and the Copeland Fund/Pittsburgh Foundation (to S.V.B.).

References

1. Zhang Y, Cook A, Kim J, Baranov SV, Jiang J, Smith K, Cormier K, Bennett E, Browser RP, Day AL, Carlisle DL, Ferrante RJ, Wang X, Friedlander RM (2013) Melatonin inhibits the caspase-1/cytochrome c/caspase-3 cell death pathway, inhibits MT1 receptor loss and delays disease progression in a mouse model of amyotrophic lateral sclerosis. Neurobiol Dis 55:26–35. https://doi.org/10.1016/j.nbd.2013.03.008 PubMed PMID: 23537713; PMCID: 3652329
2. Yano H, Baranov SV, Baranova OV, Kim J, Pan Y, Yablonska S, Carlisle DL, Ferrante RJ, Kim AH, Friedlander RM (2014) Inhibition of mitochondrial protein import by mutant huntingtin. Nat Neurosci 17:822. https://doi.org/10.1038/nn.3721 PubMed PMID: 24836077
3. Friedlander RM (2003) Apoptosis and caspases in neurodegenerative diseases. N Engl J Med 348(14):1365–1375. https://doi.org/10.1056/NEJMra022366 348/14/1365 [pii]. PubMed PMID: 12672865
4. Hensley K, Harris-White ME (2015) Redox regulation of autophagy in healthy brain and neurodegeneration. Neurobiol Dis 84:50. https://doi.org/10.1016/j.nbd.2015.03.002 PubMed PMID: 25771170
5. Wang X, Zhu S, Pei Z, Drozda M, Stavrovskaya IG, Del Signore SJ, Cormier K, Shimony EM, Wang H, Ferrante RJ, Kristal BS (2008) Friedlander RM. inhibitors of cytochrome c release with therapeutic potential for Huntington's disease. J Neurosci 28(38):9473–9485. https://doi.org/10.1523/jneurosci.1867-08.2008 Epub 2008/09/19. PubMed PMID: 18799679; PMCID: PMC2632939
6. Brown MR, Sullivan PG, Geddes JW (2006) Synaptic mitochondria are more susceptible to Ca2+overload than nonsynaptic mitochondria. J Biol Chem 281(17):11658–11668. https://doi.org/10.1074/jbc.M510303200 PubMed PMID: 16517608
7. Sims NR, Anderson MF (2008) Isolation of mitochondria from rat brain using Percoll density gradient centrifugation. Nat Protoc 3(7):1228–1239. https://doi.org/10.1038/nprot.2008.105 PubMed PMID: 18600228
8. Kristian T 2010 Isolation of mitochondria from the CNS. Curr Protoc Neurosci Chapter 7:Unit 7.22. PubMed PMID: MEDLINE:20578034

Chapter 10

Affinity Proteomics for Interactome and Phosphoproteome Screening in Synaptosomes

Kasper Engholm-Keller, Nicolai Bache, Sushma R. Rao, Jesse R. Wark, Martin R. Larsen, Phillip J. Robinson, and Mark E. Graham

Abstract

Phosphorylation is an essential regulatory protein modification, which modulates many presynaptic processes, including synaptic vesicle trafficking. Protein phosphorylation impacts neurotransmitter release and is likely to be mechanistically important for presynaptic plasticity, although only a few mechanisms have been described to date. Thus, interactomics and phosphoproteomics are important tools to determine presynaptic mechanisms. Over the last two decades, mass spectrometry-based proteomics has become the method of choice for protein-protein and protein phosphorylation analysis. This protocol is based on well-established procedures in our laboratory and will describe the application of affinity purification of presynaptic proteins via GST-fusion protein pull-downs for the study of presynaptic phospho-signalling. The pull-down is followed by on-bead proteolysis, chemical stable isotope labelling for precise quantification, enrichment of phosphorylated peptides using TiO_2, and LC-MS/MS. The method outlines the software used for peptide/protein identification and quantification, as well as some of the bioinformatics tools that are available for putting the regulated phosphorylation sites into a functional context.

Key words Synaptosome, Nerve terminal, Synaptic vesicle endocytosis, Proteomics, Phosphoproteomics, Interactomics, Protein phosphorylation, Protein pull-down, TiO_2 enrichment, Mass spectrometry

1 Introduction

1.1 Presynaptic Proteins and Vesicle Trafficking

The main function of the presynaptic terminal, or bouton, is to release neurotransmitter into the synaptic cleft via exocytosis. The rapid process of synaptic vesicle endocytosis (SVE) replenishes the pool of synaptic vesicles, supporting sustained neurotransmission. Thus, exocytic and endocytic proteins are among the most abundant proteins found in preparations of nerve terminals, i.e., synaptosomes [1]. Nevertheless, many unanswered questions remain about the protein-protein interactions of vesicle-related proteins. Neurotransmitter release is a tightly controlled process regulated by depolarization-induced signal transduction, and allows for

Kathryn M. Murphy (ed.), *Synaptosomes*, Neuromethods, vol. 141,
https://doi.org/10.1007/978-1-4939-8739-9_10,

adaptive changes to neural circuits [2, 3]. A common mode of signalling for all biological processes is protein phosphorylation: the reversible covalent modification by a phosphate group. Phosphorylation leads to changes in protein shape or ionic charge, which can change catalytic activity, protein interactions, subcellular localization, degradation, and other properties. Proteins in the presynaptic terminal mediating SVE are known to undergo phosphorylation state changes in response to depolarization [4] and these changes likely facilitate SVE [5]. This chapter describes a method to extract SVE proteins for examining the SVE interactome and to simultaneously examine changes to their phosphorylation state following depolarization.

1.2 Phosphoproteomics of the Presynaptic Terminal

Mass spectrometry-based screening of protein phosphorylation, i.e., phosphoproteomics, has become the method of choice for protein phosphorylation analysis. Several studies have contributed to mapping the presynaptic phosphoproteome over the last decade. In 2005, Collins et al. characterized protein phosphorylation in synaptosomes for the first time [6]. More than 650 phosphorylation events on 79 synaptic proteins were described, of which more than 90% were novel at the time. Two years later, Munton et al. performed subcellular fractionation of synaptosomes to detect protein phosphorylation in synaptic vesicle and postsynaptic density (PSD) fractions [7]. This study identified 974 phosphorylation sites and used absolute quantification with synthetic peptide standards to determine the stoichiometry of a few synaptic phosphorylation sites before and after depolarization. In a later study, the Burlingame lab increased the number of phosphorylation sites identified in a single study of phosphorylation and *O*-linked N-acetylglucosamine (*O*-GlcNAc) to more than 16,500, providing an in-depth global overview of the presynaptic phosphoproteome [8].

A static map of global nerve terminal protein phosphorylation sites enhances knowledge of the primary structure of presynaptic proteins. However, knowledge of dynamic changes in phosphorylation levels in response to a specific stimulus allows for ranking of phosphorylation sites in terms of potential functional importance and correlation of specific changes with specific cellular outcomes [9]. The first quantitative phosphoproteomic characterization of presynaptic proteins in response to depolarization was published in 2016 by Kohansal-Nodehi et al. [10]. Using a sustained KCl depolarizing stimulus, the study identified 252 regulated phosphorylation sites, revealing extensive activity-dependent changes to the phosphorylation state of the active zone and synaptic vesicle trafficking proteins.

Global phosphoproteomic strategies have now evolved to a stage where 10,000 s of phosphorylation sites can be identified and quantified in a single experiment. Given these technological developments, there may be limited merit in enriching a subcellular compartment or interactome prior to a phosphopeptide enrichment step. However, phosphopeptide enrichment can be opportunistically added to any proteomics workflow and so, when applied to a smaller scale project, is potentially less resource intensive and more targeted than a global analysis.

1.3 GST-Fusion Protein-Based Affinity Purification

Glutathione S-transferase (GST)-fusion protein-based affinity purification can be utilized to enrich for known and unknown interaction partners of a bait protein of choice. The changes in phosphorylation at the site level can be quantified for each interaction partner by using a quantitative approach combined with phosphopeptide enrichment. This protocol builds upon previous work where SVE proteins were enriched using their affinity for GST fusions of the adapter protein complex 2 (AP2) α-adaptin appendage ear domain and endophilin 1-SH3 domain coupled to reduced glutathione (GSH)-Sepharose beads [5]. This protein pull-down was used to simultaneously enrich AP180, synaptojanin, amphiphysin 1, dynamin 1, and synapsin 1 from ^{32}P-labelled synaptosomes, allowing for more comprehensive quantification of phosphorylation changes induced by depolarization in these proteins. This affinity purification method for endocytic proteins was the basis for several other publications [9, 11–13]. This protocol involves a crude rat brain synaptosomal preparation, which is depolarized prior to lysis. The lysate is incubated in a spin column with a Sepharose-coupled GST-fusion protein of choice to allow for binding of interaction partners (Fig. 1). The bead-bound proteins are directly digested with trypsin to rapidly provide samples for the proteomics part of the workflow.

1.4 Quantitative Phosphoproteomics

Peptide samples can be analyzed directly by LC-MS/MS for label-free protein quantification, where peptide intensities are typically summed to provide protein intensities. This would be the method of choice for a comparative analysis of presynaptic protein binding to a bait protein versus those proteins bound nonspecifically to a control (e.g., a fusion protein affinity tag). The label-free approach is sufficiently accurate to determine differences between conditions if they are three- to fourfold or higher in magnitude [14]. If smaller differences need to be distinguished, in particular when quantifying single (phospho)peptide species instead of proteins, chemical stable isotope labelling (derivatization) should be performed due to the higher precision of these approaches (e.g., [15]). Examples of this type of labelling are tandem mass tags (TMT) [16]—which are used in this protocol—iTRAQ (isobaric tags for relative and absolute quantitation) [17], or reductive dimethylation [18]. While TMT and iTRAQ rely on MS/MS-level reporter ions generated from the labelled peptide selected for fragmentation, quantification by dimethylation is achieved at the MS level. In spite of

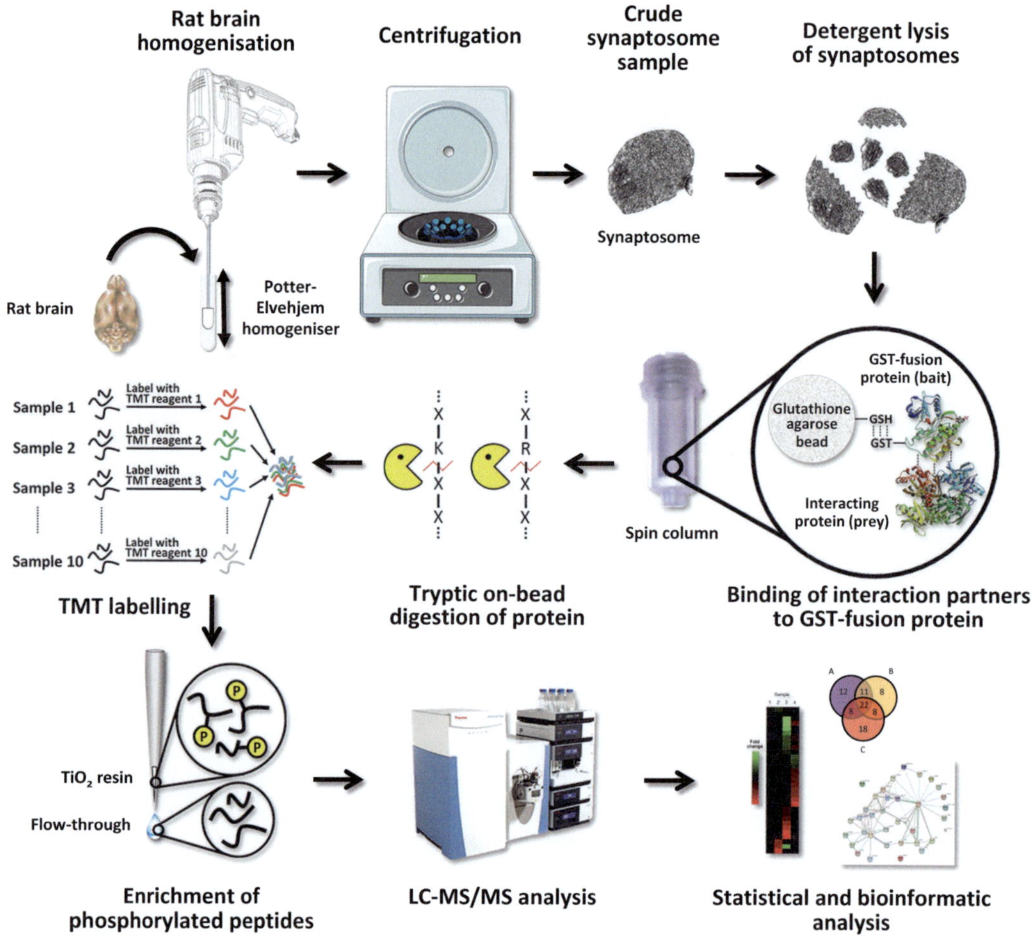

Fig. 1 General overview of the affinity purification and proteomics workflow. Firstly, rat brain tissue is homogenized in a Potter-Elvehjem homogenizer attached to a drill. Synaptosomes are enriched from the homogenate via multiple centrifugation steps. The synaptosomes are optionally stimulated under various conditions and then lysed in a solution containing physiological pH buffer, detergents, and protease and phosphatase inhibitors. The lysate is incubated in a spin column with glutathione Sepharose beads linked to a GST-fusion protein of interest. The binding proteins are separated from the non-binding proteins by centrifuging the spin column, followed by several washing steps. To allow for mass spectrometry analysis, the GST-fusion protein bait as well as interacting proteins are reduced, alkylated, and subjected to proteolysis using trypsin directly on the beads. After digestion, the resulting peptide sample is isolated by centrifugation. At this step chemical stable isotope labelling can be performed using, e.g., the TMT10plex reagents, which accommodate multiplexing of up to ten samples/conditions that can be labelled, mixed, and analyzed in a single LC-MS/MS run. Following TMT labelling, the phosphorylated peptides are enriched by TiO_2-based affinity chromatography, while non-phosphorylated peptides are not retained on the resin. The TiO_2 eluate is then analyzed by LC-MS/MS on a high-resolution/high-mass-accuracy mass spectrometer, producing data that is converted into peak lists, searched against an in silico-digested rat protein database matching the MS/MS spectra to theoretical peptide sequences from the database. These peptides are quantified based on the TMT reporter ions in the spectra and the output from the database searching/quantification software is subjected to statistical testing to identify regulated phosphosites (between the examined conditions). Bioinformatics tools can then be applied. Network analysis allows for mapping of novel phosphorylation sites/proteins onto networks of known protein-protein interactions to characterize the potential role of the protein phosphorylation in regulating interactions or other functions. Clipart from www.servier.com, www.clker.com, all-free-download.com, and www.allfreeclipart.com was used in making this figure

having varying chemistries and being based on different MS modes of quantification, chemical stable isotope labelling benefits from allowing relative quantification of isotopically distinct peptides (isotopologues) in a single spectrum. Hereby the quantitative variation can be reduced to a minimum, in contrast to label-free MS that relies on comparing intensities between different LC-MS/MS runs and therefore is prone to run-to-run variation.

Whether label-free quantification or chemical stable isotope labelling is performed, it is necessary to enrich for the phosphorylated peptides to enable identification and quantification. The phosphorylated proportion of peptides in a complex protein digest often only makes up 1–2% of the total peptide content [19]. The likelihood for a phosphopeptide to be selected for fragmentation in a typical data-dependent acquisition (DDA)-based MS analysis is therefore low. Phosphopeptides are typically enriched by immobilized metal affinity chromatography (IMAC) [20] or through metal oxide affinity chromatography using TiO_2 [21, 22] as described in this protocol. TiO_2 has high affinity for phosphorylated side chains but also attracts acidic side chains, i.e., those with carboxyl groups, as well as other acidic biomolecules [23]. Therefore, specific phosphopeptide enrichment requires very low pH buffers as well as additives to neutralize and outcompete the carboxyl groups in the acidic peptides, leaving mainly phosphopeptides bound [21]. The TiO_2 enrichment format described in this protocol is microcolumn based, using pipette tips fritted with a porous membrane plug (Empore SPE Disk) [24] retaining the TiO_2 resin, through which the sample as well as wash buffers and elution buffers can be pushed using air pressure from a plastic syringe. The enrichment can also be performed in batch format, incubating the sample with a suspension of TiO_2 beads under rotation. Handling multiple samples is more practical in batch format, and reproducibility and phosphopeptide elution efficiency can be improved compared to the microcolumn format. In either method, matching the peptide amount to the bead mass is important. In our hands, a 1:6 peptide-to-TiO_2 resin ratio (Titansphere) has been shown to produce optimal results when working with complex HeLa cell lysates [19], which is supported by reports from other laboratories [25]. For GST-fusion protein pull-down samples, the majority of peptides consist of non-phosphorylated GST-fusion protein, and not the interacting protein. Therefore, the amount of TiO_2 should be based on the peptide amount in the sample with the amount of GST-bait protein subtracted.

The choice of mass spectrometer is often a question of access and availability, but for the TMT10plex labelling performed in this protocol a high-resolution/high-mass-accuracy instrument having a resolution of 50,000 or more at m/z 150 is necessary (such as the Q Exactive Plus from Thermo Fisher Scientific). For other

MS/MS-based labelling schemes (iTRAQ or TMT 6plex) Q-TOF type (e.g., Impact II (Bruker Daltonics), 6600 TripleTOF (SCIEX), SYNAPT G2-Si (Waters), or hybrid Orbitrap instruments (LTQ-Orbitrap Elite/Velos, Thermo Fisher Scientific)) can also be used. For MS-level quantification (label-free and dimethylation) older QTOFs and ion trap-based instruments are also suitable, but will generally result in fewer phosphorylation sites being quantified compared to more current technology. Online HPLC separation prior to introduction of the peptides into the mass spectrometer is required to handle the complex samples generated by proteolysis. For achieving an adequate coverage of phosphopeptides in the sample, nanoflow HPLCs are used to perform highly sensitive peptide separation using 50–100 μm diameter capillary columns with analysis/gradient length of usually 30–180 min depending on sample complexity and speed of the MS instrument.

Various commercial and free software suites are available for MS data processing, peptide, and protein identification and quantification. The choice of software is highly dependent on the MS instrument and quantification strategy used for the study. For high-resolution data from TMT-labelled samples, as described in this protocol, software such as MaxQuant [26, 27] (free software, Max Planck Institute for Biochemistry) or Proteome Discoverer (commercial software, Thermo Fisher Scientific) can be used. Further statistical analysis of the quantified peptide identifications is needed to identify proteins or phosphorylation sites that significantly differ between sample conditions. Bioinformatics approaches, such as hierarchical clustering, gene ontology term enrichment analysis, protein kinase substrate consensus sequence analysis, and pathway enrichment analysis, will require additional software and approaches. In particular, mapping of regulated phosphoproteins onto protein-protein interaction networks can give further insight into the phospho-regulation of presynaptic interactomes [10]. For a detailed list of popular tools and protocols and their use, see [28, 29]. One software package that integrates many of the previously mentioned functionalities and is compatible with this protocol is Perseus (Max Planck Institute for Biochemistry) [30].

2 Materials

Ultrapure water and analytical grade chemicals should be used for preparing all solutions. All microcentrifuge tubes and pipette tips should be of the low-protein-binding type (*see* **Note 1**). To reduce protease/phosphatase activity in the samples, keep all buffers on ice prior to use, and perform all steps up until reduction, alkylation, and proteolysis on ice/at 4 °C using ice-cold buffers unless noted otherwise.

2.1 Synaptosome Preparation

- Adult rat.
- Rodent guillotine, sharp pair of scissors, and forceps.
- Drill capable of operating between 500 and 800 rpm.
- 15 mL Potter-Elvehjem Teflon/glass homogenizer.
- Cooled high-speed centrifuge with fixed-angle rotor capable of running at 950–23,000 g.
- 50 mL Round-bottom centrifuge tubes
- Sucrose/EDTA/Tris (SET) buffer: 0.32 M Sucrose, 1 mM EDTA/Tris (pH 7.4), 5 mM Tris/HCl (pH 7.4).
- Carbogen gas (95% oxygen (O_2), 5% carbon dioxide (CO_2), medical grade) and sintered glass gas sparger.
- HEPES-buffered Krebs-like (HBK) buffer: 143 mM NaCl, 4.7 mM KCl, 1.2 mM $CaCl_2$ (*see* **Note 2**), 1.18 mM $MgCl_2$, 20 mM Hepes/Tris (pH 7.4), 0.1 mM Na_2HPO_4, 10 mM D-glucose. This buffer should be sparged with carbogen gas (95% O_2, 5% CO_2) for 1 h on ice, prior to use, to oxygenate the buffer and adjust the pH to 7.4.
- High K^+ HBK buffer: 70.7 mM NaCl, 77 mM KCl, 1.2 mM $CaCl_2$ (*see* **Note 2**), 1.18 mM $MgCl_2$, 20 mM Hepes/Tris (pH 7.4), 0.1 mM Na_2HPO_4, 10 mM D-glucose. This buffer does not require sparging.
- Lysis buffer: 1% Triton X-100 (*see* **Note 3**), 5 mM Tris/HCl (pH 7.4), 150 mM NaCl, 1 mM EDTA/Tris (pH 7.4), 1 mM EGTA/Tris (pH 7.4), protease inhibitor cocktail, e.g., cOmplete (Roche, North Ryde, NSW, Australia), 40 μg/mL leupeptin, 2 mM phenylmethylsulfonyl fluoride (PMSF) (**CAUTION: TOXIC AND CORROSIVE!**), phosphatase inhibitor cocktails such as Phosphatase Inhibitor Cocktail Set II (Merck Millipore, Bayswater, VIC, Australia) combined with PhosSTOP (Roche, North Ryde, NSW, Australia), 5 mM NaF (**CAUTION: TOXIC!**), 2 mM beta-glycerophosphate/Tris (pH 7.4) (*see* **Notes 4** and **5**).

2.2 GST-Fusion Protein Pull-Down

- Triton lysis buffer: Same as above.
- Phosphate-buffered solution (PBS).
- Glutathione S-transferase (GST)-bait fusion protein coupled to glutathione Sepharose 4B beads (GE Healthcare, Rydalmere, NSW, Australia) (50% bead slurry) in a 1:1 glycerol/PBS solution (*see* **Note 6**).
- Bead wash solution: 20 mM Tris/HCl (pH 7.4), 1 mM EDTA/Tris (pH 7.4), 1 mM EGTA/Tris (pH 7.4), protease inhibitor cocktail, e.g., cOmplete (Roche, North Ryde, NSW, Australia), 2 mM phenylmethylsulfonyl fluoride (PMSF)

(**CAUTION: TOXIC!**), 5 mM NaF (**CAUTION: TOXIC!**), 2 mM beta-glycerophosphate/Tris (pH 7.4) (*see* **Note 5**).

- High-salt bead wash solution: 300 mM NaCl, 20 mM Tris/HCl (pH 7.4), 1 mM EDTA/Tris (pH 7.4), 1 mM EGTA/Tris (pH 7.4), protease inhibitor cocktail, e.g., cOmplete (Roche, North Ryde, NSW, Australia), 2 mM phenylmethylsulfonyl fluoride (PMSF) (**CAUTION: TOXIC!**), 5 mM NaF (**CAUTION: TOXIC!**), 2 mM beta-glycerophosphate/Tris (pH 7.4) (*see* **Note 5**).
- Digestion buffer: 50 mM HEPES/NaOH, pH 7.8.
- Fritted spin column, 0.7–2 mL capacity (e.g., Illustra Microspin columns, GE Healthcare, Rydalmere, NSW, Australia).

2.3 Reduction, Alkylation, and Proteolysis

- 0.5 M Tris(2-carboxyethyl)phosphine (TCEP) hydrochloride, pH 7 **(CAUTION: CORROSIVE!).**
- 0.5 M Iodoacetamide (IAA) (**CAUTION: TOXIC!**).
- Digestion buffer: Same as above.
- Sequencing-grade trypsin (e.g., Sequencing Grade Modified Trypsin (Promega, Alexandria, NSW, Australia)) (*see* **Notes 7** and **8**).
- Reversed-phase (RP) solid-phase extraction (SPE) cartridges (e.g., Sep-Pak C18, Waters, Rydalmere, NSW, Australia) (*see* **Note 9**).
- RP conditioning solution: 100% Acetonitrile (AcN).
- RP load/wash solution: 0.1% Trifluoroacetic acid (TFA).
- RP elution solution: 50% AcN.

2.4 Tandem Mass Tag (TMT) Labelling

- Digestion buffer: Same as above.
- TMT10plex Isobaric Label Reagent Set (0.8 mg of each) (Thermo Fisher Scientific, North Ryde, NSW, Australia) (*see* **Note 10**).
- Anhydrous AcN.

2.5 Phosphopeptide Enrichment

- RP solid-phase extraction (SPE) cartridges (e.g., Sep-Pak C18, Waters, Rydalmere, NSW, Australia) (*see* **Note 9**).
- RP conditioning solution: 100% AcN.
- RP load/wash solution: 0.1% TFA.
- RP elution solution: 50% AcN.
- 200 μL C8 StageTip and 200 μL C18 StageTip (Thermo Fisher Scientific, North Ryde, NSW, Australia) (*see* **Note 11**).
- Plastic syringe with adaptor to allow for attachment to StageTip. An adapter can be made by trimming a plastic pipette tip into

the shape of a tapered cylinder such that one end fits onto the syringe outlet and the other end fits inside the StageTip.

- Titansphere TiO 5 μm Bulk Material (GL Sciences, Tokyo, Japan).
- 100% TFA (**CAUTION: CORROSIVE!**).
- 10% TFA (**CAUTION: CORROSIVE!**).
- 100% Formic acid (FA) (**CAUTION: CORROSIVE!**).
- Glycolic acid (**CAUTION: CORROSIVE!**).
- Washing buffer 1: 80% AcN, 5% TFA, 1 M glycolic acid (**CAUTION: CORROSIVE!**).
- Washing buffer 2: 80% AcN, 5% TFA (**CAUTION: CORROSIVE!**).
- Washing buffer 3: 80% AcN, 0.1% TFA.
- Elution buffer: 1% NH_4OH (**CAUTION: CORROSIVE!**).

2.6 Analysis by Liquid Chromatography/Tandem Mass Spectrometry (LC-MS/MS)

- High-resolution/high-mass-accuracy mass spectrometer capable of acquiring MS/MS spectra at a resolution of >50,000 at m/z 150, for example a Q Exactive Plus (Thermo Fisher Scientific, North Ryde, NSW, Australia). The MS must be interfaced to a nanoflow HPLC system equipped with 50–100 μm i.d. RP capillary column for online separation of peptides, for example a Thermo Fisher Scientific UltiMate 3000 RSLCnano (Thermo Fisher Scientific, North Ryde, NSW, Australia).
- Software for LC-MS/MS data processing, TMT reporter ion quantification, and protein database searching such as MaxQuant (Max Planck Institute for Biochemistry, Planegg, Germany), Proteome Discoverer (Thermo Fisher Scientific), or similar.
- Fasta UniProt database containing all *Rattus norvegicus* protein sequences.
- Software for statistical analysis, hierarchical clustering of the data, network analysis, etc., such as Perseus (Max Planck Institute for Biochemistry, Planegg, Germany), Database for Annotation, Visualization and Integrated Discovery (DAVID), and STRING (http://string-db.org/).

3 Methods

Experiments involving animals must be carried out by trained personnel in accordance with all appropriate national and institutional regulations. A Material Safety Data Sheet (MSDS) for the reagents and the Standard Operating Procedure (SOP) for the experiment methods should be obtained where available and read before use.

3.1 P2 Synaptosome Lysate Preparation

1. Cull rat by decapitation, and rapidly remove the brain. Dissect brain to collect the region of interest—volumes described here are for whole-brain extraction. Transfer brain tissue to a beaker containing 20 mL SET buffer (approx. 1 g cerebrum per rat) and quickly rinse.
2. Transfer brain tissue to another beaker with 10 mL of SET buffer, cut tissue into several smaller pieces, and transfer the content to an ice-cold 15 mL Potter-Elvehjem Teflon/glass homogenizer. Attach the pestle to a drill running at 700 rpm and homogenize the tissue by seven slow strokes down and up.
3. Centrifuge the homogenate in a 50 mL polycarbonate tube at 950 × *g* for 10 min at 4 °C. Transfer the supernatant to a new, cold 50 mL tube.
4. Resuspend the pellet in 15 mL SET buffer by gentle pipetting. Centrifuge the solution at 950 × *g* for 10 min at 4 °C.
5. Combine the resulting supernatant with the supernatant from the first centrifugation step. At this point the preparation is known as a crude "pellet 2 (P2)" fraction. A Percoll gradient can be used to further enrich the synaptosomes from the S1 fraction if required [31], but only the crude fraction is used here (*see* **Note 12**).
6. Centrifuge the combined supernatant sample at 23,000 × *g* for 10 min at 4 °C and discard the supernatant.
7. Resuspend the pellet in 10 mL carbogen-treated HBK buffer by gentle pipetting. Centrifuge the synaptosome suspension for 10 min at 950 × *g* at 4 °C.
8. Discard the supernatant and resuspend the pellet in 1.25 mL carbogen-treated HBK buffer by gentle twirling.
9. Incubate the synaptosomes at 37 °C for at least 5 min to allow a metabolic equilibrium, although it has not been described when this occurs. A 45-min incubation is recommended if the phosphoproteomics results are intended to align with previous work using ^{32}P-labelling [5, 32, 33] (Fig. 2) (*see* **Note 13**).
10. If no stimulation is to be done, centrifuge the synaptosome suspension at 950 × *g* for 10 min at 4 °C and discard the supernatant. If a stimulation is planned, create pairs of samples, i.e., "mock stimulated" and "stimulated". Add HBK to the "mock-stimulated" sample such that the volume is doubled. Simultaneously add the same volume of high K^+ HBK to the "stimulated" samples and allow the stimulation to continue for an appropriate time (e.g., 3–10 s). Pellet the paired samples by centrifugation at >12,000 × *g* in a benchtop centrifuge for 8 s and discard the supernatant. Removal of the high K^+ concentration solution is essential for binding studies since concentrated K^+ can affect protein-protein interactions.

11. Lyse the synaptosome pellet in an appropriate volume of lysis buffer (e.g., 5 mL lysis buffer for lysing a P2 fraction pellet from one rat brain) by 30 s of vortexing.
12. Centrifuge the lysate at 14,000 × *g* for 15 min at 4 °C.
13. Collect the supernatant and store at −80 °C until required for the pull-down. Reserve a small, separate aliquot for protein concentration determination.
14. Measure protein concentration of the lysate (supernatant) using, e.g., a bicinchoninic acid (BCA) protein assay and adjust the protein concentration to ~5 mg/mL.

3.2 GST-Fusion Protein Pull-Down

1. Express the GST-fusion protein of interest in *E. coli* and couple the protein to glutathione Sepharose 4B beads as described in detail in [34].
2. The lysate should be removed from frozen storage and thawed on ice (~20 min) in preparation for steps below. If a precipitate is observed after thawing, centrifuge at 14,000 × *g* for 15 min at 4 °C and transfer the supernatant to a new ice-cold tube.
3. Pipette 400 μL PBS into a plugged micro spin column, followed by the desired amount of GST-fusion protein bead slurry. In general, the bait must be in excess to extract binding partners efficiently. Different bait-to-lysate ratios will have differential effects on binding partners [35] and this should be optimized for specific targets. Multiple baits can be used together or sequentially to extract a larger range of proteins [5] (Fig. 2) (*see* **Note 14**).
4. Remove the plug from the bottom of the spin column and place the column in a 2 mL waste collection tube.
5. Spin the liquid through the column by centrifugation at 500 × *g* for 30 s at 4 °C and discard the flow-through.
6. Dispense 400 μL PBS onto the beads and spin the liquid through the column by centrifugation at 500 × *g* for 30 s at 4 °C. Discard the flow-through.
7. Repeat these washing steps twice and discard the flow-through.
8. Remove the spin column from the waste collection tube and plug the bottom of the column.
9. Dispense 400–600 μL of lysate onto the beads, cap the top of the spin column, and incubate the spin column with end-over-end rotation for 1 h at 4 °C.
10. Place the spin column with the bottom plug attached in a 2 mL waste collection tube and centrifuge at 500 × *g* for 30 s at 4 °C to remove the liquid/beads from the cap.

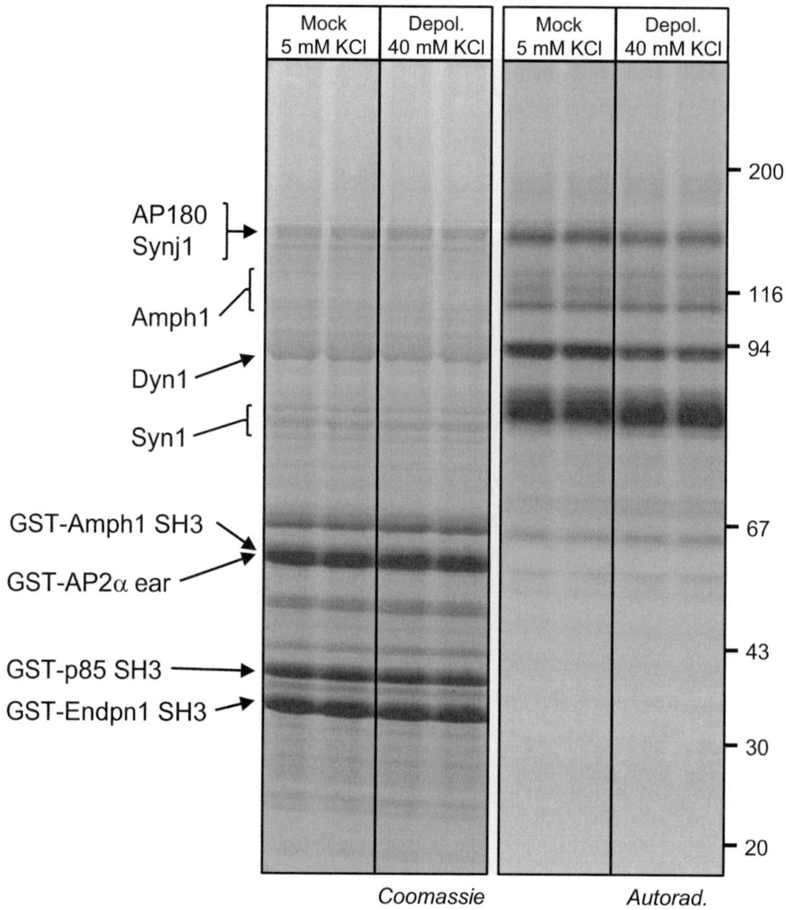

Fig. 2 SDS-PAGE of a multiple GST-fusion protein pull-down from ^{32}P-labelled synaptosomes. The four baits, GST-amphiphysin 1 SH3 domain (human, 545–695), GST-alpha adaptin (AP2) ear domain (mouse 701–938), p85 SH3 (bovine, 1–86), and endophilin 1 SH3 (human, 292–352), were used to extract synaptic vesicle endocytosis (SVE) proteins from synaptosomes. The synaptosomes were either stimulated for 3 s with 40 mM KCl or mock stimulated with a physiological level of KCl (5 mM). Coomassie-stained SDS-PAGE gel (left panel) and autoradiograph (right panel) are shown. Synaptosomes were labelled for 45 min with ^{32}P as described previously [5, 32]. Major SVE phosphoproteins dynamin 1, synaptojanin 1, AP180, and amphiphysin 1 are indicated. KCl stimulation (depolarization) causes these SVE proteins to be overall dephosphorylated [4] (right panel, last two lanes)

11. Uncap the spin column, remove the bottom plug, place in a clean 1.5 mL tube, and centrifuge at 500 × *g* for 30 s at 4 °C to collect the lysate, which now consists of proteins that did not bind the bait (*see* **Note 15**).
12. Transfer the spin column to a 2 mL waste collection tube and dispense 400 μL high-salt bead wash solution onto the beads.
13. Spin the liquid through the column by centrifugation at 500 × *g* for 30 s at 4 °C and discard the liquid.

14. Dispense 400 μL bead wash solution onto the beads, spin the liquid through the column by centrifugation at 500 × *g* for 30 s at 4 °C, and discard the liquid. Repeat the wash twice more. The proteins have now been affinity purified and can be visualized by elution and analysis by SDS-PAGE [34] (Fig. 2), but this protocol assumes that all the sample will be used for proteomics analysis.
15. Add 400 μl digestion buffer. Spin the liquid through the column by centrifugation at 500 × *g* for 30 s at 4 °C and discard the liquid.

3.3 Reduction, Alkylation, and Proteolysis

1. Plug the bottom of the spin column and add 49 μL ice-cold digestion buffer followed by 1 μL of 0.5 M TCEP. Gently mix the bead suspension and loosely cap the spin column (make sure that the cap is loose to prevent pressure buildup that may cause liquid to leak through the bottom plug). Place spin column in a clean 2 mL tube and incubate at 85 °C for 10 min.
2. Let the spin column cool to room temperature and add 2 μL of 0.5 M IAA. Gently mix the bead suspension and incubate in the dark at 23 °C for 30 min.
3. Add 1 μL of 0.5 M TCEP to quench excess IAA. Gently mix the bead suspension.
4. Add trypsin in a 1:20 (w/w) trypsin-to-GST-bait protein ratio and gently mix the bead suspension. Generally assume that the pull-down has resulted in binding of lysate proteins with ~50% of the mass of the bait protein. To ensure complete digestion, an optional second round of on-bead digestion can be performed.
5. Cap the spin column and incubate at 37 °C for 6–12 h (*see* **Note 16**).
6. Loosen the cap, unplug the column, place it in a new 1.5 mL tube, and centrifuge at 500 × *g* for 30 s.
7. Plug the column, add 50 μL digestion buffer, mix the bead suspension, unplug the column, place it in the 1.5 mL tube, and centrifuge at 500 × *g* for 30 s.
8. Discard the column and acidify the digested sample by adding 4 μL 10% trifluoroacetic acid (TFA). Mix the samples by vortexing.
9. Centrifuge the sample at 14,000 × *g* for 5 min to remove undissolved material and transfer the supernatant to a clean microcentrifuge tube.
10. For cleaning up the peptide sample prior to LC-MS/MS analysis use a reversed-phase (RP) solid-phase extraction (SPE) cartridge (*see* **Note 9**). Condition the cartridge by first passing RP conditioning solution through it, followed by RP load/wash solution (for all volumes, follow the manufacturer's instructions).

11. Dilute the sample with RP load/wash solution to a volume that follows the RP SPE manufacturer's instructions. Slowly load your sample onto the cartridge and wash the cartridge using RP load/wash solution (use a volume as recommended by the manufacturer). Apply air pressure to flush the solvent from the cartridge.
12. Elute the sample into a 1.5 mL tube using RP elution solution.
13. At this stage, measure the peptide concentration in the samples, for example using low-volume 280 nm UV spectrophotometry (Nanodrop spectrophotometer, Thermo Fisher Scientific, North Ryde, NSW, Australia) or by a colorimetric method such as the bicinchoninic acid assay (Pierce BCA Protein Assay Kit, Thermo Fisher Scientific, North Ryde, NSW, Australia) (*see* **Note 17**).
14. Reduce the volume of the sample to ~50 μL by vacuum centrifugation.
15. At this stage the sample can be subjected to LC-MS/MS directly (for label-free quantification) or chemical stable isotopic labelling (for quantification using, e.g., iTRAQ [17], reductive dimethylation [18], or tandem mass tags (TMT) [16]). Phosphopeptide enrichment can either be performed at this stage or after chemical labelling. This protocol describes the procedure of TMT labelling and TiO_2 enrichment for quantitative analysis of protein phosphorylation.

3.4 TMT Labelling

1. Adjust the volume of the samples to be labelled to 100 μL using digestion buffer. Up to ten samples can be labelled and compared in a single LC-MS/MS analysis using the TMT10plex labelling scheme. The maximum vendor-recommended amount of peptide that can be labelled with a single 0.8 mg reagent vial is 25–100 μg (*see* **Note 18**).
2. Transfer a volume equivalent to 100 μg of peptide from each sample to new tubes and adjust it to 100 μL with digestion buffer.
3. Using as many reagent vials as you have samples, let the reagent vials reach 23 °C, aspirate 41 μL of anhydrous AcN with a 100 μL gastight glass syringe, and add it to each of the required reagent vials. Allow the reagent to dissolve by vortexing a couple of times over the course of 5 min.
4. Add each of the dissolved TMT reagents to each of the samples, mix by brief vortexing, and incubate for 1 h at 23 °C.
5. Add 8 μL of 5% hydroxylamine, vortex, and incubate for 15 min at 23 °C to quench the labelling reaction.
6. Combine the labelled sample sets and mix by vortexing.
7. Dry the combined sample in a vacuum centrifuge to about 10% of the original volume.

8. Reduce the amount of excess TMT reagent in the sample by RP-SPE. Use a RP-SPE cartridge with a capacity of at least twice the peptide amount in the combined sample (*see* **Note 9**). Condition the cartridge by first passing RP conditioning solution through it, followed by RP Load/wash solution (for all volumes, follow the manufacturer's instructions).
9. Dilute the sample in RP load/wash solution to a volume that follows the RP SPE manufacturer's instructions. Slowly load your sample onto the cartridge and wash it using RP load/wash solution. Apply pressure to flush the solvent from the cartridge.
10. Elute the sample into a low-binding 1.5 mL tube using RP elution solution.
11. Dry the combined sample in a vacuum centrifuge to a final volume of ~30 μL.

3.5 Phosphopeptide Enrichment

1. Weigh out an appropriate amount (see Sect. 1.4, *Quantitative phosphoproteomics*, for more information) of TiO_2 beads and suspend the beads in AcN. Add the suspension to a C8 StageTip using air pressure applied by a 1 mL syringe.
2. Add 160 μL AcN to the sample followed by 10 μL 100% TFA. Add 15.2 mg of glycolic acid to the sample and dissolve it by vortexing.
3. Transfer the sample to the TiO_2 StageTip and slowly push it through the column by applying air pressure using the syringe (*see* **Note 19**).
4. Wash the column in 100 μL washing buffer 1, pushing it through the column by applying air pressure using the syringe.
5. Wash the column in 100 μL washing buffer 2, pushing it through the column by applying air pressure using the syringe.
6. Wash the column in 100 μL washing buffer 3, pushing it through the column by applying air pressure using the syringe.
7. Using 100 μL elution buffer, elute the peptides off the column into a 1.5 mL microcentrifuge tube, slowly pushing it through the column by applying air pressure using the syringe.
8. Some hydrophobic phosphopeptides that are released from the TiO_2 in elution buffer will bind the C8 plug (that is retaining the TiO_2 resin in the microcolumn). Thus, an additional elution is required. Elute the C8-bound phosphopeptides off the StageTip by dispensing 20 μL AcN into the tip. Slowly push it through the column by applying air pressure using the syringe. Ideally, only collect the very first droplet and combine this with the previous eluate. The first droplet is sufficient to complete

the elution and it is advantageous to keep the total AcN concentration as low as possible in subsequent steps (*see* **Note 20**).

9. Acidify the eluate by adding 10 μL 100% formic acid and 1.1 μL 10% TFA to the sample and mix by vortexing (*see* **Note 21**).
10. For cleaning up the peptide sample prior to LC-MS/MS analysis, use a C18 StageTip: Condition the StageTip by passing 50 μL RP conditioning solution through it, followed by 50 μL RP load/wash solution (*see* **Note 22**).
11. Slowly load your sample onto the StageTip and wash it using 50 μL RP load/wash solution by applying air pressure using the syringe.
12. Elute the sample into a 1.5 mL low-binding tube using 50 μL RP elution solution.
13. Dry the sample in a vacuum centrifuge to near completion (leave a few μL of liquid if possible to reduce sample losses, which can occur due to peptides either binding to the plastic surface of the tube or not being redissolved completely).
14. Adjust the volume of the sample to 5 μL using RP load/wash solution.

3.6 Mass Spectrometry and Data Processing/Analysis

1. Analyze the phosphopeptide sample using LC-MS/MS. For TMT10plex use a high-resolution mass spectrometry such as the Q Exactive Plus (Thermo Fisher Scientific) (*see* **Note 23**).
2. Process the LC-MS/MS data into peak lists using MaxQuant [26] (Max Planck Institute for Biochemistry), Proteome Discoverer (Thermo Fisher Scientific, Bremen, Germany), or similar. Perform a database search against a UniProt database containing all *Rattus norvegicus* protein sequences, and quantify the TMT reporter ion channels, which correspond to the relative quantities of (phospho-) peptides in the labelled samples. All three steps are part of the integrated workflow of each of the three software suites mentioned. For a detailed step-by-step protocol on the data analysis using MaxQuant please see [27].
3. Import the ProteinGroups.txt (for protein-level quantification) or the Phospho(STY)Sites.txt (for phosphorylation site-level quantification) result output file from MaxQuant into Perseus [30]. Log2-transform the reporter intensities and filter rows for valid values, removing all sites, which have quantitative values from less than three replicates. Filter the rows to remove reverse hits resulting from the forward-reverse decoy database search. Create categorical annotation rows to define the column-replicate and column-condition relationship. Normalize the data to account for differences in sample amounts in each replicate by subtracting the column log2 intensity median from each of the protein/phosphosite intensities. Perform two-sample Student's *t*-test with Benjamini-Hochberg

Phosphopeptide	Protein Name	Site	Av. Log$_2$ (Depol./Mock)
LQRpSPVLK	RLTPR	Ser-1243	-0.44
SPpSQLR	Amphiphysin 2	Ser-334	-0.42
pSPSQLR	Amphiphysin 2	Ser-332	-0.41
pSPSQTR	Amphiphysin 1	Ser-293	-0.36
pSPAPARK	Synaptojanin 1	Ser-1131	-0.34
QSRPVAGGPGAPPAARPPApSPpSPQR	Synapsin 1	Ser-549+551	-0.31
QLpSPESLGTLQFGELNLGKEER	MAP1A	Ser-1206	0.21
VAEQTFIKDDYLETLSpSPK	SGIP1	Ser-398	0.24
SSEGpSGK	Phospholipase C β-1	Ser-477	0.25
HSSpSAPPPPPPGR	WIPF2	Ser175	0.31
RpTPSEADR	Numb homolog	Ser-437	0.44
IGpSLDNITHVPGGGNK	Tau	Ser-667	0.48
QASIpSGPAPPK	Synapsin 1	Ser-568	0.63

Fig. 3 Changes to phosphorylation sites in response to depolarization. The SVE protein network was enriched using the baits shown in Fig. 2. The unbound protein was then used in an additional pull-down with a mixture of GST-amphiphysin 1 CLAP (clathrin and AP2 binding) domain (human, 308–435) and GST-dynamin 1 proline-rich domain (rat, 746–864). The bound proteins from each pull-down were digested with trypsin. The tryptic peptides from the sequential pull-downs were pooled. Mock-stimulated and 40 mM KCl-stimulated conditions were produced in four independent experiments and labelled with TMT10plex reagents. Phosphopeptides were enriched and analyzed by LC-MS/MS. The phosphopeptides/sites were identified and quantified by MaxQuant and the data further processed in Graphpad Prism 4. Results with $p > 0.05$ were discarded after statistical analysis (one-way ANOVA and Dunnett's post hoc test, $n = 3$, 4 or 5). The average log2-transformed ratio, relative amount of phosphopeptide after depolarization divided by the amount after mock stimulation, is shown. Regulated phosphosites from SVE proteins amphiphysin 1 and 2, synaptojanin 1, and SGIP1 were among those detected

correction for multiple testing (*see* **Note 24**). At the protein level, comparing the GST-only controls to the bait samples will facilitate identification of specific bait interactors. Performing the *t*-test at the phosphosite-level comparing control to stimulated samples allows for detecting stimulus-induced protein phosphorylation changes. Filter the phosphorylation site localization probability to remove dubious phosphorylation sites (e.g., sites with probability <0.75) (*see* **Note 25**). The same process of log2 transformation, normalization, statistical analysis, and filtering can be applied using various combinations of software to produce a list of up/downregulated phosphorylation sites (Fig. 3).

4. Use other features of the Perseus software to identify enriched gene ontology terms (Fig. 4), do hierarchical clustering of proteins and samples, determine enriched phosphorylation site motifs, etc. Export the data from Perseus to, optionally, perform soft clustering to identify co-regulated phosphorylation sites or proteins, e.g., GProX [36]. The data can also be imported into bioinformatics resources such as DAVID (Database for Annotation, Visualization and Integrated Discovery) [37] or network analysis tools such as STRING, which can be used to map protein-protein interaction networks (Fig. 5) [38].

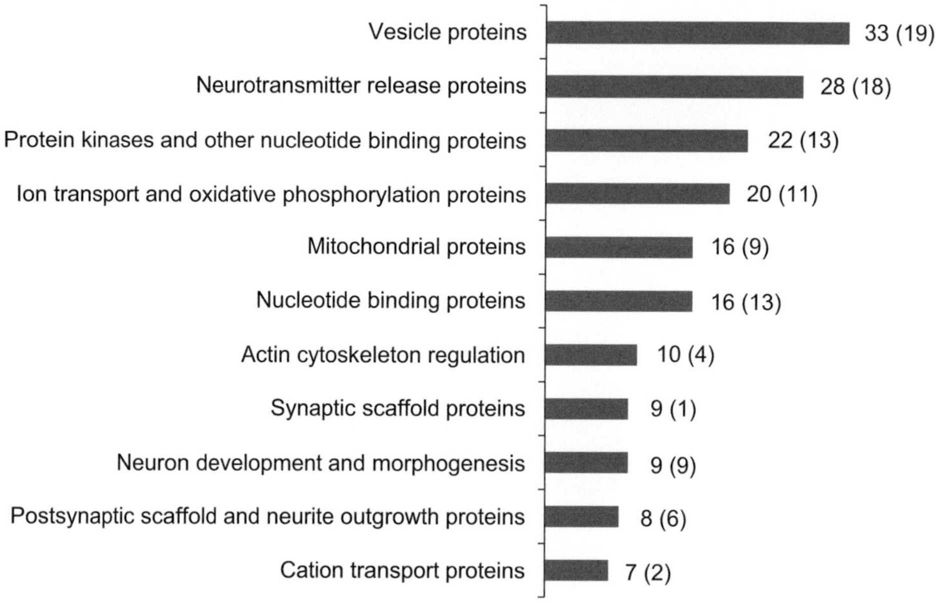

Fig. 4 Classification of enriched synaptic proteins according to function. The proteins detected by LC-MS analysis after the sequential pull-down from synaptosomes (described in Figs. 2 and 3) were functionally clustered using Database for Annotation, Visualization and Integrated Discovery (DAVID) [37]. The number of proteins in each category is shown. Category descriptions have been manually edited to produce classes familiar to the field of presynaptic vesicle trafficking. The number of proteins assigned to more than one class is shown in parentheses. Proteins determined to be contaminants (bound to GST only) were not included. A successful enrichment of the SVE machinery is indicated by the two top classes being "vesicle proteins" and "neurotransmitter release proteins"

4 Conclusions

In the example experiment presented in this chapter, a mixture of GST-p85 SH3, GST-amph1 SH3, GST-endpn1 SH3, and GST-AP2α-adaptin appendage ear domain, followed by a mixture of GST-amph1 CLAP domain and GST-dyn1 PRD domain, was used to enrich for the SVE proteome from four biological replicates of P2 synaptosomal lysates. Prior to lysis, each synaptosome sample was subjected to 3 s of either mock stimulation (5 mM KCl) or 40 mM KCl depolarization. After sample preparation and LC-MS/MS analysis, 280 proteins were identified with at least two unique peptides, after disregarding nonspecific/GST-binding proteins. Functional annotation based on gene ontology in DAVID revealed that vesicle proteins were the largest group, which included endocytosis and exocytosis proteins (Fig. 4), indicating a successful affinity purification. In support of this, 15 of the proteins/protein families with genetic evidence for their involvement in SVE known from literature were detected as specific binders in the pull-down (clathrin, AP2, EPS15, intersectin, AP180/CALM, epsin, amphiphysin, dynamin, auxilin, HSC70, synaptojanin, endophilin, syn-

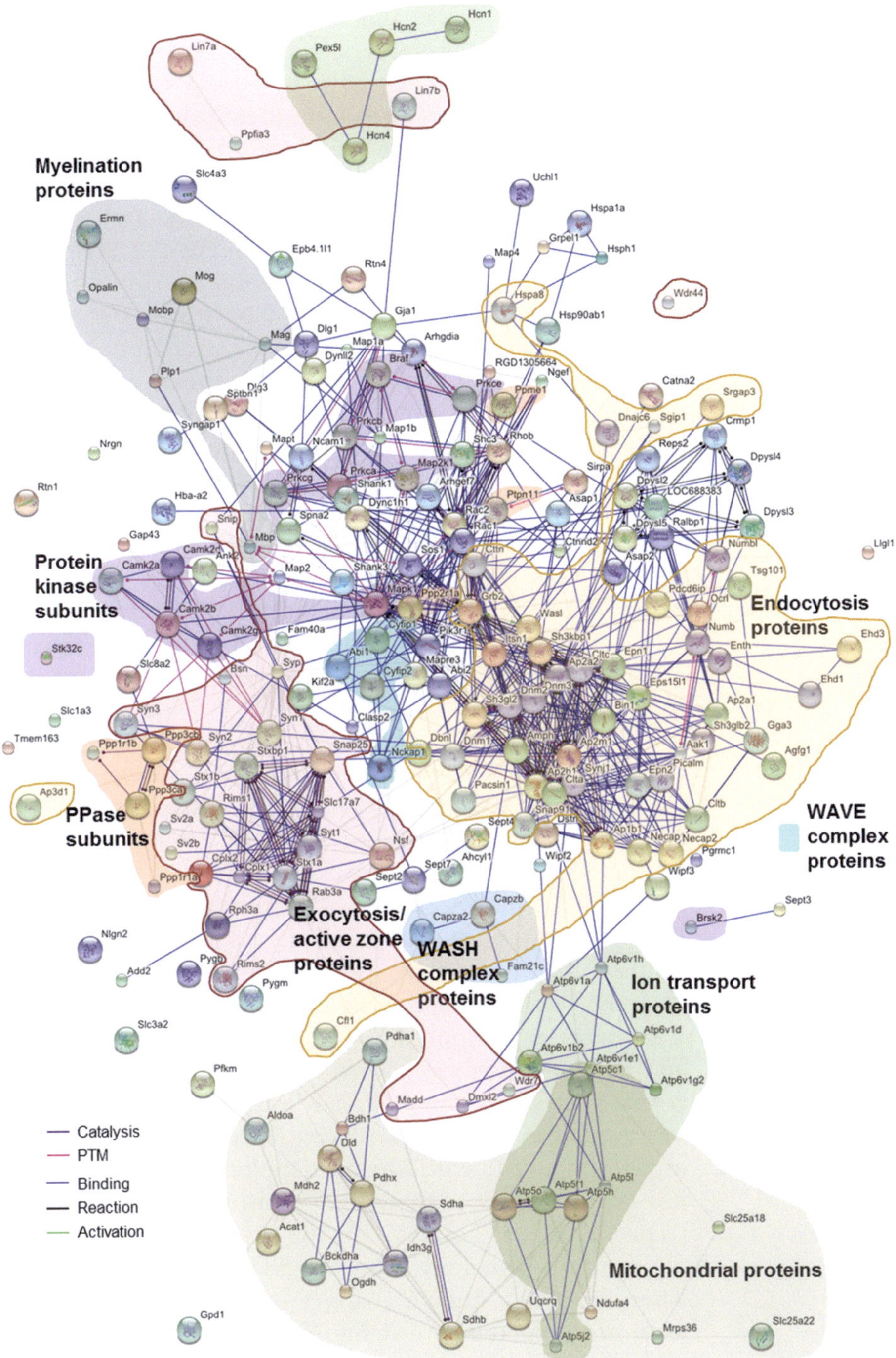

Fig. 5 Network of protein interactions for the GST-fusion protein affinity pull-downs. STRING network of confidently identified proteins enriched in the sequential pull-down from synaptosomes described in Figs. 2 and 3. Unconnected proteins are not shown. Proteins (nodes), labelled by their gene names, are connected by lines (edges) that correlate with association scores calculated by STRING. Purple, blue, pink, and black lines indicate catalysis, binding, posttranslational modification (PTM), and reaction, respectively. Node colors and sizes have no meaning, but indicate if structures are known for the protein or protein domains. Endocytic proteins were highly clustered in the network

dapin, nervous wreck, and PIPK1γ [39]). Additional or more appropriate bait proteins could be used to improve this enrichment.

To complement the functional clustering analysis in DAVID, an interaction network of the identified proteins was generated using STRING [38] as a visual aid to identify binding proteins, functionally linked proteins, and proteins with many network connections (Fig. 5). As expected from their status as hub proteins, four AP2 complex subunits and clathrin (Cltc) had a large number of network connections. The dynamin gene family (Dnm1, Dnm2, and Dnm3), endophilin 1 (Sh3gl2), growth factor receptor-bound protein 2/GRB2 (Grb2), and SH3 domain-containing kinase-binding protein 1/CIN85 (Sh3kbp1) also had multiple network connections. The combined DAVID and STRING output clearly identified proteins associated with endocytosis as a major group enriched by the nerve terminal pull-down and these proteins demonstrated connectivity within the network. Tools such as STRING rely on curation of data in the scientific literature and data mining. Thus, there is the possibility for inappropriate use; however, resources such as STRING are likely to improve with time.

A modest set of phosphorylation sites were significantly up/downregulated in the example phosphoproteomic data shown in Fig. 3. Synapsin 1 increased phosphorylation at S568 and was dephosphorylated at S549 (calcineurin substrate site 6) and S551 [40]. Phospho-S549 and S551 were detected on the same phosphopeptide. Nevertheless, the established response from synapsin 1 site 6 serves as an internal control to validate the method of stimulation, quantification, and data analysis. Fast dephosphorylation was observed for synaptojanin 1 (S1131), amphiphysin 1 (S293), and amphiphysin 2 (S332 and S334), which are known calcineurin substrates [4]. SH3-containing GRB2-like protein 3-interacting protein 1 (SGIP1), a known phosphoprotein [12] implicated in SVE [41], increased in phosphorylation following depolarization (Fig. 3). We did not explore whether changes in phosphorylation affect binding to protein baits. This was not the case for one published example, i.e., the dynamin 1-Amph1 SH3 domain interaction [9]. There are other known changes to SVE phosphoproteins that were not detected or not significant, e.g., the dephosphorylation of dynamin 1 [9] was not significant. There can be several reasons for lack of detection or significance. The sample may have been too complex for the chromatography used, in which case fractionation prior to LC-MS/MS is advised [19]. Alternatively, variation in the sample preparation or LC-MS analysis can affect the standard deviation in the data analysis. Sources of variation need to be carefully considered at the earliest opportunity. Another consideration is the statistical power of the test.

In the field of proteomics, a low number of experimental replicates is commonly used (e.g., $n = 3$). However, a larger sample size will increase the statistical power of the significance test, i.e., a difference is more likely to be detected, if it exists. Using power analysis, the sample size necessary for detecting a difference between two conditions with a certain power and significance level can be calculated if the standard deviation of the normal distributed data is known. Multiple Web-based power analysis tools are available online (such as www.powerandsamplesize.com [42]). For example, consider an experiment investigating whether a specific stimulation of synaptosomes has an effect compared to control, with a standard deviation of the two populations of 0.25. If we need to distinguish differences of 50% (1.5-fold change) nine of out ten times (power = 0.9) at a 5% false-positive rate, we need six replicates. While analyzing so many replicates previously resulted in extensive analysis time, multiplexing the mass spectrometry analysis of replicates, using for example TMT labelling, reduces the length of the analysis markedly.

The experiment outlined in this chapter facilitates identification of binding partners of GST-bait fusion proteins and detection of phosphorylation-site changes. Preparation of the synaptosomes, the pull-down itself, and overnight tryptic on-bead digestion can be completed on day 1. The sample cleanup and chemical stable isotope labelling using TMT can be performed on day 2. Phosphopeptide enrichment and desalting of the sample require another 6 h on day 3. The duration of mass spectrometry analysis of the samples depends on the number of samples, but will via TMT10plex multiplexing often require less than 3 h per ten samples (conditions x replicates). The entire procedure, including data acquisition, takes approximately 4 days, while data processing and bioinformatic analysis require day- to week-long analysis.

5 Notes

1. By using low-protein-binding microcentrifuge tubes and pipette tips (such as Eppendorf LoBind (Eppendorf North Ryde, NSW, Australia) or Axygen MAXYmum recovery (Pacific Laboratory Products, Blackburn, Victoria, Australia)), protein and peptide adsorption to plastic surfaces is reduced. The use of this type of plasticware is particularly important when handling the small amount of phosphopeptides present after enrichment.

2. Earlier work used 0.1 mM $CaCl_2$ to align with ^{32}P-labelling conditions [5, 32, 33]). The choice of Ca^{2+} concentration in the final buffer is determined by the experimental goals.
3. Some active-zone proteins cannot be adequately solubilized in the Triton X-100 buffer used in this protocol [43]. To maximize the solubilization efficiency of active zone targets, use a lysis buffer containing a stronger (ionic) detergent such as 1% deoxycholate. In this case, minimize the lysis buffer volume. This will allow for later dilution of deoxycholate concentration in the lysate at least twofold with 1% Triton X-100 lysis buffer prior to the GST pull-down, as ionic detergents like deoxycholate can interfere with protein-protein interactions.
4. As serine/threonine phosphatases such as protein phosphatase 1 (PP1), protein phosphatase 2A (PP2A), and protein phosphatase 2B (PP2B, also known as calcineurin) are abundant proteins in the nerve terminal, including phosphatase inhibitors targeting these phosphatases in the lysis and pull-down buffers is crucial to reduce the phosphatase activity and retain the phosphorylation status of the sample.
5. Add all protease and phosphatase inhibitors to the lysis buffer immediately before use as some of them are unstable; for example PMSF has a half-life of ~1 h in aqueous solutions at pH 7.4.
6. For a detailed protocol describing the procedure of bacterial transfection of the GST-fusion protein construct, cloning, bacterial expressing, and coupling of the GST-bait fusion protein to glutathione Sepharose 4B beads please see [34].
7. Dissolve the trypsin in ice-cold digestion buffer immediately before use to limit autolysis.
8. A cheaper alternative for digesting larger amounts of protein is TrypZean (Sigma-Aldrich, Castle Hill, NSW, Australia), a recombinant bovine trypsin expressed in corn. This product has been shown to have similar efficiency and specificity as for example Promega Sequencing Grade Modified Trypsin [44].
9. Reversed-phase (RP) solid-phase extraction (SPE) cartridges generally have a peptide binding capacity of ~5% of the sorbent weight. Use a cartridge with a capacity of at least fivefold the GST-bait protein weight.
10. The TMT10plex reagents are highly sensitive to moisture and should be stored at −20 °C until use. Let the reagent vials equilibrate to room temperature before opening to avoid condensation that might lead to reagent hydrolysis.
11. As a more cost-effective alternative, the StageTips can be prepared in-house. For a detailed protocol on how to produce these please see [24].

12. The crude P2 synaptosome preparation can be particularly useful for dynamic metabolic studies such as stimulus-dependent protein phosphorylation, without the need for further purification on Percoll gradients. This is because respiration and ATP production occur only within sealed viable synaptosomes, which are able to maintain a membrane potential and thus exhibit stimulus-responsive changes in phosphorylation. Any phosphorylation events not within the sealed synaptosomes are not expected to be dynamic or stimulus responsive. Phosphorylation changes therefore only occur within the synaptosomes. In contrast, the Percoll-purified synaptosomes may increase the specificity of protein-protein interaction studies.
13. At this stage the synaptosomes can be subjected to chemical depolarization, using for example KCl or 4-aminopyridine, or activation or inhibition of specific kinases, phosphatases, or other enzymes depending on the aim of the study.
14. Apart from the GST-fusion protein bead samples, remember to include GST-only bead control sample(s) to allow for distinguishing between specific and nonspecific interactors.
15. The cap should always be loosened before unplugging the column to equalize the pressure in the column. Unequal pressure can cause the liquid to escape unexpectedly.
16. To avoid condensation in the cap of the column, which could cause the beads to dry out, the column should be heated evenly from both the bottom and top. This can be achieved by using a heated block with an enclosed lid or a water bath with lid.
17. Determination of protein/peptide concentration by UV spectrophotometry at 280 nm on a microliter scale is fast, does not waste sample, and is sufficiently accurate. However, this method is not appropriate for lysates, due to chemical interference, and should therefore only be used for highly purified samples. Many alternative quantification methods exist and should be assessed for appropriate use for particular sample types, chemical compatibility, sensitivity, economics, and ease of use. For cell lysates, the bicinchoninic acid (BCA) protein assay is the preferred protein concentration assay in our laboratory.
18. We have successfully labelled up to 250 μg desalted peptide with >95% efficiency. To reduce the reagent costs, samples from multiple experiments/replicates should be labelled simultaneously, as the TMT reagents will quickly degrade over time, even when stored at −80 °C. According to the vendor, the reagents are stable for 1 week at −20 °C after being dissolved in anhydrous AcN.
19. Avoid letting the column dry out in between the individual loading/washing steps as this can reduce the efficacy of the procedure.

20. Do not elute all 20 μL into the tube—only the first droplet that appears at the tip of the microcolumn; it is important to avoid a too high AcN concentration in the combined eluate to allow for direct desalting by RP-SPE after acidification of the sample. If the concentration of AcN is >5%, some hydrophilic phosphopeptides will not be retained on the RP column. Another way to ensure ending up with a low concentration of AcN in the combined eluate is to dispense only 2–5 μL of AcN into the microcolumn.
21. The low pH increases the hydrophobicity of the peptides, while TFA acts as an ion-pairing reagent, both of which leads to increased retention of hydrophilic peptides.
22. The C18 StageTip has a capacity of up to 10 μg peptide, which roughly corresponds to the phosphorylated peptides from the 10 × 100 μg cell lysate peptide recommended as starting material for a TMT10plex labelling experiment. If higher phosphopeptide amounts are expected to be present or if losses should be limited to a minimum, load the flow-through from the StageTip onto a second StageTip, repeat the washing and elution steps, and combine the eluates from the two StageTips. The binding capacity can also be increased by adding, e.g., OLIGO R3 RP resin (Thermo Fisher Scientific, North Ryde, NSW, Australia) on top of the C18 StageTip membrane, or, if manufacturing the StageTip in-house, by inserting multiple plugs of SPE membrane into the pipette tip.
23. Make sure to analyze the sample using an LC-MS/MS method acquiring fragment scans at >50,000 at m/z 150 to ensure adequate resolving power for each reporter ion in the MS/MS spectra.
24. Ensure that the statistical test is suitable for the experimental setup regarding distribution of the data, number of variables, dependence, etc. For further information on how to choose the correct statistical test see [45].
25. Notice that the phosphorylation-site localization in a peptide is often difficult to pinpoint from MS/MS data, especially when the potential phosphosite resides in a peptide sequence with multiple neighboring serine and/or threonine residues. Thus phosphoproteomics software will regularly report erroneous identifications of protein phosphorylation on sites adjacent to well-established sites if stringent filtering of the localization probability is not applied. The example data in this chapter was filtered to remove phosphosites with a localization probability of less than 0.9.

Acknowledgements

Mass spectrometry data was obtained and analyzed at the Biomedical Proteomics and Australian Cancer Research Foundation Centre for Kinomics, Children's Medical Research Institute, Westmead, NSW 2145, Australia. This work was supported by National Health and Medical Research Council of Australia, the Rebecca L Cooper Medical Research Foundation, the Cancer Institute New South Wales, the Ian Potter Foundation, the Lundbeck Foundation (M.R.L., Lundbeck Junior Group Leader Fellowship and K.E.-K., Postdoctoral Fellowship), the Danish Council for Independent Research (N.B., Postdoc Fellowship)/ the European Union FP7 Marie Curie Actions—COFUND programme (K.E.-K., MOBILEX Postdoc Fellowship, grant ID DFF–1325-00154), and the Carlsberg Foundation (K.E.-K., Internationalization Fellowship). We thank Lauren E. Smith for proofreading the manuscript.

References

1. Wilhelm BG, Mandad S, Truckenbrodt S, Krohnert K, Schafer C, Rammner B, Koo SJ, Classen GA, Krauss M, Haucke V, Urlaub H, Rizzoli SO (2014) Composition of isolated synaptic boutons reveals the amounts of vesicle trafficking proteins. Science 344(6187):1023–1028. https://doi.org/10.1126/science.1252884
2. Castillo PE (2012) Presynaptic LTP and LTD of excitatory and inhibitory synapses. Cold Spring Harb Perspect Biol 4(2):a005728. https://doi.org/10.1101/cshperspect.a005728
3. Regehr WG (2012) Short-term presynaptic plasticity. Cold Spring Harb Perspect Biol 4(7):a005702. https://doi.org/10.1101/cshperspect.a005702
4. Cousin MA, Robinson PJ (2001) The dephosphins: dephosphorylation by calcineurin triggers synaptic vesicle endocytosis. Trends Neurosci 24(11):659–665
5. Tan TC, Valova VA, Malladi CS, Graham ME, Berven LA, Jupp OJ, Hansra G, McClure SJ, Sarcevic B, Boadle RA, Larsen MR, Cousin MA, Robinson PJ (2003) Cdk5 is essential for synaptic vesicle endocytosis. Nat Cell Biol 5(8):701–710. https://doi.org/10.1038/ncb1020
6. Collins MO, Yu L, Coba MP, Husi H, Campuzano I, Blackstock WP, Choudhary JS, Grant SG (2005) Proteomic analysis of in vivo phosphorylated synaptic proteins. J Biol Chem 280(7):5972–5982. https://doi.org/10.1074/jbc.M411220200
7. Munton RP, Tweedie-Cullen R, Livingstone-Zatchej M, Weinandy F, Waidelich M, Longo D, Gehrig P, Potthast F, Rutishauser D, Gerrits B, Panse C, Schlapbach R, Mansuy IM (2007) Qualitative and quantitative analyses of protein phosphorylation in naive and stimulated mouse synaptosomal preparations. Mol Cell Proteomics 6(2):283–293. https://doi.org/10.1074/mcp.M600046-MCP200
8. Trinidad JC, Barkan DT, Gulledge BF, Thalhammer A, Sali A, Schoepfer R, Burlingame AL (2012) Global identification and characterization of both O-GlcNAcylation and phosphorylation at the murine synapse. Mol Cell Proteomics 11(8):215–229. https://doi.org/10.1074/mcp.O112.018366
9. Graham ME, Anggono V, Bache N, Larsen MR, Craft GE, Robinson PJ (2007) The in vivo phosphorylation sites of rat brain dynamin I. J Biol Chem 282(20):14695–14707. https://doi.org/10.1074/jbc.M609713200
10. Kohansal-Nodehi M, Chua JJ, Urlaub H, Jahn R, Czernik D (2016) Analysis of protein phosphorylation in nerve terminal reveals extensive changes in active zone proteins upon exocytosis. eLife 5:e14530. https://doi.org/10.7554/eLife.14530
11. Chan LS, Hansra G, Robinson PJ, Graham ME (2010) Differential phosphorylation of dynamin I isoforms in subcellular compartments demonstrates the hidden complexity of phos-

phoproteomes. J Proteome Res 9(8):4028–4037. https://doi.org/10.1021/pr100223n

12. Craft GE, Graham ME, Bache N, Larsen MR, Robinson PJ (2008) The in vivo phosphorylation sites in multiple isoforms of amphiphysin I from rat brain nerve terminals. Mol Cell Proteomics 7(6):1146–1161. https://doi.org/10.1074/mcp.M700351-MCP200
13. Xue J, Graham ME, Novelle AE, Sue N, Gray N, McNiven MA, Smillie KJ, Cousin MA, Robinson PJ (2011) Calcineurin selectively docks with the dynamin Ixb splice variant to regulate activity-dependent bulk endocytosis. J Biol Chem 286(35):30295–30303. https://doi.org/10.1074/jbc.M111.273110
14. Merl J, Ueffing M, Hauck SM, von Toerne C (2012) Direct comparison of MS-based label-free and SILAC quantitative proteome profiling strategies in primary retinal Muller cells. Proteomics 12(12):1902–1911. https://doi.org/10.1002/pmic.201100549
15. Li Z, Adams RM, Chourey K, Hurst GB, Hettich RL, Pan C (2012) Systematic comparison of label-free, metabolic labeling, and isobaric chemical labeling for quantitative proteomics on LTQ Orbitrap Velos. J Proteome Res 11(3):1582–1590. https://doi.org/10.1021/pr200748h
16. Thompson A, Schafer J, Kuhn K, Kienle S, Schwarz J, Schmidt G, Neumann T, Johnstone R, Mohammed AK, Hamon C (2003) Tandem mass tags: a novel quantification strategy for comparative analysis of complex protein mixtures by MS/MS. Anal Chem 75(8):1895–1904
17. Ross PL, Huang YN, Marchese JN, Williamson B, Parker K, Hattan S, Khainovski N, Pillai S, Dey S, Daniels S, Purkayastha S, Juhasz P, Martin S, Bartlet-Jones M, He F, Jacobson A, Pappin DJ (2004) Multiplexed protein quantitation in Saccharomyces cerevisiae using amine-reactive isobaric tagging reagents. Mol Cell Proteomics 3(12):1154–1169. https://doi.org/10.1074/mcp.M400129-MCP200
18. Hsu JL, Huang SY, Chow NH, Chen SH (2003) Stable-isotope dimethyl labeling for quantitative proteomics. Anal Chem 75(24):6843–6852. https://doi.org/10.1021/ac0348625
19. Engholm-Keller K, Birck P, Storling J, Pociot F, Mandrup-Poulsen T, Larsen MR (2012) TiSH--a robust and sensitive global phosphoproteomics strategy employing a combination of TiO_2, SIMAC, and HILIC. J Proteomics 75(18):5749–5761. https://doi.org/10.1016/j.jprot.2012.08.007
20. Andersson L, Porath J (1986) Isolation of phosphoproteins by immobilized metal (Fe3+) affinity chromatography. Anal Biochem 154(1):250–254
21. Larsen MR, Thingholm TE, Jensen ON, Roepstorff P, Jorgensen TJ (2005) Highly selective enrichment of phosphorylated peptides from peptide mixtures using titanium dioxide microcolumns. Mol Cell Proteomics 4(7):873–886. https://doi.org/10.1074/mcp.T500007-MCP200
22. Pinkse MW, Uitto PM, Hilhorst MJ, Ooms B, Heck AJ (2004) Selective isolation at the femtomole level of phosphopeptides from proteolytic digests using 2D-NanoLC-ESI-MS/MS and titanium oxide precolumns. Anal Chem 76(14):3935–3943. https://doi.org/10.1021/ac0498617
23. Engholm-Keller K, Larsen MR (2011) Titanium dioxide as chemo-affinity chromatographic sorbent of biomolecular compounds--applications in acidic modification-specific proteomics. J Proteome 75(2):317–328. https://doi.org/10.1016/j.jprot.2011.07.024
24. Rappsilber J, Mann M, Ishihama Y (2007) Protocol for micro-purification, enrichment, pre-fractionation and storage of peptides for proteomics using StageTips. Nat Protoc 2(8):1896–1906. https://doi.org/10.1038/nprot.2007.261
25. Li QR, Ning ZB, Tang JS, Nie S, Zeng R (2009) Effect of peptide-to-TiO_2 beads ratio on phosphopeptide enrichment selectivity. J Proteome Res 8(11):5375–5381. https://doi.org/10.1021/pr900659n
26. Cox J, Mann M (2008) MaxQuant enables high peptide identification rates, individualized p.p.b.-range mass accuracies and proteome-wide protein quantification. Nat Biotechnol 26(12):1367–1372. https://doi.org/10.1038/nbt.1511
27. Tyanova S, Temu T, Cox J (2016) The MaxQuant computational platform for mass spectrometry-based shotgun proteomics. Nat Protoc 11(12):2301–2319. https://doi.org/10.1038/nprot.2016.136
28. Munk S, Refsgaard JC, Olsen JV (2016) Systems analysis for interpretation of phosphoproteomics data. Methods Mol Biol 1355:341–360. https://doi.org/10.1007/978-1-4939-3049-4_23
29. Munk S, Refsgaard JC, Olsen JV, Jensen LJ (2016) From phosphosites to kinases. Methods Mol Biol 1355:307–321. https://doi.org/10.1007/978-1-4939-3049-4_21
30. Tyanova S, Temu T, Sinitcyn P, Carlson A, Hein MY, Geiger T, Mann M, Cox J (2016)

The Perseus computational platform for comprehensive analysis of (prote)omics data. Nat Methods 13(9):731–740. https://doi.org/10.1038/nmeth.3901

31. Dunkley PR, Jarvie PE, Robinson PJ (2008) A rapid Percoll gradient procedure for preparation of synaptosomes. Nat Protoc 3(11):1718–1728. https://doi.org/10.1038/nprot.2008.171
32. Robinson PJ, Dunkley PR (1983) Depolarisation-dependent protein phosphorylation in rat cortical synaptosomes: factors determining the magnitude of the response. J Neurochem 41(4):909–918
33. Robinson PJ, Dunkley PR (1985) Depolarisation-dependent protein phosphorylation and dephosphorylation in rat cortical synaptosomes is modulated by calcium. J Neurochem 44(2):338–348
34. Brymora A, Valova VA, Robinson PJ (2004) Protein-protein interactions identified by pull-down experiments and mass spectrometry. Curr Protoc Cell Biol Chapter 17:Unit 17:15. https://doi.org/10.1002/0471143030.cb1705s22
35. Praefcke GJ, Ford MG, Schmid EM, Olesen LE, Gallop JL, Peak-Chew SY, Vallis Y, Babu MM, Mills IG, McMahon HT (2004) Evolving nature of the AP2 alpha-appendage hub during clathrin-coated vesicle endocytosis. EMBO J 23(22):4371–4383. https://doi.org/10.1038/sj.emboj.7600445
36. Rigbolt KT, Vanselow JT, Blagoev B (2011) GProX, a user-friendly platform for bioinformatics analysis and visualization of quantitative proteomics data. Mol Cell Proteomics 10(8):O110.007450. https://doi.org/10.1074/mcp.O110.007450
37. Huang d W, Sherman BT, Lempicki RA (2009) Systematic and integrative analysis of large gene lists using DAVID bioinformatics resources. Nat Protoc 4(1):44–57. https://doi.org/10.1038/nprot.2008.211
38. Franceschini A, Szklarczyk D, Frankild S, Kuhn M, Simonovic M, Roth A, Lin J, Minguez P, Bork P, von Mering C, Jensen LJ (2013) STRING v9.1: protein-protein interaction networks, with increased coverage and integration. Nucleic Acids Res 41(Database issue):D808–D815. https://doi.org/10.1093/nar/gks1094
39. Saheki Y, De Camilli P (2012) Synaptic vesicle endocytosis. Cold Spring Harb Perspect Biol 4(9):a005645. https://doi.org/10.1101/cshperspect.a005645
40. Jovanovic JN, Sihra TS, Nairn AC, Hemmings HC Jr, Greengard P, Czernik AJ (2001) Opposing changes in phosphorylation of specific sites in synapsin I during Ca2+−dependent glutamate release in isolated nerve terminals. J Neurosci 21(20):7944–7953
41. McMahon HT, Boucrot E (2011) Molecular mechanism and physiological functions of clathrin-mediated endocytosis. Nat Rev Mol Cell Biol 12(8):517–533. https://doi.org/10.1038/nrm3151
42. Chow S-C, Wang H, Shao J (2007) Sample size calculations in clinical research. CRC Press, Boca Raton
43. Sudhof TC (2012) The presynaptic active zone. Neuron 75(1):11–25. https://doi.org/10.1016/j.neuron.2012.06.012
44. Bunkenborg J, Espadas G, Molina H (2013) Cutting edge proteomics: benchmarking of six commercial trypsins. J Proteome Res 12(8):3631–3641. https://doi.org/10.1021/pr4001465
45. Jung K (2011) Statistics in experimental design, preprocessing, and analysis of proteomics data. Methods Mol Biol 696:259–272. https://doi.org/10.1007/978-1-60761-987-1_16

Part V

Studying Synaptic Proteomics, Function, and Plasticity Using Synaptosomes

Chapter 11

Manipulating Synapse-Enriched Biochemical Fractions to Detect a Purified Synaptic Protein's Effect on Neurotransmitter Release

Marianne Cilluffo, Theodore A. Sarafian, and Joseph B. Watson

Abstract

The availability of tractable biochemical fractions enriched in synaptic terminals is essential to complement a variety of other experimental synaptic approaches utilizing electrophysiology, neuropharmacology, proteomics, and neural imaging. This review focuses on a rapid, convenient, scaled-down synaptoneurosome (SN) method to pre-internalize a given exogenous purified synaptic protein in presynaptic terminals for in vitro neurotransmitter assays. The chapter builds on previous publications that targeted the presynaptic role of the Parkinson's disease-related protein, α-synuclein, in glutamate neurotransmitter release. Additional detailed methods are described for transmission electron microscopy combined with immunohistochemistry to localize α-synuclein in SN synaptic terminals. A recently reported enzyme-based, fluorescent assay to measure glutamate release quantitatively in a 96-well plate format is also described. It is envisioned that the SN-centric methods can be easily extrapolated to synaptosome-driven approaches to synaptic biology.

Key words Synaptosome, Synaptoneurosome, Recombinant α-synuclein, Transmission electron microscopy, Immunohistochemistry, Glutamate assay, Amplex® Red, Fluorescence, 96-Well plates

1 Introduction

There are a variety of ways to measure a synaptic protein's role in presynaptic neurotransmitter release including neurochemical assays of synapse-enriched biochemical fractions such as synaptosomes [1]. A key step is to know that a target protein is localized and present in sufficient amounts in a synaptosome, due to either endogenous expression or heterologous expression by transgenic mice [2, 3]. However, while pre-assembly and/or association of a purified synaptic protein in a reconstituted lipid bilayer or in a presynaptic complex is quite feasible [4, 5], methods for protein pre-internalization in synaptosomes enriched in resealed, presynaptic membranous terminals are lacking.

Kathryn M. Murphy (ed.), *Synaptosomes*, Neuromethods, vol. 141,
https://doi.org/10.1007/978-1-4939-8739-9_11,

Previous reports showed that overexpression of monomeric human α-synuclein can have either positive or negative effects on presynaptic neurotransmitter release, specifically glutamate from corticostriatal terminals [6, 7]. More recently, recombinant human α-synuclein proteoforms (monomers, fibrillated) were pre-internalized in synaptoneurosomes (SNs) [8–10] containing both pre- and postsynaptic endings to measure their effects on presynaptic glutamate release [11]. Here a detailed protocol captures many of these same approaches so that the role of other synaptic proteins in neurotransmitter release can be studied as internalized proteins in a desired synaptosome preparation, enriched in mainly presynaptic terminals.

2 Materials

2.1 Laboratory Equipment

1. Microliter pipettors and sterile tips.
2. Several pairs of Dumont nonmagnetic curved #7 Forceps with clamping rings (clamping rings can be purchased from an EM supplier or made from pieces of Tygon tubing cut down into 3–4 mm rings).
3. 130 mm Glass petri dishes.
4. Glow discharge unit.
5. Electron microscope.
6. Centrifuge for microcentrifuge tubes (200–2000 × *g*):
   (a) 95% O_2/CO_2 pressurized tank.
   (b) Freezers: −20 °C/−80 °C.
   (c) pH meter; strips.
   (d) Fluorescent plate reader, e.g., Fluoroskan FL Ascent (Thermo Scientific).

2.2 Supplies

1. Carbon/formvar 200–400 mesh copper and/or nickel grids.
2. Whatman #1 filter paper circles, 25 mm diameter (larger diameter can be cut into wedges of similar size).
3. Parafilm.
4. Syringes.
5. PVDF syringe filters (0.2 um, 13 mm diameter).
6. Gelatin capsules and capsule rack.
7. Microcentrifuge tubes.
8. SN fractionation-related supplies [10]:
   (a) Swinnex filter holders.
   (b) Nylon filters (100 μM).

(c) Fine 5 μM filters.
(d) Microcentrifuge tube pestle.
(e) 1 mL Tuberculin syringes.

9. 96-Well plates.

2.3 Chemicals/Reagents

1. Purified proteins: Human α-synuclein proteins (full-length 1-140aa ± NH_2-terminal acetylation; truncated 1-95aa ± NH_2-terminal acetylation).
2. Synaptosomes/synaptoneurosome fractions.
3. 2% Uranyl acetate solution in ddH_2O.
4. Saturated uranyl acetate solution in ddH_2O.
5. 2% Glutaraldehyde/0.1 M phosphate buffer.
6. 4% Paraformaldhehyde/0.1 M phosphate buffer.
7. Anti-alpha synuclein antisera.
8. Donkey anti-rabbit IgG coupled to 10–20 nm gold.
9. Nonimmune rabbit IgG.
10. LR White resin.
11. 100% Ethanol.
12. PBS (0.1 M phosphate buffer, 0.9% sodium chloride).
13. PBS with 0.06% sodium azide for storage of pellets.
14. 2% Bovine serum albumin in PBS.
15. Reynold's lead citrate.
16. Ca^{2+}-free, modified Krebs-Henseleit (mKRBS) buffer pH 7.4.
17. Amplex® Red Glutamic Acid/Glutamate Oxidase Kit.
18. Krebs ringer buffer (KR).
19. 0.1 M Tris buffer (pH 7.5).

3 Methods

We will first describe the basic technique to pre-internalize a chosen purified protein within synaptic terminals enriched in a desired synaptosome fraction. Note that there are a multitude of detailed methods for preparing synaptosome fractions (see relevant chapters within in this book). For illustration, we will describe in detail an example of how a purified recombinant human α-synuclein protein can be preincubated with an SN fraction prepared from mouse brain. Methods for SN fractionation were described previously [9, 10] and will not be detailed here. Studies were carried out according to guidelines of the National Institutes of Health Guide for Care and Use of Laboratory Animals (NIH Publications No.80–23), "Guidelines for the Use of Animals in Neuroscience Research"

(Society for Neuroscience), and with approval from the Institutional Animal Care and Use Committee at UCLA. Both wild-type (WT) mice and α-synuclein knockout (KO) mice were used [2, 11].

3.1 Preincubation of Purified Synaptic Protein with a Synaptosome Fraction (Fig. 1)

1. Purified proteins are simply added to the starting brain tissue on ice (or at 4 °C, for example cold room or refrigerated container) prior to the homogenization protocol used in the initial step of the synaptosome preparation of choice. An SN fractionation scheme [10] was used in the example shown here (Fig.1) (see Note c).
2. Homogenize sample vigorously with a pestle or dounce in starting synaptosome buffer to facilitate internalization of proteins during membrane disruption and random resealing within vesicularized structures (see Note d).
3. Perform remaining synaptosome prep in normal fashion as desired.
   Notes
   (a) Recombinant proteins were produced in the UCLA-DOE Protein Expression Technology Center as described previously [11].
   (b) For recombinant human α-synuclein protein, on average 15–25 μg of protein works well when added to 50–100 mg wet weight of starting mouse forebrain tissue contained in a microfuge tube for SN preps (see outline in Fig. 1).
   (c) Place about 50 μL of protein sample against the wall of the microfuge tube (2.0 mL) and allow to run down into 200 μL Ca^{2+}-free mKREBS or KR buffer preincubated with brain tissue for about 10 s.
   (d) For SNs, at least 20 vigorous turns of the pestle sufficiently homogenizes the brain tissue in buffer with minimal insoluble tissue remaining at the bottom of the tube. Lastly add another 800 μL of buffer and mix gently by tapping the outside of the tube. The homogenate is now ready for the remainder of the SN or synaptosome prep.
   (e) Homogenization conditions may need to be modified depending on the source of brain tissue and the nature of the synaptosome preps.
   (f) *Caution:* Working with purified α-synuclein and other folding proteins can be hazardous and appropriate safety precautions should be followed [12].

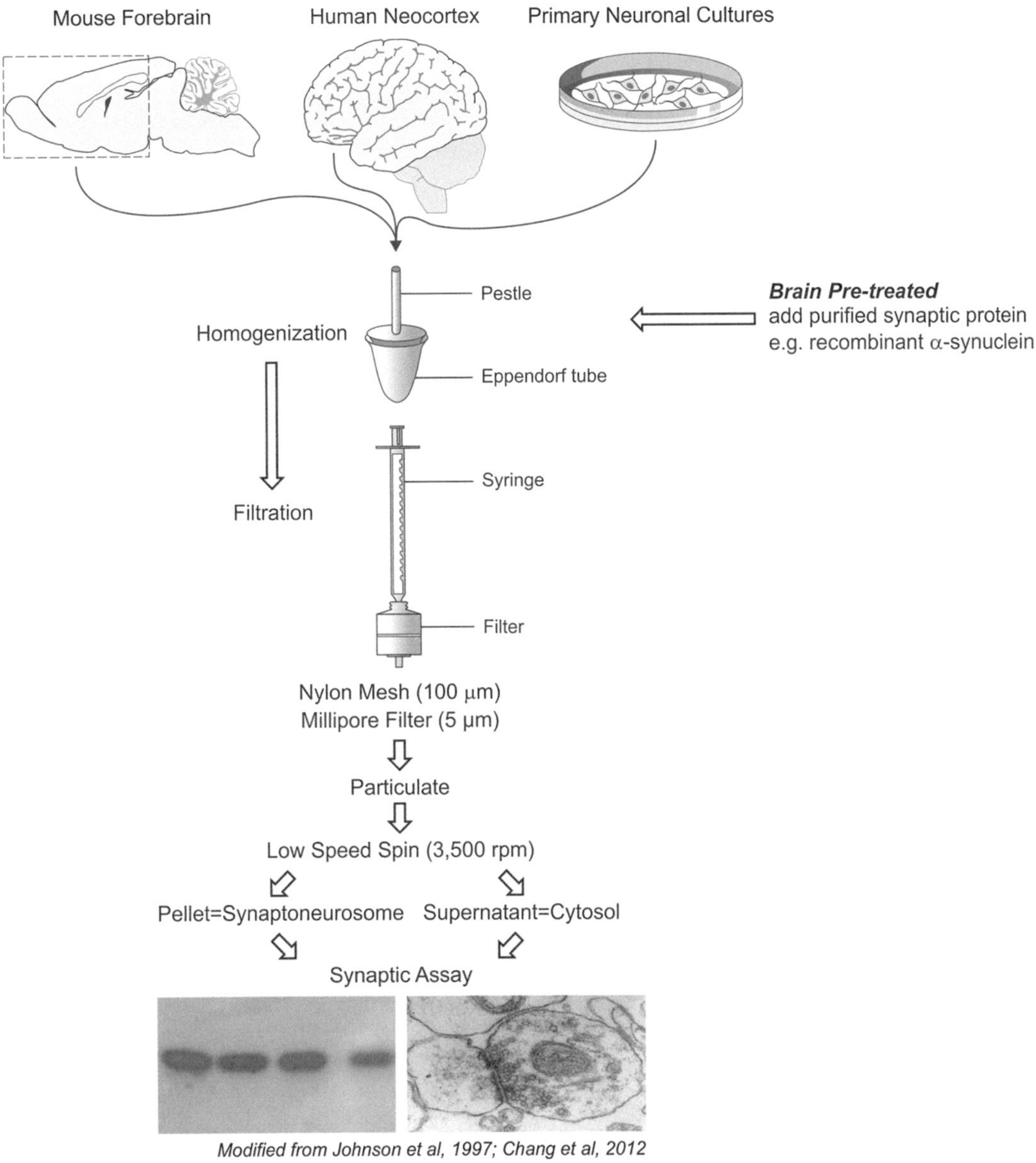

Fig. 1 Pre-internalization of purified synaptic protein in SNs. SNs were freshly prepared from forebrains of either WT or KO mice as described [11]. Recombinant monomeric, full-length human α-synuclein was preincubated with brain tissue (brain pretreated) prior to the homogenization step and subsequent SN fractionation. Note: The figure illustration was modified from previously published figures [see 9, 10]

3.2 Confirm Pre-internalization of Synaptic Protein in Synaptic Terminals Within Synaptosome Fractions by Microscopic Approaches (Fig. 2)

The key to making an approach such as this work is to first confirm that given purified synaptic protein was pre-internalized, ideally by a microscopic approach. Transmission electron microscopy (TEM) and immunohistochemistry (IHC) can be co-utilized to examine the synaptosome population as well as the isolated recombinant proteins. Isolated proteins can be deposited directly on grids, stained, and examined in the electron microscope. Synaptosomes should be embedded, sectioned, and stained to allow for visualization of internal structures as previously described for SN preps [11]. They can be identified by the presynaptic vesicles and/or postsynaptic densities.

3.2.1 TEM Immunogold on Purified Protein Preparations (Fig. 2a)

Initially it is helpful to test the primary antisera before proceeding to IHC of synaptosome preparations [13]. Appropriate controls should be run to check for nonspecific binding of the primary antisera. These should include a no primary (incubation solution) and a nonimmune IgG. The nonimmune serum should match the concentration and species of the primary antisera as closely as possible. The following conditions were used for analyzing purified recombinant human α-synuclein protein and SNs prepared from forebrain KO mice, but can easily be adapted to other synaptic proteins and synaptosome fractions.

1. Prepare and isolate purified proteins as desired.
2. Absorb the protein to freshly glow-discharged 200 mesh formvar/carbon nickel grids by clamping a grid between the tips of a pair of forceps and placing the forceps in large petri dish lined with Parafilm (see notes below).
3. Deposit 2–5 μL of sample onto each grid. Absorb for 3–10 min (see notes below).
4. Wick solution off the grid by touching a piece of filter paper to the juncture between the grid and forceps. Quickly add 5 μL of 4% paraformaldehyde. Fix for 5 min. Do not allow the grid to dry at this point. Wick and add fix to each grid one at a time to avoid drying.
5. Wick solution off the grid by touching a piece of filter paper to the juncture between the grid and forceps. Quickly rinse with PBS three times. Keep the grid moist on a drop of PBS (see Note c below).
6. Place a fresh strip of Parafilm in a large petri dish. If stored overnight, give the grids one additional PBS wash. Invert the grids on 20 μL drops of 2% BSA/PBS. Cover the dish and incubate for 20 min.
7. Wick solution off the grid by touching a piece of filter paper to the juncture between the grid and forceps. Invert the grids onto 10–20 μL of primary antisera diluted in 1% BSA/PBS, cover the dish, and incubate for 1 h. Use separate forceps for

anti-synuclein antibody, nonimmune rabbit IgG and BSA controls.

8. Wick solution off the grid and invert the grid onto a drop of PBS. Wash six times, 5 min each, wicking the solution from the grid between changes.
9. Wick solution off the grid and invert the grid onto a drop of anti-rabbit IgG-10 nm gold diluted 1:20 in 1% BSA/PBS. Cover and incubate for 1 h.
10. Repeat washing as in step 9.
11. Wick the solution off the grid and place on a drop of 2% glutaraldehyde. Fix for 3 min.
12. Wick and wash the grid three times with PBS and one time with ddH_2O.
13. Negative stain the grid by wicking the solution off the grid. Stain grid with 5 μL of 1–2% uranyl acetate for 3 min.
14. Wick the grid dry and let dry covered for a minimum of 10 min.

Notes

(a) The grid should be clamped so that it is flat and the solution deposits evenly over the surface. If the grid is tilted the protein could absorb unevenly.

(b) Protein also can be absorbed to the grid by placing 5–10 μL droplets of solution on a strip of Parafilm. The freshly glow-discharged grid is inverted onto the droplet and incubated for 5–60 min, covered.

(c) The appropriate concentration of the protein solution often needs to be determined empirically. If the grid is completely electron-opaque or the film breaks when being examined, there is too much protein present and the sample would need to be diluted. Alternatively, if the grid does not appear to have any sample, the absorption time can be increased.

(d) All solutions used for IHC should be passed through a 0.2 μm filter before use.

(e) Do not allow the grids to dry during the IHC procedure.

(f) The grids can be stored overnight at 4 °C in PBS.

3.2.2 Immuno-gold TEM of SNs Plus Recombinant Proteins (Figs. 2b–d)

Post-embedding immuno-electron microscopy can be used to examine the presence of recombinant proteins in synaptosomes or in this example SN-pelleted fractions from α-synuclein KO mice. Pellets are embedded in acrylic resin and sectioned. The sections are then stained for the protein of interest [14, 15]. Instructional pamphlets from manufacturers of resins and antibodies can also be of assistance in determining the best IHC procedure.

1. Pre-internalize protein of interest in synaptosome or SNs as described above (Sect. 3.1).
2. After protein pretreatment, pellet the synaptosome or SNs by microcentrifugation at 1000 × *g* for 5–10 min at 4 °C.
3. Remove the supernatant and fix in 0.1% glutaraldehyde/4% paraformaldehyde in KR buffer for 1–2 h at room temperature. See Note g below. Use a volume of fixative equal to 10–20 times the pellet volume.
4. Remove the fixative and store the pellets in PBS plus 0.06% sodium azide at 4 °C until they can be embedded in resin.
5. To embed the pellets: gently wash the pellets two times in PBS (see Note b).
6. Dehydrate the pellets in 50% pure ethanol for 15 min. Follow with two 30-min incubations in 70% ethanol. Use 10–20 times the sample volume in each step.
7. Incubate the pellet with a 1:2 mixture of 80% ethanol:LR White resin for 1 h.
8. Transfer the pellets to pure resin for 1 h. Change the resin and incubate for an additional hour.
9. Transfer to fresh resin and incubate overnight at 4 °C.
10. The following day, transfer to fresh resin for an additional hour and embed in gelatin capsules. Polymerize at 55 °C overnight.
11. Section the LR White blocks at 70–80 nm (gold interference color) on an ultramicrotome and deposit the sections on 200–300 mesh nickel grids.
12. Perform IHC on the grids as described in Sect. 3.2.1 (*steps 6–12*).
13. Counterstain the grids with saturated uranyl acetate and Reynold's lead citrate.

Notes

(a) α-Synuclein KO mice were used for SN preps to minimize interference from the endogenous mouse protein.

(b) If the pellets are not solid, they can be embedded in 4% low-melt agarose before dehydration.

(c) Various acrylic resins are available for post-embedding IHC. We find LR White to be simple to use.

(d) All dehydration and incubation steps for embedding should be done on a rotator or with gentle agitation.

(e) The primary antisera will typically need to be run at a lower dilution (higher concentration) than for purified proteins.

(f) Incubation times can be increased in the primary and secondary antisera to assist in penetration of the antibody into the section.

(g) The concentration and type of fixative should be checked using dot blots, or staining of purified proteins on grids.

(h) Many pretreatments can be performed on the grids to enhance the immunoreactivity of the embedded sections.

3.3 Functional Neurotransmitter Assay of Synaptosome Fractions

Once a given synaptic protein's pre-internalization in synaptic terminals has been confirmed by TEM as described above (Fig. 2) or another preferred microscopic technique, the next step is up to the investigator to choose the appropriate functional neurotransmitter assay. To provide an example, glutamate was measured neurochemically in mouse forebrain SNs by glutamate oxidase activity coupled to Amplex® Red fluorescence using a modification of the manufacturer's protocol (Thermo Fisher Scientific/Invitrogen) [11]. With a slight modification of the SN protocol using WT mice, we were able to establish a model for measuring sustained (15–30 min) glutamate neurotransmitter release.

3.3.1 Glutamate Assay (Fig. 3)

1. SNs (or synaptosomes) are first washed in a Ca^{2+}-free KR buffer and suspended in about 2 mL of the same buffer (see Note c below).
2. A volume of 100 μL of SN suspension is then aliquoted into a series of 0.5 mL tubes and treated with various concentrations of proteins of interest for 10–15 min at room temperature.
3. Paired samples are treated with 2 μL of H_2O (basal) or 2 μL of 2 M KCl (40 μM final; stimulated) and triplicate 25 μL aliquots of SNs are transferred to a 96-well flat-bottom plate.
4. When all SN samples are transferred, an enzyme cocktail is prepared just before starting the assay by a modification of the kit protocol for the glutamate assay from Thermo Fisher Scientific/ Invitrogen.
5. The enzyme cocktail for 90 assays consisted of 2.2 mL reaction buffer (0.1 M Tris pH 7.5), 20 μL Amplex® *R*ed (2.6 mg/mL in DMSO), 5 μL of horseradish peroxidase (100 U/mL), 21 μL glutamate oxidase (2.5 U/mL), 10 μL glutamate pyruvate transanion (100 U/mL), and 2 μL of 200 mM alanine (see Note d).
6. Aliquots (25 μL) of the enzyme cocktail are added to wells pre-aliquoted with equal volumes of SNs (25 μL).
7. After 10 s of shaking, the plate is incubated for 30 min at 37 °C in the dark.
8. Red fluorescence is measured using a Fluoroskan Fl Ascent (Thermo Scientific) plate reader (E_m = 535, E_m = 585).
9. Glutamate concentrations are determined based on glutamate standards in separate wells. For most experiments, absolute values for SN glutamate release are expressed as pmoles per μg protein.

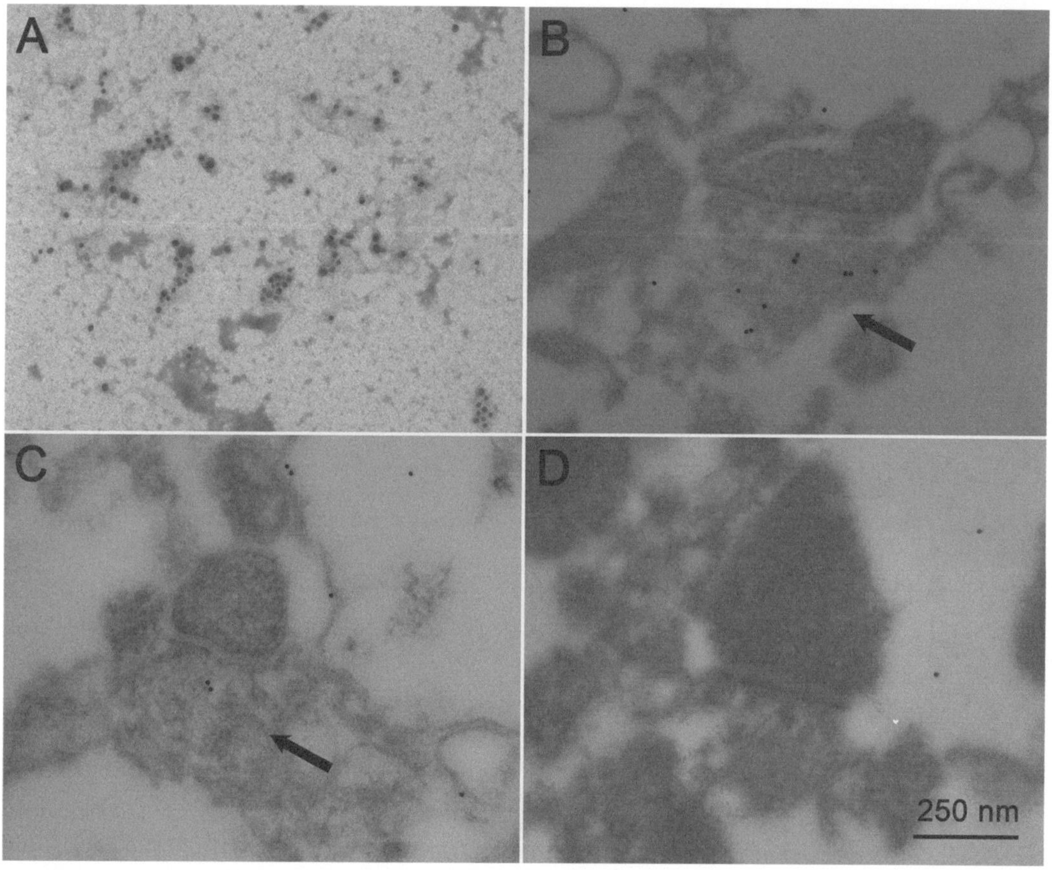

Fig. 2 Confirmation of pre-internalization of α-synuclein in SN synaptic terminals by TEM and IHC. (**a**) An initial control study examined purified α-synuclein monomer directly deposited on carbon/formvar film and stained with α-synuclein antisera, followed by anti-rabbit IgG coupled to 15 nm colloidal gold. The grid was negatively stained with 2% uranyl acetate after the IHC procedure to allow for visualization of the deposited proteins. Gold particles can be seen covering deposited protein. (**b**) LR White section of forebrain SNs from α-synuclein KO mice that were pretreated with recombinant α-synuclein protein (see Fig.1) and then stained with α-synuclein antisera. Numerous gold particles are found covering the presynaptic terminal (arrow). (**c**) LR White section from WT mouse forebrain stained for α-synuclein. Endogenous α-synuclein, identified by the presence of gold particles, is found in the presynaptic terminal (arrow) (also see section: Functional Neurotransmitter Assay and Figs. 3 and 4). (**d**) LR White section from KO mouse forebrain stained for α-synuclein. There are no gold particles associated with the synaptic terminals. Scale bar for all figures is shown in **d**. Note: Figure panels 2**b** and **c** were modified from a previous report [11]

10. Reliable measurements of SN glutamate were achievable with the glutamate oxidase assay and a 96-well plate format (see Fig. 3). The addition of 0.5 mM $CaCl_2$ (final conc.) and 40 mM KCl to SNs to induce membrane depolarization substantially increased glutamate levels over the same time period relative to controls (KR buffer, EGTA to quench divalent cations).

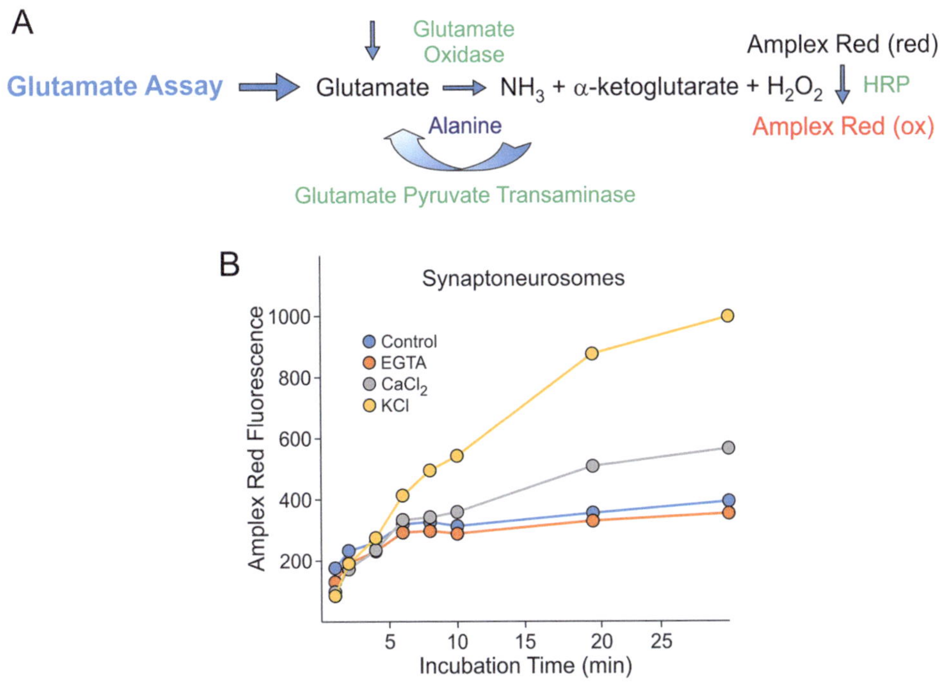

Fig. 3 Glutamate release assay using SNs. (**a**) The basic enzymology for glutamate oxidase activity in SNs (arrow) is shown. Enzyme activity is coupled to horseradish peroxidase (HRP)-mediated Amplex® Red fluorescence, using the Amplex® Red Glutamic Acid/Glutamate Oxidase Assay Kit (Thermo Fisher). (**b**) Enhanced fluorescence was observed for WT forebrain SNs over 30 min in response to both 0.5 mM $CaCl_2$ and 40 mM KCl; baseline values were observed for both H_2O control (con) and EGTA (2 mM). Note: The figure illustration was modified from previous published figures [11].

Notes

(a) It is important to use synaptic fractions (synaptosomes or SNs used here) prepared from freshly dissected brains or harvested tissue cultures on the same day.

(b) Synaptic preparations should ideally not sit on ice longer than 2 h prior to the glutamate release assay.

(c) For washes, SNs were microcentrifuged at $1000 \times g$ (5 min) in the cold room (4 °C) and pellets were resuspended in fresh cold Ca^{2+}-free KR buffer.

(d) It is prudent to have each ingredient of the enzyme cocktail (e.g., reaction buffer, Amplex® Red) prepared ahead of time and handy on the lab bench top prior to step 5.

(e) Conditions (e.g., 96-well plates, instrumentation) may need to be optimized for alternative synaptosome fractions and neurotransmitter assays.

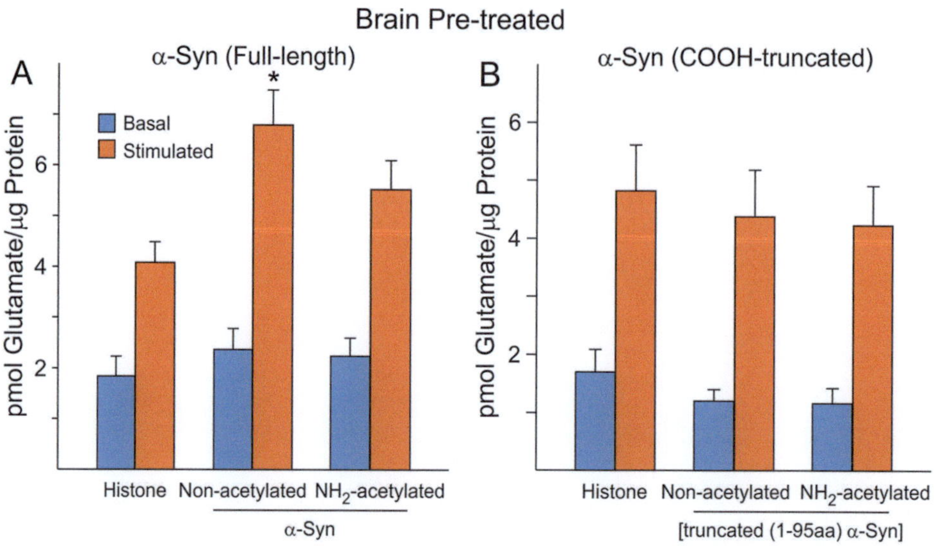

Fig. 4 Enhancement of SN glutamate release by brain pretreatment with monomeric full-length α-synuclein (α-Syn). (**a**) Both native (non-acetylated) and NH_2-terminally acetylated, full-length human α-Syn enhanced the stimulated component of (i.e., Ca^{2+}/KCl) glutamate release from SNs prepared from pretreated WT forebrain (see Fig.1) (histone 4.09 ± 0.393, non-acetylated α-Syn 6.81 ± 0.67; acetylated α-Syn 5.54 ± 0.56, $N = 5$ mice) ($P < 0.01$, One Way ANOVA for all groups; $^*P < 0.05$, post hoc Bonferroni t-test for stimulated-non-acetylated vs. stimulated-histone control). Under basal conditions, glutamate release was not significantly affected. (**b**) Both native and acetylated forms of COOH-truncated (1-95aa) α-Syn failed to enhance SN glutamate release under stimulated conditions (control 4.13 ± 0.68, $N = 20$, 9 mice; truncated/non-acetylated α-Syn 3.75 ± 0.69, $N = 10$, 9 mice; truncated/acetylated α-Syn 3.62 ± 0.58, $N = 10$, 4 mice) ($P > 0.05$, one-way ANOVA). Data suggested lack of interaction with SNARE's synaptobrevin. Note: Data appeared previously within text [11]

3.3.2 Glutamate Release Assay of SNs in Response to Pre-internalized Monomeric α-Synuclein Protein (Fig. 4)

To demonstrate the utility of the glutamate assay in a synapse-enriched fraction, VAMP2/synaptobrevin of the SNARE complex was examined as a likely presynaptic target site of monomeric α-synuclein in WT mouse forebrain SN fractions [5]. To test this hypothesis, either full-length monomeric α-synuclein or truncated (1-95aa) α-synuclein protein lacking the carboxyl terminal domain that interacts with synaptobrevin [16] was pre-internalized using the protocols described above and SN glutamate release was examined (Fig. 4). Unlike the control full-length protein, truncated forms failed to enhance stimulated glutamate release beyond the histone control, consistent with the need for some α-synuclein interaction with the SNARE complex ($P > 0.05$, one-way ANOVA).

4 Conclusion

We have described in detail how a variety of synapse-enriched biochemical fractions such as synaptosomes or SNs can be manipulated to (1) pre-internalize a desired purified protein, (2) confirm synaptic terminal localization by electron microscopy, and (3) measure glutamate neurotransmitter release by an enzyme-based fluorescence assay. We anticipate that the combined experimental methods will have broad applications for studying synaptic biology in many areas of neuroscience.

Acknowledgements

Recombinant α-synuclein protein was prepared by the UCLA-DOE Protein Expression Technology Center [supported by the US Department of Energy, Office of Biological and Environmental Research (BER) program under Award Number DE-FC02-02ER63421]. Dr. Michael Levine's UCLA lab provided wild-type (WT) and α-synuclein knockout (KO) mice. Special thanks to Donna Crandall for artwork and illustrations. Some figures were modified from previous publications and are denoted in legends.

References

1. Whittaker VP, Michaelson IA, Kirkland RJA (1964) The separation of synaptic vesicles from nerve-ending particles ('synaptosomes'). Biochem J 90(2):293–305
2. Cabin DE, Shimazu K, Murphy D, Cole NB, Gottschalk W, McIlwain KL, Orrison B, Chen A, Ellis CE, Paylor R, Lu B, Nussbaum RL (2002) Synaptic vesicle depletion correlates with attenuated synaptic responses to prolonged repetitive stimulation in mice lacking alpha-synuclein. J Neurosci 22(20): 8797–8807
3. Wislet-Gendebien S, D'Souza C, Kawarai T, St George-Hyslop P, Westaway D, Fraser P, Tandon A (2006) Cytosolic proteins regulate alpha-synuclein dissociation from presynaptic membranes. J Biol Chem 28(43): 32148–32155
4. DeWitt DC, Rhoades E (2013) Alpha-Synuclein can inhibit SNARE-mediated vesicle fusion through direct interactions with lipid bilayers. Biochemistry 52(14):2385–2387
5. Burre J, Sharma M, Tsetsenis T, Buchman V, Etherton MR, Sudhof TC (2010) Alpha-Synuclein promotes SNARE-complex assembly in vivo and in vitro. Science 329(5999): 1663–1667
6. Watson JB, Hatami A, David H, Masliah E, Roberts K, Evans CE, Levine MS (2009) Alterations in corticostriatal synaptic plasticity in mice overexpressing human alpha-synuclein. Neuroscience 159(2):501–513
7. Bendor JT, Logan TP, Edwards RH (2013) The function of alpha-synuclein. Neuron 79(6):1044–1066
8. Hollingsworth EB, McNeal ET, Burton JL, Williams RJ, Daly JW, Creveling CR (1985) Biochemical characterization of a filtered synaptoneuorosome preparation from Guinea pig cerebral cortex: cyclic adenosine 3%:5%-monophosphate-generating systems, receptors, and enzymes. J Neurosci 5(8):2240–2253
9. Johnson MW, Chotiner JK, Watson JB (1997) Isolation and characterization of synaptoneurosomes from single rat hippocampal slices. J Neurosci Methods 77(2):151–156
10. Chang JW, Arnold MM, Rozenbaum A, Caputo A, Schweizer FE, Huynh M, Mathern GW, Sarafian TA, Watson JB (2012) Synaptoneurosome micromethod for fractionation of mouse and human brain, and primary neuronal cultures. J Neurosci Methods 211(2):289–295
11. Sarafian TA, Littlejohn K, Yuan S, Fernandez C, Cilluffo M, Koo B-Y, Whitelegge JP, Watson JB (2017) Stimulation of synaptoneurosome glutamate release by monomeric and fibrillated α-synuclein. J Neurosci Res 95:1871. https://doi.org/10.1002/jnr.24024
12. Bousset L, Brundin P, Böckmann A, Meier B, Melki R (2016) An efficient procedure for removal and inactivation of alpha-synuclein assemblies from laboratory materials.

J Parkinson's Dis 6(1):143–151. https://doi.org/10.3233/JPD-150691

13. Roychaudhuri R, Yang M, Deshpande A, Cole GM, Frautschy S, Lomakin A, Benedek GB, Teplow DB (2013) C-terminal turn stability determines assembly differences between Abeta40 and Abeta42. J Mol Biol 425(2):292–308. https://doi.org/10.1016/j.jmb2012.11.006
14. Zhang N, Wei W, Mody I, Houser CR (2007) Altered localization of GABA(A) receptor subunits on dentate granule cell dendrites influences tonic and phasic inhibition in a mouse model of epilepsy. J Neurosci 27(28):7520–7531
15. Gangalum RK, Bhat AM, Kohan SA, Bhat SP (2016) Inhibition of the expression of the small heat shock protein αB-srystallin inhibits exosome secretion in human retinal pigment epithelial cells in culture. J Biol Chem 291(25):12930–12942. https://doi.org/10.1074/jbc.M115.698530 Epub 2016 Apr 27
16. Burre J, Sharma M, Sudhof TC (2012) Systematic mutagenesis of alpha-synuclein reveals distinct sequence requirements for physiological and pathological activities. J Neurosci 32(43):15227–15242

Chapter 12

Porosomes: Supramolecular Structures at the Synaptosome Membrane Involved in Vesicle Docking, Fusion, and Neurotransmitter Release

Bhanu P. Jena

Abstract

Cup-shaped supramolecular structures at the cell plasma membrane called *porosomes* mediate the fractional discharge of vesicular contents from cells during secretion. Membrane-bound secretory vesicles transiently dock and fuse at the porosome base via SNARE proteins, to expel vesicular contents to the outside during cell secretion. Porosomes range in size from 15 nm in neurons and astrocytes to 180 nm in endocrine and exocrine cells. Neuronal porosomes are composed of nearly 40 proteins, compared to the 120 nm nuclear pore composed of nearly 1000 protein molecules. Determination of the presence of porosomes in all cells examined and an understanding of their structure, composition, and functional reconstitution into artificial lipid membrane and in live cells provide a molecular understanding of cell secretion. In this chapter, the discovery and structural characterization of the neuronal porosome at the synaptosome membrane using atomic force microscopy, electron microscopy, and small-angle X-ray solution scattering; its isolation and both structural and functional reconstitution into artificial lipid membrane; and the possible molecular mechanism of its involvement in neurotransmitter release are presented. The assembly of neuronal t/v-SNARE complexes in a rosette or ring when t-SNAREs (present at the porosome base) in one membrane interact with v-SNAREs (secretory vesicle) in an opposing membrane and the molecular mechanism of synaptic vesicle swelling required for the regulated fractional release of vesicle contents are also presented in this chapter.

Key words Synaptosome membrane, Neuronal porosome complex, t/v-SNARE ring complex, Synaptic vesicle volume regulation, Neurotransmitter release

1 Introduction and Background

While membrane-bound secretory vesicles may completely collapse at the cell plasma membrane resulting in the diffusion of intravesicular contents to the cell exterior and the compensatory retrieval of the excess membrane by endocytosis, it had failed to justify the generation of partially empty secretory vesicles observed in electron micrographs of cells following secretion. This suggested the involvement of an additional cellular mechanism, enabling the release of a portion of the vesicle content

Kathryn M. Murphy (ed.), *Synaptosomes*, Neuromethods, vol. 141,
https://doi.org/10.1007/978-1-4939-8739-9_12,

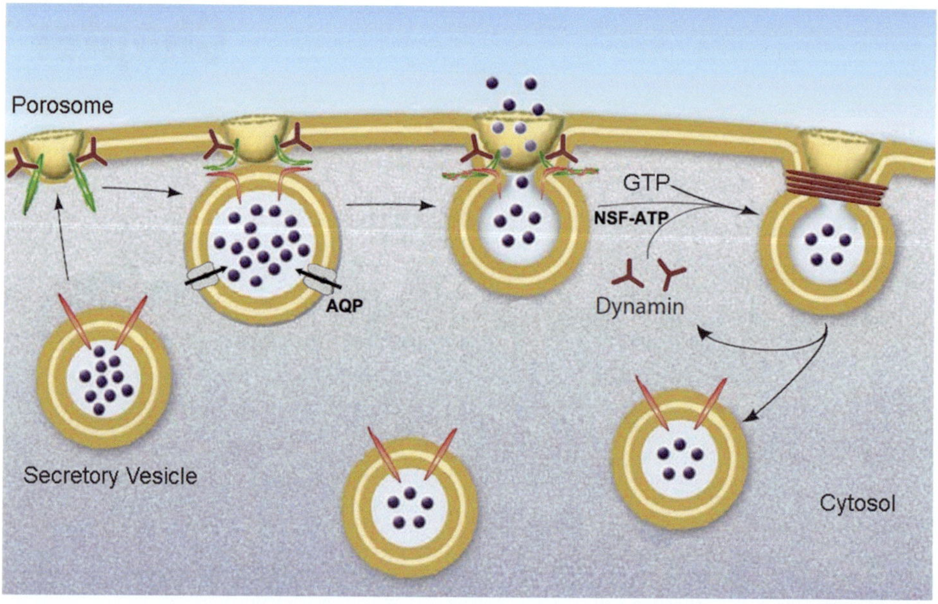

Fig. 1 Schematic drawing depicting the presence and increased association of dynamin with the porosome complex following stimulation of neurotransmitter release. Following stimulation of secretion, synaptic vesicles would dock at the porosome base, develop intravesicular pressure via active transport of water through water channels or aquaporins (AQP) at the vesicle membrane, transiently fuse at the porosome base via SNAREs and calcium, and expel neurotransmitters. After secretion, NSF, an ATPase, and dynamin, a GTPase, would work synchronously to disassemble t/v-SNARE complexes and fission the neck of fused vesicles at the porosome base, respectively. By this mechanism, partially empty vesicles could go through multiple rounds of docking-fusion-expulsion-dissociation. Unlike protein- and peptide-containing vesicles, synaptic vesicles have neurotransmitter transporters at the vesicle membrane to rapidly refill vesicles [14]. ©Bhanu Jena

during secretion. The partial emptying of vesicles during secretion would require the secretory vesicle to transiently establish continuity with the cell plasma membrane, expel a portion of its contents (exocytosis), then detach, reseal, and withdraw from the plasma membrane into the cytosol (endocytosis) (Fig. 1). Utilizing this mechanism, secretory vesicle could be reused for subsequent rounds of exo-endocytosis, until complete emptying of contents. As opposed to cells secreting large molecules such as proteins or glycoproteins, synaptic vesicles have the additional advantage of rapidly refilling spent vesicles utilizing neurotransmitter transporters present at the synaptic vesicle membrane. Consequently, in 1973, Bruno Ceccarelli coined the term "kiss-and-run" or the "transient" mechanism of secretory vesicle fusion at the cell plasma membrane [1] enabling fractional discharge of intravesicular contents during secretion. To accomplish fractional release of intravesicular contents from cells, it was hypothesized in 1990 that a continuity established between the vesicle membrane and the cell plasma membrane or a "fusion pore," results from a "preassembled ion channel-like structure that could open and close" [2]. In a 1992 review article Monck and Fernandez [3] opined that observing these structures had been the lack of adequate imaging tools to directly observe their presence and study their activity in live cells. Then in 1997 [4], a cup-shaped supramolecular secretory portal later named "porosome" was discovered at the plasma membrane of exocrine pancreas employing atomic force microscopy (AFM) [5], and subsequently confirmed using electron microscopy (EM) [6–9], and in neurons (Fig. 2) using AFM, EM, and small-angle X-ray solution scattering (SAXS) [10–16]. Neuronal porosomes measure 15 nm, and are composed of nearly 40 proteins [14] and several lipids [13] (Tables 1 and 2), compared to the 120 nm nuclear pore complex composed of 1000 protein molecules [17–19]. In the past decade, a large number of studies implicate neuronal porosome proteins in neurotransmission and their altered states in neurological disorders [20–66].

Membrane fusion in cells is mediated via a specialized set of proteins present in opposing bilayers [67–70]. Target membrane proteins, SNAP-25 and syntaxin (t-SNAREs), and secretory vesicle-associated protein (v-SNARE) are part of the conserved protein complex involved in fusion of opposing lipid membranes. The structure and arrangement of the membrane-associated full-length neuronal SNARE complex were first determined using AFM [71]. Results from the study demonstrate that neuronal t-SNAREs and v-SNARE, when present in opposing bilayers, interact in a circular array (Fig. 3) to form supramolecular ring complexes, each measuring a few nanometers [71]. In the presence of calcium, the ring complex helps in establishing continuity or the fusion pore between the opposing bilayers [71]. In contrast, in the absence of membrane, v- and t-SNAREs fail to assemble in a ring or rosette, nor establish continuity between the opposing membranes [71]. Size of the t/v-SNARE ring complex is dictated by the curvature of SNARE-associated membranes [72]. The higher the membrane curvature, the smaller the SNARE ring complex [72] (Fig. 3). At the porosome base facing the cytosol, the t-SNARE syntaxin directly interacts with L-type calcium channels [73, 74], helping vesicles dock and transiently fuse to release intravesicular contents during cell secretion. Secretory vesicle volume increase then regulates the precise amount of the fractional expulsion of intravesicular contents from cells through the fusion pore established at the porosome base [75–82] (Fig. 4).

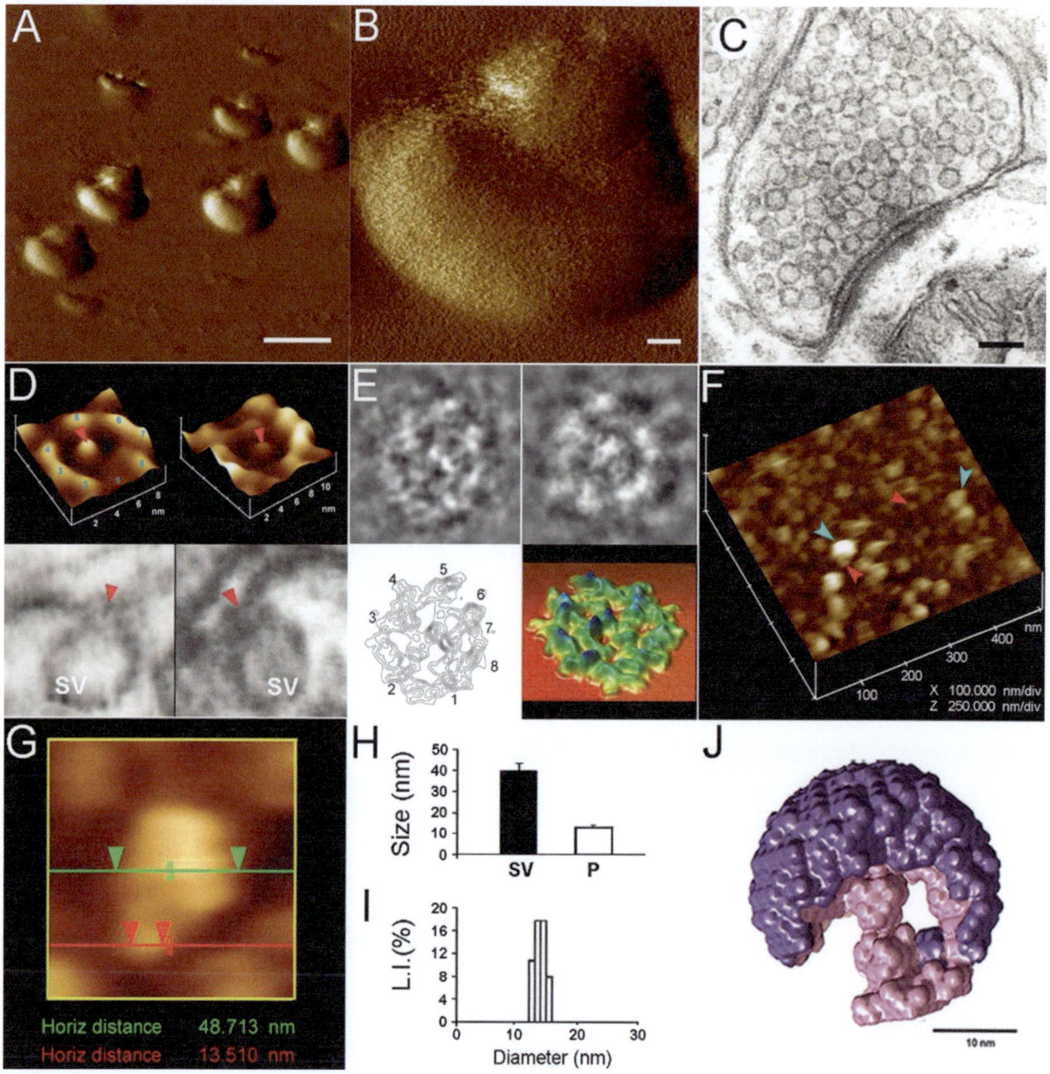

Fig. 2 Neuronal porosome structure and organization [8–13]. (**a**) Low- [scale = 1 μm] (**b**) and high-resolution [scale = 100 nm] atomic force microscope (AFM) micrographs of isolated rat-brain nerve terminals or synaptosomes in buffer. (**c**) Electron microscope (EM) picture of a bouton [scale = 100 nm]. (**d**) AFM micrographs of native neuronal porosome complex at the presynaptic membrane (Fig. **d**, top left), and of an isolated porosome complex reconstituted into lipid membrane (Fig. **d**, top right). Note the native and reconstituted porosomes to be morphologically identical. In the electron micrographs below, lower panels are two EM micrographs demonstrating synaptic vesicles (SV) docked at the base of cup-shaped porosome, with a central plug (red arrowhead). (**e**) EM, electron density, and 3D contour maps provide at nanometer resolution of protein arrangement in the porosome. (**f**) AFM micrograph of the cytosolic compartment of an isolated bouton or synaptosome, demonstrating SV (blue arrowhead) docked at the base of porosomes (red arrowhead). (**g**) AFM micrograph of a SV docked at a porosome. (**h**) Measurements ($n = 15$) using AFM of SV (SV, 40.15 ± 3.14 nm) and porosomes (*P*, 13.05 ± 0.91 nm) at the presynaptic membrane. (**i**) Photon correlation spectroscopy performed on isolated neuronal porosomes measures 12–17 nm. (**j**) Small-angle X-ray solution scattering (SAXS) of averaged 3D structure of a SV (purple) docked at the base of a native neuronal porosome (pink) [10, 12–15]. ©Bhanu Jena

Table 1
Proteins identified in purified neuronal porosomes in multiple experiments

Gene symbol	MW	Protein name	Found in earlier studies
ACTB	42 kDa	Actin, cytoplasmic 1	x, *
AT1A3	112 kDa	Sodium/potassium-transporting ATPase subunit alpha-3	
AT2B1	139 kDa	Plasma membrane calcium-transporting ATPase 1	
AT2B2	137 kDa	Plasma membrane calcium-transporting ATPase 2	
BASP1	22 kDa	Brain acid-soluble protein 1	
CAP1	52 kDa	Adenylyl cyclase-associated protein 1	
CN37	47 kDa	2′,3′-Cyclic-nucleotide 3′-phosphodiesterase	
DPYL2	62 kDa	Dihydropyrimidinase-related protein 2	
DPYL3	62 kDa	Dihydropyrimidinase-related protein 3	
DPYL5	62 kDa	Dihydropyrimidinase-related protein 5	
GLNA	42 kDa	Glutamine synthetase	
GNAO	40 kDa	Guanine nucleotide-binding protein G(o) subunit alpha	x, *
NCAM1	95 kDa	Neural cell adhesion molecule 1	
NSF	83 kDa	Vesicle-fusing ATPase	*
RAB3A	25 kDa	Ras-related protein Rab-3A	
RTN3	102 kDa	Reticulon-3	
RTN4	126 kDa	Reticulon-4	
SNP25	25 kDa	Synaptosomal associated protein 25	x, *
STX1A	33 kDa	Syntaxin-1A	*
STX1B	33 kDa	Syntaxin-1B	*
STXB1	68 kDa	Syntaxin-binding protein 1	
SYN2	63 kDa	Synapsin-2	
SYPH	33 kDa	Synaptophysin	
SYT1	47 kDa	Synaptotagmin-1	*
TBA1A	50 kDa	Tubulin alpha-1A chain	x
VAMP1	13 kDa	Vesicle-associated membrane protein 1	
VAMP2	13 kDa	Vesicle-associated membrane protein 2	
VATB2	57 kDa	V-type proton ATPase subunit B, brain isoform	

Purified rat brain porosomes from two separate experiments were analyzed by LC-MS/MS on both LTQ and QSTAR XL. Only proteins identified in both samples are reported here, all of which had protein confidence ≥95% with at least two unique peptides, each having 95% confidence or above. Proteins also found in earlier immunoisolation studies are marked in a separate column, x indicating proteins identified using MALDI-TOF/TOF; * indicates proteins identified using immunoblot analysis

Table 2
Major lipid species associated with the neuronal porosome complex identified using mass spectrometry

Lipid species	Porosome	Synaptosome	AQP1
Cer (ceramides)	**4.82**	0.00	0.00
Cer-P (ceramide phosphates)	**0.48**	0.00	0.00
LPC (lysophosphatidylcholines)	0.00	0.20	0.00
PC (phosphatidylcholines)	4.94	18.50	13.62
SM (sphingomyelins)	0.00	13.78	0.07
LPG (lysophosphatidylglycerols)	0.00	0.50	0.00
LPIP (lysophosphatidylinositol phosphates)	**23.61**	2.81	15.45
LPIP2 (lysophosphatidylinositol diphosphates)	**0.34**	0.14	0.00
LPS (lysophosphatidylserines)	0.00	0.32	0.00
PA (phosphatidic acids)	0.40	2.53	0.00
PE (phosphatidylethanolamines)	0.00	0.94	0.00
PG (phosphatidylglycerols)	0.34	5.19	0.00
PI (phosphatidylinositols)	0.74	6.86	0.00
PIP (phosphatidylinositol phosphates)	1.61	3.23	0.00
PIP2 (phosphatidylinositol diphosphates)	2.98	5.16	0.00
PS (phosphatidylserines)	2.31	20.36	2.21
DAG (diacylglycerols)	**393.74**	2.12	675.89
MADAG (1-alkyl-2-acylglycerols)	**3.00**	0.51	0.00
TAG (triacylglycerols)	0.00	0.19	0.00

Both the porosome and AQP1 (control) were purified from isolated synaptosome preparation. Note the enriched presence of Cer, Cer-P, LPIP, LPIP2, DAG, and MADAG in the isolated neuronal porosome complex. The specific association of PI, PIP, and PIP2 with the neuronal porosome complex and their absence in the AQP1 immunoisolates are further noted. Protein-normalized lipid data is presented as ng/μg protein [13]

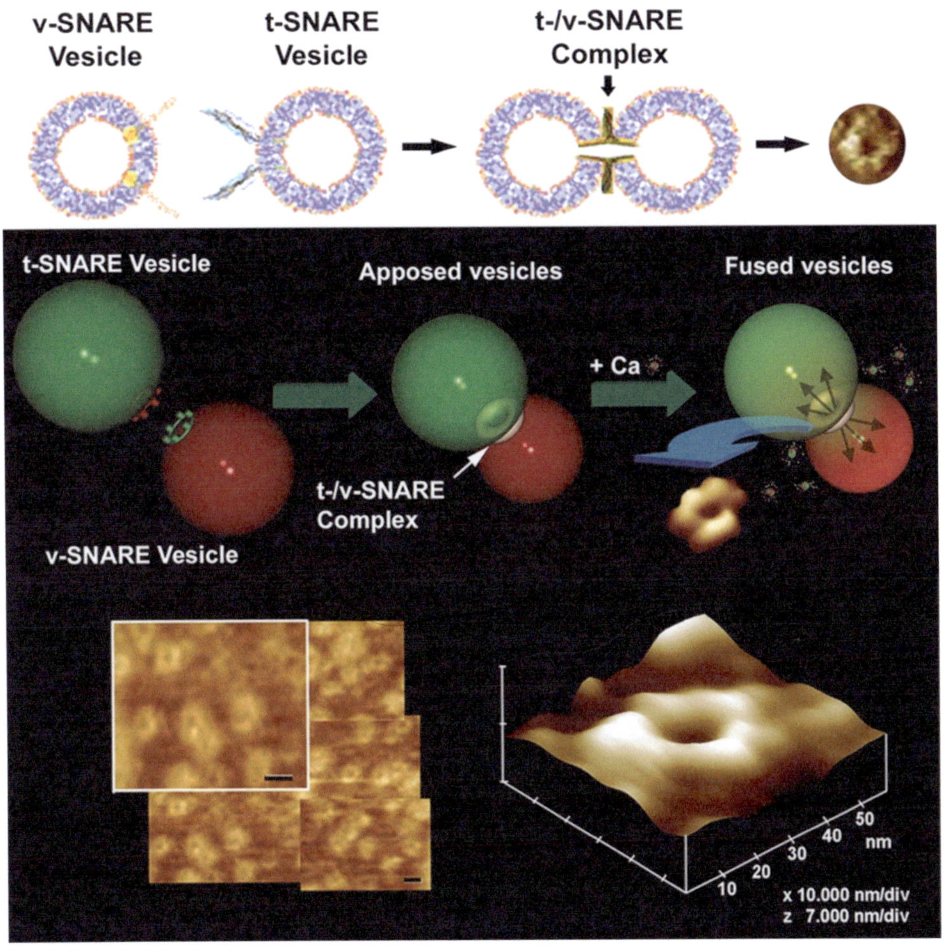

Fig. 3 Interaction of t-SNARE-associated lipid vesicle or lipid membrane with v-SNARE-associated liposomes establishes SNARE ring complexes. t/v-SNARE ring complexes or rosettes established following interactions between t-SNAREs and v-SNAREs reconstituted lipid membranes. Top and center panels are schematic drawing of the interaction between t-SNARE liposomes and v-SNARE liposome in opposing vesicles. Atomic force micrographs of actual SNARE rosettes formed, whose size is proportional to the vesicle size and hence vesicle curvature. Besides vesicle size, vesicle composition is also known to influence the size of SNARE rosette [71, 72]. ©Bhanu Jena

2 Methods

2.1 Synaptosomes, Synaptosome Membrane, and Synaptic Vesicle Isolation

Rat brain synaptosomes, synaptosomal membrane, and synaptic vesicles were prepared using whole-brain tissue from Sprague-Dawley rats. Brains from Sprague-Dawley rats weighing 100–150 g were isolated and placed in an ice-cold buffered sucrose solution (5 mM Hepes pH 7.4, 0.32 M sucrose) supplemented with protease inhibitor cocktail (Sigma, St. Louis, MO) and homogenized using a Teflon-glass homogenizer (ten strokes). The total homogenate was centrifuged for 3 min at 2500 × *g* and the supernatant fraction was further centrifuged for 15 min at 14,500 × *g*. The

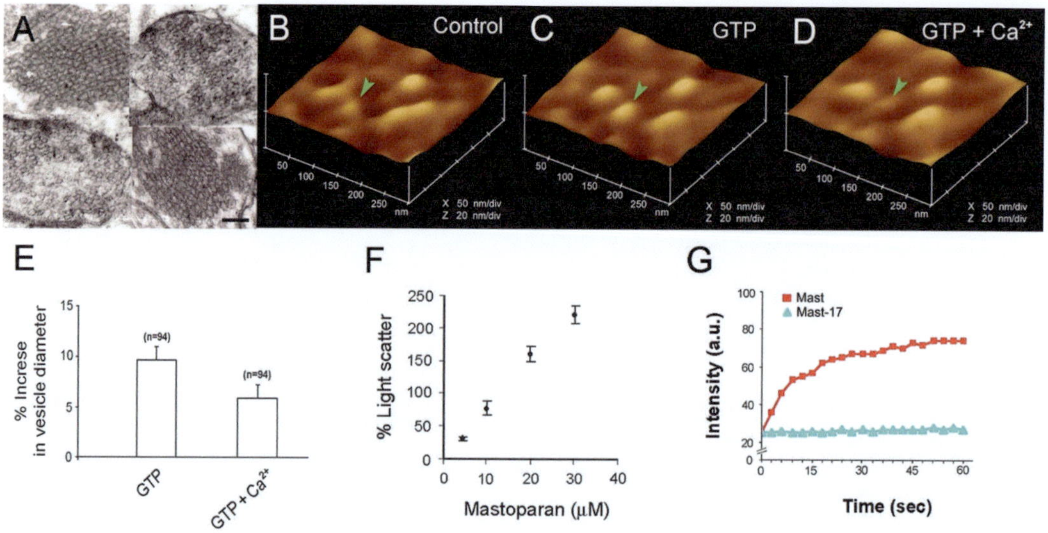

Fig. 4 Synaptic vesicles swell in response to GTP and mastoparan, and vesicle swelling is required for neurotransmitter release. (**a**) Electron micrographs of brain synaptosomes, demonstrating the presence of 40–50 nm synaptic vesicles within. Bar = 200 nm. (**b**) AFM micrographs of synaptosomal membrane, demonstrating the presence of 40–50 nm synaptic vesicles docked to the cytosolic face of the presynaptic membrane. (**c**) Exposure of the synaptic vesicles to 20 μM GTP results in vesicle swelling (blue arrowhead). (**d**, **e**) Further, addition of calcium results in the transient fusion of the synaptic vesicles at porosomes in the presynaptic membrane of the nerve terminal, and expulsion of intravesicular contents. Note the decrease in size of the synaptic vesicle following content expulsion. (**f**) Light scattering assays on isolated synaptic vesicles demonstrate the mastoparan dose-dependent increase in vesicle swelling ($n = 5$), and further confirm the AFM results. (**g**) Exposure of isolated synaptic vesicles to 20 μM mastoparan demonstrates a time-dependent increase in synaptic swelling via water and ion transport [78]. Note the control peptide mast-17 has little or no effect on synaptic vesicle swelling [75, 80–82]. ©Bhanu Jena

resultant pellet was resuspended in buffered sucrose solution, which was loaded onto 3–10–23% Percoll gradients for synaptosome isolation. After centrifugation at 28,000 × *g* for 6 min, the enriched synaptosomal fraction was collected at the 10–23% Percoll gradient interface. To isolate synaptic vesicles and synaptosomal membrane, isolated synaptosomes were diluted with nine volumes of ice-cold H_2O (hypotonic lysis) and immediately homogenized with three strokes in Dounce homogenizer, followed by 30-min incubation on ice. The homogenate was centrifuged for 20 min at 25,500 × *g*, and the resultant pellet enriched in synaptosomal membrane and the supernatant fraction enriched in synaptic vesicles were used in studies.

2.2 Isolation of the Neuronal Porosome Complex

To isolate the neuronal porosome complex, SNAP-25-specific antibody conjugated to protein A-sepharose was used. Five milligram of synaptosome membrane preparation was solubilized in Triton/Lubrol solubilization buffer (0.5% Lubrol; 1 mM benzamidine; 5 mM Mg-ATP; 5 mM EDTA; 0.5% Triton X-100, in

PBS) supplemented with protease inhibitor mix (Sigma, St. Louis, MO). SNAP-25 antibody conjugated to protein A-sepharose was incubated with the solubilized homogenate for 1 h on ice, followed by washing with wash buffer (500 mM NaCl, 10 mM Tris, 2 mM EDTA, pH 7.5). The immunopulldown porosomes associated with the immunosepharose beads were eluted using a low pH (pH 3) buffer.

2.3 Neuronal Porosome Reconstitution into Lipid Membrane for AFM Imaging

Isolated neuronal porosome complexes were reconstituted into lipid membrane on mica in buffered solution. To establish lipid membrane on mica for AFM studies, freshly cleaved mica disks were placed in a fluid chamber. Bilayer bath solution (200 μL), containing 140 mM NaCl, 10 mM HEPES, and 1 mM $CaCl_2$, was placed at the center of the cleaved mica disk. Brain lipid vesicles (10 μL) were added to the above bath solution. The mixture was then allowed to incubate for 60 min at room temperature, before washing (×10), using 100 μL bath solution/wash. The lipid membrane on mica was imaged by the AFM before and following addition of the isolated porosomes.

2.4 Neuronal Porosome Reconstitution into Lipid Membrane for Electrophysiological Measurements

Isolated neuronal porosome complexes were reconstituted into the brushed lipid membrane of the EPC9 bilayer apparatus. Lipid bilayers were established using brain phosphatidylethanolamine (PE), phosphatidylcholine (PC), dioleoylphosphatidylcholine (DOPC), and dioleoylphosphatidylserine (DOPS) obtained from Avanti Lipids (Alabaster, AL). A suspension of PE:PC in a ratio of 7:3, and at a concentration of 10 mg/mL, was prepared. Lipid suspension in 100 μL aliquots was dried under nitrogen gas and resuspended in 50 μL decane. To reconstitute the bilayer with neuronal porosomes, the preparation was added to the lipid suspension and brushed onto a 200 μm hole between the cis and the trans bilayer chamber until a stable bilayer with a capacitance between 100 and 250 pF is obtained.

2.5 Atomic Force Microscopy

Synaptosomes, synaptic vesicle preparations, isolated porosomes, and t/v-SNARE complexes adhered to a flat mica sheet were used for AFM imaging. The BioAFM or BAFM (Digital Instruments, Santa Barbara, CA) mounted on an inverted optical microscope (Olympus IX70) was used to image the sample. Sample images were obtained by the BAFM, working in both the "contact" and "tapping" mode in fluid, using a very low vertical imaging force <1 to 3 nN. Silicon and silicon nitride tips were used with spring constants of 0.25 and 0.06 $N \cdot m^{-1}$, respectively. Control experiments were performed, where a scanning force of several nN over a 60-min period was performed to determine any significant changes to the sample and consequently to the acquired images.

2.6 Photon Correlation Spectroscopy to Determine Porosome Size Distribution

The size of neuronal porosomes was also estimated using photon correlation spectroscopy (PCS). PCS measurements were performed using a Zetasizer Nano ZS (Malvern Instruments, UK). In a typical experiment, the size distribution of isolated porosomes was determined using built-in software provided by Malvern Instruments. Prior to determination of porosome size, calibration of the instrument was performed using latex spheres of known size. In PCS, subtle fluctuations in the sample scattering intensity are correlated across microsecond timescales. The correlation function is calculated, from which the diffusion coefficient is determined by the instrument. Using Stokes-Einstein equation, hydrodynamics radius can be calculated from diffusion coefficient. The intensity size distribution, which is obtained as a plot of the relative intensity of light scattered by particles in various size classes, is calculated from correlation function using the built-in software. The particle scattering intensity is proportional to the molecular weight squared. Volume distribution, which assigns more realistic weights to both small and big particles, is calculated from the intensity distribution using Mie theory. The transforms of the PCS intensity distribution to volume distributions are obtained using the provided software by Malvern Instruments.

2.7 Transmission Electron Microscopy of Synaptosome and Synaptic Vesicle Preparation

Isolated rat brain synaptic vesicles and synaptosome preparation were fixed in phosphate buffer (pH 7.4) containing 2.5% glutaraldehyde. The sample is then postfixed in phosphate buffer containing 1.5% osmium tetroxide, dehydrated in graded ethanol and acetone, and embedded in araldite. Tissue blocks were appropriately trimmed and the 40–50 nm sections obtained were stained with lead citrate, and examined using a JEOL JEM-100C transmission electron microscope.

2.8 Electron Microscopy and Contour Mapping of Isolated Neuronal Porosome Complex

Aliquots (2.5 μL) of immunoisolated porosomes from rat brain synaptosome membrane were adhered to carbon-coated 400 mesh copper grids (Cu-400CN, Pacific Grid-Tech, San Francisco, CA) previously rendered hydrophilic by glow discharge for 10 s. The grids were washed with drops of deionized water and then exposed to drops of 2% (w/v) uranyl nitrate. Images at 80,000× magnification were obtained at low defocus on 4 × 4 K Gatan UltraScan CCD under low-electron-dose conditions using a Tecnai 20 electron microscope (Philips Electron Optics/FEI, Eindhoven, The Netherlands) operating at 200 kV. Each pixel of the micrographs corresponds to 1.4 Å at the level of the specimen. Particles in micrographs were selected and windowed using the EMAN software. Each window size was 160 pixels, corresponding to 225 Å in specimen.

To display the protein structure at greater detail, contour map of proteins within the neuronal porosome complex was determined using the Spider software. All particles are first normalized using a mean density of 0 and standard deviation of 10, and then contoured by the lowest ring at 0.5. Outside the protein complex, the background is flattened and smoothed, and the protein boundary is defined by the darkest shadow around the particles, which results from the contrast transfer function of electrons. Heavy metal staining of the porosome complex enabled estimation of the relative size and arrangement of proteins in electron micrographs of the complex. Density of negative-stained proteins in electron micrographs reflects the dimension, concentration, and arrangement of protein at the various locations within the porosome complex. Higher density corresponds to greater amount of protein at that location. Using Chimera software, protein density distributions were determined and revealed in three dimension at nm resolution.

2.9 Proteome of the Isolated Neuronal Porosome Complex

In studies using in-solution digestion, purified neuronal porosomes were solubilized in lysis buffer, followed by precipitation in cold acetone. The resulting protein pellets (25 μg) were solubilized in 8 M urea, 0.4% SDS, and 250 mM TEAB, a potent buffer system used in our previous studies for membrane and membrane-associated proteins. After reduction and cysteine blocking with 55 mM iodoacetamide, trypsin was added to the diluted sample solution (1:10 w/w) for overnight digestion at 37 °C. Mass spectrometry was performed using the Applied Biosystems (ABI, Carlsbad, CA) 4700 Proteomics Analyzer (TOF/TOF) in positive ion mode. A fraction of the tryptic peptides from each sample was spotted onto a MALDI target plate for mass spectrometric analysis. Peptide mass fingerprints were collected for each well and the four most intense peaks above S/N of 60 were selected for MS/MS analysis. After the MS and MS/MS, spectra were processed using 4700 Explorer™ software (v2.0, Applied Biosystems). The monoisotopic peak lists generated in ABI's GPS Explorer™ v2.0 were submitted to the GPS Explorer™ v2.0 search tool (based on MASCOT) for protein identities. The Non-redundant Protein Database, NCBInr, was searched using the following parameters for 0 or 1 missed cleavage by trypsin, carboxyamidomethylation of cysteines as fixed modification, and methionine oxidations, N-terminal protein acetylation, Pyro-glu (N-term E), and Pyro-glu (N-term Q) as variable modifications. After a detergent removal procedure, tryptic peptides were separated by reverse-phase chromatography using Magic C18 column (Michrom, Auburn, CA), followed by ionization with the ADVANCE ion source (Michrom, Auburn, CA), and then analyzed in an LTQ-XL mass spectrometer (Thermo Scientific, Barrington, IL). Six abundant species were fragmented with collision-induced dissociation. Data analysis was performed using Proteome Discoverer 1.1 (Thermo), which incorporated the Mascot algorithm (Matrix Science, Boston, MA). The NCBI database was used against rat protein sequences and a reverse decoy protein database was run simultaneously for false discovery rate (FDR) determination. A duplicate porosome sample was also analyzed by nanoLC-MS/MS. In this case, the tryptic peptides were separated on a reversed-phase C18 column with a 90-min gradient using the Dionex Ultimate™ HPLC system (Thermo Scientific, Barrington, IL). Then the MS and MS/MS spectra were acquired on an Applied Biosystems QSTAR XL mass analyzer using information-dependent acquisition mode. A MS scan was performed from m/z 400–1500 for 1 s followed by product ion scans on two most intense multiply charged ions. Peak lists were submitted to Mascot server to search against the NCBInr database for rat sequences with carbamidomethyl (C) used as a fixed modification and oxidation (M), and N-acetylation (protein N terminus) as variable modifications. Secondary analysis of both the LTO XL and QSTAR XL was next performed using Scaffold (Proteome Software, Portland, OR). A fixed modification of +57 on cysteine (carbamidomethylation) and variable modifications of +16 on methionine (oxidation) and +42 on protein N-terminus (acetylation) were included in the database search. Minimum protein identification probability was set at ≥95% with two unique peptides at 95% minimum peptide identification probability.

2.10 Lipidome of Isolated Neuronal Porosome Complex, Synaptosome Membrane, and Synaptic Vesicle Preparation

Lipids were extracted with methanol and methyl-*tert*-butyl ether (MTBE). Briefly, methanol (1.5 mL) containing 100 ng each of internal standards (diheptadecanoyl PC, diheptadecanoyl PE, diheptadecanoyl PS, diheptadecanoyl PA, diheptadecanoyl PG, diheptadecanoyl glycerol-d5, 1,3-diheptadecanoyl-2-(10Z)heptadecenoyl glycerol-d5, 1-palmitoyl(d31)-2-oleoyl-sn-glycero-3-phosphoinositol, N-heptadecanoyl C-18 ceramide, N-heptadecanoyl C-18 sphingomyelin, and PAF-C16-d4) was added to the extracts (20 μL) followed by MTBE (5 mL), and mixed well. The mixture was left for 1 h at room temperature with occasional mixing. Water (1.5 mL) was added to the mixture, mixed thoroughly, and centrifuged (1000 × *g*) for 5 min to assist the separation of phases. The upper organic phase was collected to a clean glass tube. The lower aqueous phase was extracted twice (2 mL each time) with MTBE saturated with methanol and water (10:3:2.5 v/v) and the extracts were combined. The MTBE extracts were evaporated to dryness under a gentle stream of nitrogen and the residue was dissolved in LC-MS-grade isopropanol-hexane-100 mM aqueous ammonium acetate (58:40:2 v/v). The reconstituted lipid extract was analyzed for lipids by mass spectrometry. Lipid extracts were directly infused into the TurboVion source by a syringe pump at 10 μL/min and analyzed by QTRAP5500 mass spectrometer (ABSCIEX). Multiple precursor ion and neutral loss scanning methods were used for information-dependent acquisition of ms/ms data to detect and quantify the lipid classes as described earlier. Mass analyzer conditions in the positive ion mode are as follows: ionization potential: 5500 V, declustering potential: 120 V, entrance potential: 9 V, and collision cell exit potential: 9 V. Collision energy for the survey scan was 10 and 45 eV for enhanced product ion scans. In each scan, three ions with highest intensity were chosen for dependent product ion acquisition and the detected ions were excluded for the rest of the experiment after three occurrences. Data were analyzed for the identification of lipid species using LipidView software (ABSCIEX), and lipids were quantified against internal standards and normalized against protein values.

2.11 Expression and Purification of Neuronal NSF and SNAREs

N-terminal 6 × His-tag constructs for SNAP25 and NSF and C-terminal 6 × His-tag constructs for syntaxin 1A and VAMP2 were generated. All four proteins were expressed with 6 × His at full length in *E. coli* (*BL21DE3*) and isolated by Ni-NTA (nickel-nitrilotriacetic acid) affinity chromatography (Qiagen, Valencia, CA). Protein concentration was determined by BCA assay.

2.12 Dynamic Light Scattering to Determine Volume Dynamics of Isolated Synaptic Vesicles

Kinetics of the synaptic vesicle volume was monitored using 90° scattering of light, in a Hitachi F-2000 spectrophotometer. Synaptic vesicle suspension was injected into the thermoregulated cuvette containing 700 μL of a buffer solution (NaCl 140 mM; KCl 2.5 mM; NaH_2PO_4 0.25 mM; KH_2PO_4 0.25 mM; pH 7.4) at 37°C. Light scattering was monitored for 1 min following addition of vesicles to either buffer alone (control) or buffer in the presence and absence of GTP and the heterotrimeric G_i-protein stimulant mastoparan (experimental) or its non-stimulatory peptide, Mast-17 (control). Values are expressed as percent increase in light scattered over control values.

3 Summary

In summary, neuronal porosomes are 15 nm cup-shaped lipoprotein structures composed of nearly 40 proteins [14], and enriched in specific lipids compared to the synaptosome [13]. Recent studies on the lipidome of the synaptosome membrane and synaptic vesicles demonstrate the unique composition of lipids in the two compartments, supporting their tight regulation [83]. In further agreement of the existence of transient or kiss-and-run fusion of synaptic vesicles at the synaptosome membrane, it has been demonstrated that single synaptic vesicles fuse transiently and successively without loss of identity [84]. Additionally, the presence of neurotransmitter transporters to rapidly refill spent synaptic vesicles would be of little use if synaptic vesicles would collapse as an all-or-none mechanism of the release of neurotransmitters. To further determine the molecular mechanism of the involvement of the neuronal porosome in neurotransmission, an understanding of its atomic structure will be required, which is no trivial matter given the 40 proteins composing the structure, a great many of which are membrane proteins. Although STRING maps generated on the neuronal porosome proteome identify associated proteins and their predicted interactions [14], further chemical cross-linking studies combined with mass spectrometry are required to confirm these interactions, and to determine the stoichiometry of the proteins present within the complex. This will allow in the building of a working model of the neuronal porosome complex. The great number of studies that implicate various porosome proteins in neurotransmission and in various neurological disorders further lend support that this secretory portal at the nerve terminal is critical to neurotransmission, and that its altered state leads to various neurological disorders and diseases, requiring further molecular understanding of its structure-function.

Acknowledgements

Work presented in this chapter was supported in part by the National Institutes of Health grants DK56212 and NS39918 and the National Science Foundation grants EB00303 and CBET1066661 (BPJ).

References

1. Ceccarelli B, Hurlbut WP, Mauro A (1973) Turnover of transmitter and synaptic vesicles at the frog neuromuscular junction. J Cell Biol 57:499–524
2. Almers W, Tse FW (1990) Transmitter release from synapses: does a preassembled fusion pore initiate exocytosis? Neuron 4:813–818
3. Monck JR, Fernandez JM (1992) The exocytotic fusion pore. J Cell Biol 119:1395–1404
4. Schneider SW, Sritharan KC, Geibel JP, Oberleithner H, Jena BP (1997) Surface dynamics in living acinar cells imaged by atomic force microscopy: identification of plasma membrane structures involved in exocytosis. Proc Natl Acad Sci U S A 94:316–321
5. Alexander S, Hellemans L, Marti O, Schneir J, Elings V, Hansma PK (1989) An atomic resolution atomic force microscope implemented using an optical lever. J Appl Phys 65:164–167
6. Jena BP, Cho S-J, Jeremic A, Stromer MH, Abu-Hamdah R (2003) Structure and composition of the fusion pore. Biophys J 84:1337–1343
7. Jeremic A, Kelly M, Cho S-J, Stromer MH, Jena BP (2003) Reconstituted fusion pore. Biophys J 85:2035–2043
8. Craciun C, Barbu-Tudoran L (2013) Identification of new structural elements within 'porosomes' of the exocrine pancreas: a detailed study using high-resolution electron microscopy. Micron 44:137–142
9. Paknikar KM, Jeremic A (2007) (2007) Discovery of the cell secretion machinery. J Biomed Nanotechnol 3:218–222
10. Cho W-J, Jeremic A, Rognlien KT, Zhvania MG, Lazrishvili I, Tamar B, Jena BP (2004) Structure, isolation, composition and reconstitution of the neuronal fusion pore. Cell Biol Int 28:699–708
11. Cho WJ, Ren G, Jena BP (2008) EM 3D contour maps provide protein assembly at the nanoscale within the neuronal porosome complex. J Microsc 232:106–111
12. Kovari LC, Brunzelle JS, Lewis KT, Cho WJ, Lee J-S, Taatjes DJ, Jena BP (2014) X-ray solution structure of the native neuronal porosome-synaptic vesicle complex: implication in neurotransmitter release. Micron 56:37–43
13. Lewis KT, Maddipati KR, Taatjes DJ, Jena BP (2014) Neuronal Porosome Lipidome. J Cell Mol Med 18:1927–1937
14. Lee J-S, Jeremic A, Shin L, Cho WJ, Chen X, Jena BP (2012) Neuronal porosome proteome: molecular dynamics and architecture. J Proteome 75:3952–3962

15. Cho WJ, Jeremic A, Jin H, Ren G, Jena BP (2007) Neuronal fusion pore assembly requires membrane cholesterol. Cell Biol Int 31:1301–1308
16. Okuneva VG, Japaridze ND, Kotaria NT, Zhvania MG (2012) Neuronal porosome in the rat and cat brain. Cell Tissue Biol 6:69–72
17. Lin DH, Hoelz A (2016) Nuclear pores come into sharper focus. The Scientist 1:2016
18. Lin DH et al (2016) Architecture of the symmetric core of the nuclear pore. Science 352:aaf1015
19. Hoelz A et al (2011) The structure of the nuclear pore complex. Annu Rev Biochem 80:613–643
20. Nemhauser I, Goldberg DJ (1985) Structural effects in axoplasm of DNase I, an actin depolymerizer that blocks fast axonal transport. Brain Res 334:47–58
21. Cole JC, Villa BR, Wilkinson RS (2000) Disruption of actin impedes transmitter release in snake motor terminals. J Physiol 525(3):579–586
22. Klein ME, Younts TJ, Castillo PE, Jordan BA (2013) RNA-binding protein Sam68 controls synapse number and local beta-actin mRNA metabolism in dendrites. Proc Natl Acad Sci U S A 110:3125–3130
23. Khanna R, Zougman A, Stanley EF (2007) A proteomic screen for presynaptic terminal N-type calcium channel (CaV2.2) binding partners. J Biochem Mol Biol 40:302–314
24. Khanna R, Li Q, Bewersdorf J, Stanley EF (2007) The presynaptic CaV2.2 channel-transmitter release site core complex. Eur J Neurosci 26:547–559
25. Balestrino M, Young J, Aitken P (1999) Block of (Na+, K+)ATPase with ouabain induces spreading depression-like depolarization in hippocampal slices. Brain Res 838:37–44
26. Li KC, Zhang FX, Li CL, Wang F, Yu MY, Zhong YQ, Zhang KH, Lu YJ, Wang Q, Ma XL, Yao JR, Wang JY, Lin LB, Han M, Zhang YQ, Kuner R, Xiao HS, Bao L, Gao X, Zhang X (2011) Follistatin-like 1 suppresses sensory afferent transmission by activating Na+, K+-ATPase. Neuron 69:974–987
27. Scuri R, Lombardo P, Cataldo E, Ristori C, Brunelli M (2007) Inhibition of Na+/K+ ATPase potentiates synaptic transmission in tactile sensory neurons of the leech. Eur J Neurosci 25:59–167
28. Kim JH, Sizov I, Dobretsov M, von Gersdorff H (2007) Presynaptic Ca2+ buffers control the strength of a fast post-tetanic hyperpolarization mediated by the alpha3 Na(+)/K(+)-ATPase. Nat Neurosci 10:196–205
29. Jensen TP, Filoteo AG, Knopfel T, Empson RM (2007) Presynaptic plasma membrane Ca2+ ATPase isoform 2a regulates excitatory synaptic transmission in rat hippocampal CA3. J Physiol 579:85–99
30. Garside ML, Turner PR, Austen B, Strehler EE, Beesley PW, Empson RM (2009) Molecular interactions of the plasma membrane calcium ATPase 2 at pre- and post-synaptic sites in rat cerebellum. Neuroscience 162:383–395
31. Geerlings A, Nunez E, Lopez-Corcuera B, Aragon C (2001) Calcium- and syntaxin 1-mediated trafficking of the neuronal glycine transporter GLYT2. J Biol Chem 276:17584–17590
32. de Juan-Sanz J, Núñez E, Villarejo-López L, Pérez-Hernández D, Rodriguez-Fraticelli AE, López-Corcuera B, Vázquez J, Aragón C (2013) Na+/K+-ATPase is a new interacting partner for the neuronal glycine transporter GlyT2 that downregulates its expression in vitro and in vivo. J Neurosci 33:14269–14281
33. Empson RM, Akemann W, Knopfel T (2010) The role of the calcium transporter protein plasma membrane calcium ATPase PMCA2 in cerebellar Purkinje neuron function. Funct Neuronal 25:53–158
34. Dodson HC, Charalabapoulou M (2001) PMCA2 mutation causes structural changes in the auditory system in deafwaddler mice. J Neurocytol 30:281–292
35. Iino S, Kobayashi S, Maekawa S (1999) Immunohistochemical localization of a novel acidic calmodulin-binding protein, NAP-22, in the rat brain. Neuroscience 91:1435–4436
36. IinoS,MaekawaS(1999)Immunohistochemical demonstration of a neuronal calmodulin-binding protein, NAP-22, in the rat spinal cord. Brain Res 834:66–73
37. Yamamoto Y, Sokawa Y, Maekawa S (1997) Biochemical evidence for the presence of NAP-22, a novel acidic calmodulin binding protein, in the synaptic vesicles of rat brain. Neurosci Lett 224:127–130
38. Freeman NL, Field J (2000) Mammalian homolog of the yeast cyclase associated protein, CAP/Srv2p, regulates actin filament assembly. Cell Motil Cytoskeleton 45:106–120
39. Zhang H, Ghai P, Wu H, Wang C, Field J, Zhou G-L (2013) Mammalian adenylyl cyclase-associated protein 1 (CAP1) regulates cofilin function, the actin cytoskeleton, and cell adhesion. J Biol Chem 288:20966–20977
40. Sultana R, Boyd-Kimball D, Cai J, Pierce WM, Klein JB, Merchant M, Butterfield DA (2007)

Proteomics analysis of the Alzheimer's disease hippocampal proteome. J Alzheimers Dis 11:153–164
41. Vlkolinsky R, Cairns N, Fountoulakis M, Lubec G (2001) Decreased brain levels of 2′,3′-cyclic nucleotide-3′-phosphodiesterase in down syndrome and Alzheimer's disease. Neurobiol Aging 22:547–553
42. Reinikainen KJ, Pitkanen A, Riekkinen PJ (1989) 2′,3′-cyclic nucleotide-3′-phosphodiesterase activity as an index of myelin in the post-mortem brains of patients with Alzheimer's disease. Neurosci Lett 106:229–232
43. Flynn SW, Lang DJ, Mackay AL, Goghari V, Vavasour IM, Whittall KP, Smith GN, Arango V, Mann JJ, Dwork AJ, Falkai P, Honer WG (2003) Abnormalities of myelination in schizophrenia detected in vivo with MRI, and post-mortem with analysis of oligodendrocyte proteins. Mol Psychiatry 8:811–820
44. Peirce TR, Bray NJ, Williams NM, Norton N, Moskvina V, Preece A, Haroutunian V, Buxbaum JD, Owen MJ, O'Donovan MC (2006) Convergent evidence for 2′,3′-cyclic nucleotide 3′-phosphodiesterase as a possible susceptibility gene for schizophrenia. Arch Gen Psychiatry 63:18–24
45. Mikoshiba K, Aoki E, Tsukada Y (1980) 2′-3′-cyclic nucleotide 3′-phosphohydrolase activity in the central nervous system of a myelin deficient mutant (Shiverer). Brain Res 192:195–204
46. Zhao YY, Shi XY, Qiu X, Zhang L, Lu W, Yang S, Li C, Cheng GH, Yang ZW, Tang Y (2011) Enriched environment increases the total number of CNPase positive cells in the corpus callosum of middle-aged rats. Acta Neurobiol Exp (Wars) 71:322–330
47. Sinclair LI, Tayler HM, Love S (2015) Synaptic protein levels altered in vascular dementia. Neuropathol Appl Neurobiol 41:533–543
48. Mukaetova-Ladinska EB, Xuereb JH, Garcia-Sierra F, Hurt J, Gertz HJ, Hills R, Brayne C, Huppert FA, Paykel ES, McGee MA, Jakes R, Honer WG, Harrington CR, Wischik CM (2009) Lewy body variant of Alzheimer's disease: selective neocortical loss of t-SNARE proteins and loss of MAP2 and alpha-synuclein in medial temporal lobe. Sci World J 9:1463–1475
49. Greber S, Lubec G, Cairns N, Fountoulakis M (1999) Decreased levels of synaptosomal associated protein 25 in the brain of patients with down syndrome and Alzheimer's disease. Electrophoresis 20:928–934
50. Corradini I, Donzelli A, Antonucci F, Welzl H, Loos M, Martucci R, De Astis S, Pattini L, Inverardi F, Wolfer D, Caleo M, Bozzi Y, Verderio C, Frassoni C, Braida D, Clerici M, Lipp HP, Sala M, Matteoli M (2014) Epileptiform activity and cognitive deficits in SNAP-25 (+/−) mice are normalized by antiepileptic drugs. Cereb Cortex 24:364–376
51. McKee AG, Loscher JS, O'Sullivan NC, Chadderton N, Palfi A, Batti L, Sheridan GK, O'Shea S, Moran M, McCabe O, Fernandez AB, Pangalos MN, O'Connor JJ, Regan CM, O'Connor WT, Humphries P, Farrar GJ, Murphy KJ (2010) AAV-mediated chronic over-expression of SNAP-25 in adult rat dorsal hippocampus impairs memory-associated synaptic plasticity. J Neurochem 112:991–1004
52. Smith R, Klein P, Koc-Schmitz Y, Waldvogel HJ, Faull RL, Brundin P, Plomann M, Li JY (2007) Loss of SNAP-25 and rabphilin 3a in sensory-motor cortex in Huntington's disease. J Neurochem 103:115–123
53. Jeans AF, Oliver PL, Johnson R, Capogna M, Vikman J, Molnar Z, Babbs A, Partridge CJ, Salehi A, Bengtsson M, Eliasson L, Rorsman P, Davies KE (2007) A dominant mutation in Snap25 causes impaired vesicle trafficking, sensorimotor gating, and ataxia in the blind-drunk mouse. Proc Natl Acad Sci U S A 104:2431–2436
54. Scarr E, Gray L, Keriakous D, Robinson PJ, Dean B (2006) Increased levels of SNAP-25 and synaptophysin in the dorsolateral prefrontal cortex in bipolar I disorder. Bipolar Disord 8:133–143
55. Cao F, Hata R, Zhu P, Niinobe M, Sakanaka M (2009) Up-regulation of syntaxin1 in ischemic cortex after permanent focal ischemia in rats. Brain Res 1272:52–61
56. Wu MN, Fergestad T, Lloyd TE, He Y, Broadie K, Bellen HJ (1999) Syntaxin 1A interacts with multiple exocytic proteins to regulate neurotransmitter release in vivo. Neuron 23:593–605
57. Lagow RD, Bao H, Cohen EN, Daniels RW, Zuzek A, Williams WH, Macleod GT, Sutton RB, Zhang B (2007) Modification of a hydrophobic layer by a point mutation in syntaxin 1A regulates the rate of synaptic vesicle fusion. PLoS Biol 5:e72
58. Chapman ER, An S, Edwardson JM, Jahn R (1996) A novel function for the second C2 domain of synaptotagmin. Ca2+−triggered dimerization. J Biol Chem 271:5844–5849
59. Zhang X, Kim-Miller MJ, Fukuda M, Kowalchyk JA, Martin TF (2002) Ca2+-dependent synaptotagmin binding to SNAP-25 is essential for Ca2+-triggered exocytosis. Neuron 34:599–611

60. Japaridze NJ, Okuneva VG, Qsovreli MG, Surmava AG, Lordkipanidze TG, Kiladze M, Zhvania MG (2012) Hypokineitic stress and neuronal porosome complex in the rat brain: the electron microscopic study. Micron 43:948–963
61. Okuneva VG, Japaridze NJ, Kotaria NT, Zhvania MG (2012) Neuronal porosome in the rat and cat brain. Tsitoloiya 54:210–215
62. Zhvania MG, Japaridze NJ, Qsovreli MG, Okuneva VG, Surmava AG, Lordkipanidze TG (2014) The neuronal porosome complex in mammalian brain: a study using electron microscope. In: Jena BP, Taajes DJ (eds) NanoCellBiology, Multimodal imaging in biology and medicine. Pan Stanford Publishing Pte. Ltd., Verlag
63. Zhvania MG, Japaridze NJ, Ksovreli M (2014) Effect of different forms of hypokinesia on the ultrastructure of limbic, extrapyramidal and neocortical areas of the rat brain: electron microscopic study. Proceedings of Physics, Springer International Publishing Switzerland 15:21–16
64. Zhvania MG, Bikashvili TZ, Japaridze NJ, Lazrishvili II, Ksovreli M (2014) White noise and neuronal porosome complex: transmission electron microscopic study. *Discoveries* 2(3):e25
65. Eggermont JJ (2008) The role of sound in adult and developmental auditory cortical plasticity. Ear Hear 29(6):819–829
66. Zhu X, Wang F, Hu H, Sun X, Kilgard MP, Merzenich MM et al (2014) Environmental acoustic enrichment promotes recovery from developmentally degraded auditory cortical processing. J Neurosci 34:5406–5415
67. Trimble WS, Cowan DW, Scheller RH (1988) VAMP-1: a synaptic vesicle-associated integral membrane protein. Proc Natl Acad Sci U S A 85:4538–4542
68. Oyler GA, Higgins GA, Hart RA, Battenberg E, Billingsley M, Bloom FE, Wilson MC (1989) The identification of a novel synaptosomal-associated protein, SNAP-25, differentially expressed by neuronal subpopulations. J Cell Biol 109:3039–3052
69. Bennett MK, Calakos N, Schller RH (1992) Syntaxin: a synaptic protein implicated in docking of synaptic vesicles at presynaptic active zones. Science 257:255–259
70. Weber T, Zemelman BV, McNew JA, Westerman B, Gmachi M, Parlati F, Söllner TH, Rothman JE (1998) SNAREpins: minimal machinery for membrane fusion. Cell 92:759–772
71. Cho S-J, Kelly M, Rognlien KT, Cho J, Hörber JK, Jena BP (2002) SNAREs in opposing bilayers interact in a circular array to form conducting pores. Biophys J 83:2522–2527
72. Cho WJ, Jeremic A, Jena BP (2005) Size of supramolecular SNARE complex: membrane-directed self-assembly. J Am Chem Soc 127:10156–10157
73. Cho WJ, Jeremic A, Jena BP (2005) Direct interaction between SNAP-23 and L-type Ca2+ channel. J Cell Mol Med 9(2):380–386
74. Jena BP, Cho S-J, Jeremic A, Stromer MH, Abu-Hamdah R (2003) Structure and composition of the fusion pore. Biophys J 84:1–7
75. Kelly M, Cho WJ, Jeremic A, Abu-Hamdah R, Jena BP (2004) Vesicle swelling regulates content expulsion during secretion. Cell Biol Int 28:709–716
76. Jena BP, Schneider SW, Geibel JP, Webster P, Oberleithner H, Sritharan KC (1997) G_i regulation of secretory vesicle swelling examined by atomic force microscopy. Proc Natl Acad Sci U S A 94:13317–13322
77. Cho S–J, Sattar AK, Jeong EH, Satchi M, Cho J, Dash S, Mayes MS, Stromer MH, Jena BP (2002) Aquaporin 1 regulates GTP-induced rapid gating of water in secretory vesicles. Proc Natl Acad Sci U S A 99:4720–4724
78. Jeremic A, Cho W-J, Jena BP (2005) Involvement of water channels in synaptic vesicle swelling. Exp Biol Med 230:674–680
79. Abu-Hamdah R, Cho W-J, Cho S-J, Jeremic A, Kelly M, Ilie AE, Jena BP (2004) Regulation of the water channel aquaporin-1: isolation and reconstitution of the regulatory complex. Cell Biol Int 28:7–17
80. Shin L, Basi N, Lee J-S, Cho W-J, Chen Z, Abu-Hamdah R, Oupicky D, Jena BP (2010) Involvement of vH^+-ATPase in synaptic vesicle swelling. J Neurosci Res 88:95–101
81. Lee J-S, Cho W-J, Shin L, Jena BP (2010) Involvement of cholesterol in synaptic vesicle swelling. Exp Biol Med 235:470–477
82. Chen Z-H, Lee J-S, Shin L, Cho W-J, Jena BP (2010) Involvement of β-adrenergic receptor in synaptic vesicle swelling and implication in neurotransmitter release. J Cell Mol Med 15:572–576
83. Lewis KT, Maddipati KR, Naik AR, Jena BP (2017) Unique lipid chemistry of synaptic vesicle and synaptosome membrane revealed using mass spectrometry. ACS Chem Neurosci 8:1163. https://doi.org/10.1021/acschemneuro.7b00030
84. Aravanis AM, Pyle JL, Tsien RW (2003) Single synaptic vesicles fusing transiently and successively without loss of identity. Nature 423:643–647

Chapter 13

SMPL Synaptic Membranes: Nanodisc-Mediated Synaptic Membrane Mimetics Expand the Toolkit for Drug Discovery and the Molecular Cell Biology of Synapses

Riana Lo Bu, Jeffrey Clark, Thomas DiChiara, Stephen G. Sligar, and William L. Klein

Abstract

We describe in detail how proteins of synaptic membranes can be extracted and reconstituted within "Nanodiscs" which renders them soluble in aqueous solutions. Each Nanodisc is a nanometer-scaled lipid bilayer containing an individual membrane protein. The method can provide a *soluble membrane protein library* (SMPL) from any tissue fraction, notably from the standpoint of this treatise, from synaptosome plasma membranes (*SMPL synaptic membranes*). Nanodiscs facilitate investigations of membrane proteins that are difficult to tease apart from protein-protein complexes while at the same time maintaining conformational integrity. The procedure captures virtually all membrane proteins in a manner that preserves their activities in soluble form. The usefulness of SMPL synaptic membranes is exemplified by experiments that concern the specific binding of Alzheimer's-related amyloid β oligomers (AβOs). AβO binding to SMPL synaptic membranes is saturable, has the same high affinity found in cell-based assays, and is mediated adventitiously by particular proteins. Immunoaffinity isolation experiments have identified one AβO toxin "receptor" as the NaK ATPase α3 (NKAα3). For drug discovery, recombinant membrane proteins in Nanodiscs can be readily coupled to high-throughput screening platforms; SMPL synaptic membranes can also be readily coupled and used for unbiased high-throughput screening of unknown targets. SMPL synaptic membranes, and analogous organelle libraries, offer new resources for the CNS membrane protein toolkit with a wide array of applications to synapse molecular cell biology and drug discovery.

Key words Abeta oligomers, Alzheimer's, Nanodisc, Synapse, Synaptic membrane, Receptor

Abbreviations

AD	Alzheimer's disease
AβOs	Amyloid beta oligomers
CNS	Central nervous system
MP	Membrane protein
MSP	Membrane scaffold protein
MW	Molecular weight
NaKAα3	Sodium potassium ATPase alpha3

Kathryn M. Murphy (ed.), *Synaptosomes*, Neuromethods, vol. 141,
https://doi.org/10.1007/978-1-4939-8739-9_13,

PL	Phospholipid
PSD	Postsynaptic density
SEC	Size-exclusion chromatography
SMPL	Soluble membrane protein library

1 Introduction

Synaptosomes provide a prepackaged collection of proteins essential for CNS communication and valuable for CNS drug discovery [1, 2]. A high percentage of the current as well as emerging synaptosome therapeutic targets are membrane or membrane-associated proteins. These include G protein-coupled receptors (GPCRs), transduction proteins, ion channels, machinery for neurotransmitter release, various transporters, and other membrane proteins that often are key drug targets. Synaptosomes also include as-yet unidentified molecules that may be potential new drug targets. These include, e.g., receptors for exosomes or for the pathological oligomeric forms of synuclein, tau, or amyloid beta (AβOs) associated with neurodegenerative disease [3–9].

It would be advantageous for CNS drug discovery if synaptic membrane proteins could be investigated using powerful technologies that have been developed to study soluble proteins. This would open the door to advanced analytics and molecular modeling, including structure-activity relationships, with increased potential for discovering and developing investigational new drugs. Complex membrane systems, however, include proteins that are recalcitrant to such analyses. Even when membrane proteins have been solubilized by adding detergent or exogenous lipids, their conformations are uncertain, often inactive and, most importantly, not readily amenable to high-throughput analysis for discovery of lead compounds for new therapeutics.

There have been many attempts to surmount this difficulty, including incorporation into liposomes or micellar detergent aggregates [10]. Liposomes, however, are inherently unstable and prone to fusion, while placing a single membrane protein target into a ~100 nm liposome is a challenge and overloads an assay with excess lipids. Proteins solubilized in micelles most often have an altered activity profile. Thus, difficulties in obtaining membrane proteins for molecular level studies under physiological conditions leave us with a fundamental shortcoming in our experimental toolkit. These difficulties can be overcome via use of Nanodiscs.

As noted, the general problem in revealing the structure and function of membrane proteins is that removing them from their lipid bilayer environment results in inactivation, unfolding, or critical modifications of activity. Membrane proteins require the presence of a phospholipid bilayer to maintain their functional structure. Remove a membrane protein from the bilayer and it often aggregates and becomes inactive. To solve these solubility problems, a method based on nanotechnology has been developed that opens the door to powerful biophysical analysis of membrane proteins as well as robust exploitation of membrane proteins for high-throughput screening and drug discovery. The methodology is based upon the Nanodisc system.

A relatively recent technology [10, 11], the Nanodisc system self-assembles a membrane protein into a nanometer-scale phospholipid bilayer, ~10 nm in diameter. The entire entity is stabilized by two encircling amphipathic helical protein belts termed membrane scaffold proteins (MSP). Following self-assembly into Nanodiscs, the membrane protein finds itself in a native-like environment and displays full functionality, and the entire entity is soluble in aqueous media. The molecular engineering of the MSP makes it possible to carry out separation protocols useful for protein isolation and for high-throughput screening. The Nanodisc-incorporated proteins are stable to isolation, sorting, and assay without worrying about poisoning of a sensor surface or inactivation of the target. Extensive work over the past 5 years has finely honed the ability to analyze Nanodiscs by mass spectrometry and identify the incorporated target protein.

The Nanodisc technology is now being used in over 1300 laboratories worldwide (https://www.addgene.org/Stephen_Sligar/) and has been described in an exhaustive 2017 review [10].

The Nanodisc system has found three broad applications: The first is the direct assembly of a purified membrane protein into the Nanodisc bilayer where the incorporated target protein finds itself in a native-like environment and maintains full functionality. The second is as a 6–18 nm diameter membrane surface where the lipid composition can be precisely controlled, allowing one to understand the mechanisms whereby entities such as signaling proteins bind to membrane surfaces. Both have been extensively discussed in several reviews (for an inroad to the literature and comparison to other methods see the Chemical Review article cited earlier). The third application is the subject of this review: Moving the entire membrane proteome from a tissue into Nanodiscs such that each membrane protein present in the starting material is incorporated into an individual 10–12 nm Nanodisc.

Once in the Nanodisc, the targets are stable and functional and can be assayed by a wide variety of analytical methods. Importantly, since the resultant soluble membrane protein library (SMPL) can be sorted, derivatized with luminescent compounds or pull-down labels, and tethered to surfaces without loss of function, a new avenue for exploring the synaptic membrane is opened. While numerous studies have established the value of applying Nanodisc technology to recombinant membrane proteins, Nanodisc technology only recently has been introduced as a significant new resource for investigations of the synaptic membrane proteome [12].

Synaptosomes can be studied readily as organelles or as membrane fragments, but ideally one would like to study membrane proteins in a simple reconstituted system that can operate at the single-molecule level. Difficulties for investigating synaptic membrane proteins include the following: [1] gentle detergents yield gross ill-defined complexes due to scaffolding (e.g., the postsynaptic density (PSD) scaffold alone appears to have more than 1000 constituent proteins) and [2] strong detergents that release scaffolding cause altered structure. These difficulties can be overcome by capturing the synaptic membrane proteome as a soluble membrane protein library (a SMPL solution), a newly introduced Nanodisc application that is ready for use by neuroscientists.

Figure 1 illustrates the two fundamental Nanodisc approaches, one to recombinant membrane proteins (Fig. 1b) and the other to the unbiased collection of all members of a solubilized membrane proteome (Fig. 1a). For the solubilized membrane proteome, a

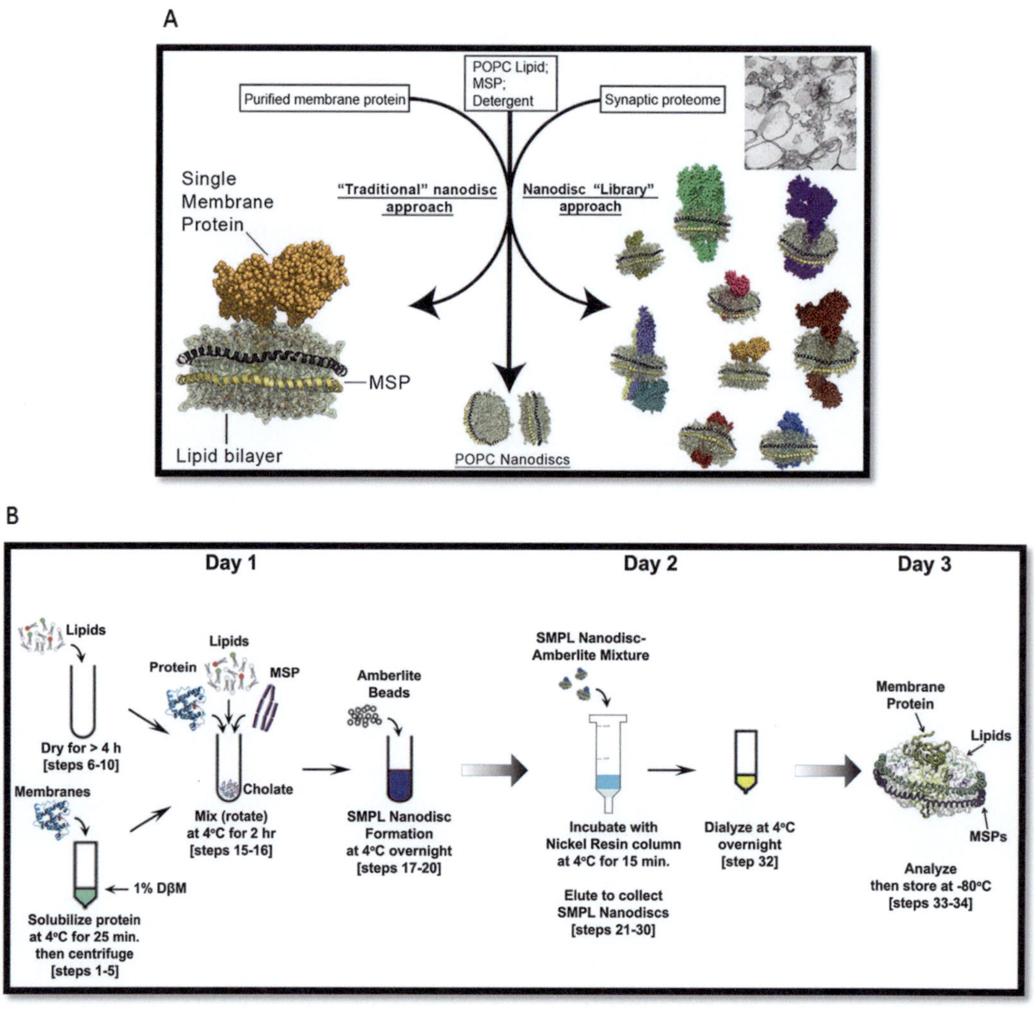

Fig. 1 SMPL synaptic membrane production

starting biological membrane is detergent solubilized and mixed with a membrane scaffold protein (MSP) and phospholipids. With excess MSP and lipid, upon removal of the detergent, self-assembly distributes each membrane protein into an individual Nanodisc wherein it is soluble and fully active. This results in the formation of a soluble membrane protein library (SMPL) which faithfully replicates the composition of the starting membrane population. SDS-PAGE analysis shows nearly identical bands in the SMPL synaptosome and the starting synaptosome membrane. Functional activities have been confirmed, and numerous expected individual components identified (Fig. 2a,b) [13, 14].

In this chapter, we provide a tutorial for use of Nanodiscs in the study of synaptic membrane proteins. In the first section, we

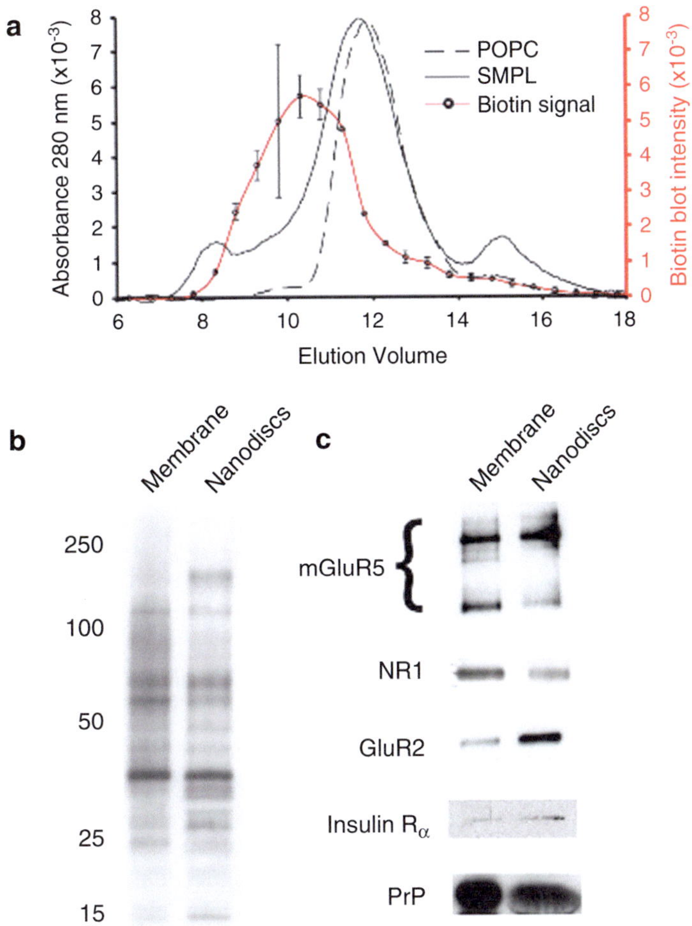

Fig. 2 SMPL synaptic membrane preserves synaptic protein composition and structure

describe the self-assembly process whereby phospholipids and the Nanodisc encircling protein belt (membrane scaffold protein or MSP) come together to form the soluble discoidal bilayer. We then show how a purified membrane protein can be added to the protocol to result in the incorporation of the target into the Nanodisc bilayer. We then describe the general procedures for moving membrane proteins from a tissue, such as the synaptosome, into Nanodiscs to form the soluble membrane protein library.

2 The Tutorial: Incorporating Membrane Proteins into Nanodiscs

The methodologies and detailed protocols for generating empty Nanodiscs as well as incorporating membrane proteins into these bilayer nanostructures have been described in detail in numerous published reviews [15–17] as well as on the web site for the Sligar laboratory (http://sligarlab.life.illinois.edu). We briefly summarize the salient points involved in order to point out some important parameters.

2.1 Assembly of Bare Nanodiscs

Before attempting to insert membrane proteins into Nanodiscs, either from purified proteins or from an entire membrane, it is strongly advised to begin by assembling bare Nanodiscs and assaying the results by size-exclusion chromatography (SEC). This analytical technique is simple and can readily reveal the generation of a homogenous Nanodisc peak verses aggregates of lipid or the membrane scaffold protein. The assembly process involves detergent solubilizing the membrane scaffold proteins (MSP) together with desired synthetic phospholipid (PL) or a mixture of lipid types. In starting with a single lipid, we recommend using DMPC as the interaction of the saturated acyl chains favors formation. Much work has also been successful using unsaturated phospholipids, notable POPC, as well as lipid mixtures. In assembling bare Nanodiscs a detergent, such as cholate, is desirable as it readily will solubilize both MSP and PL. The assembly process is optimized at a temperature near the phase transition temperature of lipids being used. For DMPC this allows work at room temperature. For unsaturated lipids such as POPC, with a bulk-phase transition temperature below 0 °C, working on ice at 4 °C is ideal. These lower temperatures can also be used with mixtures of lipids, and is often an advantage in stabilizing the target protein. Perhaps the most important parameter in realizing successful self-assembly is having the correct stoichiometric ratio of MSP to lipid. A table describing the optimum stoichiometry for various PLs has been presented [15]. If one is exact in matching the number of lipids needed to fill out a bare Nanodisc, there will essentially be only fully formed disks in solution. If the lipid ratio is close, there may appear lipid or MSP aggregates in the void volume or small peaks near that of the fully formed Nanodisc. Getting the stoichiometry right requires knowing the concentration of the lipid stocks one is using. This mandates determining the lipid concentration, such as with a phosphate assay. It is not sufficient to rely on a manufacturer's stated quantities! The detergent is most easily removed by passage over or batch mixing with hydrophobic beads, such as Biobeads® or Amberlite resin.

Various sizes of Nanodiscs can be generated, using longer or shorter membrane scaffold proteins (MSP) and the corresponding number of lipids. Most common are the "MSP1D1" family that results in ~10 nm disks and the "MSP1E3D1" family that form ~12 nm disks. Nanodiscs from ~7 to ~17 nm have been formed [10]. MSPs with various tags, including His, FLAG, and biotin, are also available.

2.2 Inserting Purified Membrane Proteins into Nanodiscs

The fundamental discovery underlying the Nanodisc technology is that membrane proteins, if added to the detergent-solubilized mixture of MSP and PL, will self-assemble into the resultant discoidal bilayer. Importantly, the target protein finds itself in a native-like bilayer environment and displays its native conformation and activity. A surprise was that membrane proteins of many different topologies, including multi-protein and multi-subunit complexes, will also self-assemble in the correct configuration. A detailed presentation of the vast number of successful uses of Nanodisc technology has recently appeared [10] and contains over 500 references citing work from numerous laboratories worldwide.

One aspect of this self-assembly process needs to be explicitly mentioned. As detergent is desorbed from the various surfaces there is a branch point: target protein finding lipids and assembling into a bilayer, ultimately stabilized by the encircling MSP versus aggregation of the target protein. In the latter case, the aggregate could be soluble, but not in a membrane environment, and hence often inactive. While this undesired product can be readily separated from that incorporated into the Nanodisc, it reduces the overall yield of successful product. This yield is dependent on the protein target, lipid detergent used, and amount of hydrophobic beads used. A full discussion of this has been provided [10, 15], and one can expect success to range from 25 to 95%. Hence one should not be discouraged with less than full incorporation of a particular target.

In adding a membrane protein target to the assembly mixture it is clear that when it is in the bilayer, less PL will be needed to completely fill out the disk. In addition, larger targets will require fewer lipids for a given size Nanodisc. Again, the PL:MSP ratio must be closely matched. This requires some experimentation in trial assembly runs with different phospholipid ratios and using size-exclusion chromatography to assay for correct Nanodisc formation. There is an accumulating database as to how many lipids various anchor domains of a membrane protein target will displace. For instance, a GPCR monomer can be assembled into a 10 nm disc with reduced PL content [15, 18–20].

2.3 Assembling Targets Directly from a Membrane

Perhaps most relevant to the study of synaptosomal function is the ability to directly incorporate all membrane proteins from a given tissue into Nanodiscs without pre-purifying the proteins. This process was first described in 2003 [21] wherein a membrane protein target was overexpressed in insect cells and moved from cell membrane directly into Nanodiscs. These authors also demonstrated that the fatty acid and phospholipid content of the starting membrane was faithfully replicated in the Nanodisc population. This approach as a general method for studying the membrane proteome was verified by Marty et al. [13] who showed that the protein content of a starting *E. coli* membrane was fully (>90%) incorporated into Nanodiscs and that the assayed proteins were fully active in the nanoscale bilayer. Importantly, Marty et al. documented the optimum phospholipid ratio necessary for the production of Nanodiscs, necessary as additional phospholipid needs to be added to fill in the encircling MSP belts containing a single target protein.

The ability to generate a soluble membrane protein library (SMPL) in Nanodiscs that faithfully represents the starting tissue composition is a major enablement to understanding the interaction of external factors with cellular receptors. Each membrane protein (MP) is in an individual Nanodisc, where it displays its native conformation, thus preserving any binding epitope. Given that multi-protein complexes can also be incorporated into Nanodiscs by the same procedure means that complex cellular machines such as ion pumps and transporters can be assayed by the same methodology.

The SMPL technology has become a major factor in the design of experiments to reveal the association of external peptide components, such as the amyloid beta oligomer (AβO), to synaptosomes. Since each Nanodisc is robust to a variety of sorting procedures, including FACS, fluidics, and pull-down assays, the identity of the specific receptor or receptors that bind AβO can be identified. Alternatively, an assay can be formulated that monitors AβO binding to the entire library and this can be used in a high-throughput screening format to identify small molecule or peptide modulators. This work has been described in detail in Wilcox et al. [12].

2.4 Mass Spectrometry Identification of Membrane Proteins in Nanodiscs

The ability to directly analyze a membrane protein library (SMPL) in Nanodiscs by mass spectrometry means that it is straightforward to identify the set of proteins that are present in any isolate from a selective assay [22, 23]. Indeed, it has been recently shown that the entire Nanodisc, together with incorporated target protein, can be flown on a mass spectrometer [24, 25]. We thus have an immensely enabling approach for neuroscientists in dissecting the function of synaptosomal proteins and their interaction with signaling partners.

3 Preparation of a Soluble Membrane Protein Library (SMPL) from Synapse Membranes (SMPL Synapse Membranes)

3.1 Materials and Preparation of Stock

SMPL Synaptic Membrane Production

1. 1-Palmitoyl-2-oleoyl-sn-glycero-3-phosphocholine (POPC) *(Avanti Polar Lipids) (product #: 850457C)*:
   (a) Used to mimic mammalian phospholipid composition:
   - Note: Glass syringes are used to measure volumes of organic solutions as plastic pipette tips cannot accurately dispense nonaqueous solutions and organic solvents will partially dissolve the plastic tip and contaminate the lipid.

   (b) Store at −20 °C in 4 mL glass vials with Teflon-lined screw caps.
2. Phosphate-buffered saline (PBS):
   (a) (155 mM NaCl, 1 mM potassium phosphate monobasic, 3 mM sodium phosphate dibasic).
3. 1% N-dodecyl-β-D-maltoside (DβM) [in PBS] *(Sigma Aldrich) (product #: D4641)*:
   (a) DβM is a relatively gentle, nonionic detergent. It is the detergent of choice for protein solubilization because it can be removed through dialysis and/or hydrophobic bead treatment as a soluble aggregated protein will not self-assemble into the SMPL synaptic membrane.
   (b) Filter and store at 4 °C.
4. Rat cerebrum.
5. Membrane scaffold protein (MSP1D1) *(Sigma Aldrich) (product #: M6574)*:
   (a) Protein that forms the amphipathic helical protein belt that encircles the phospholipid bilayer (contains a 6-His tag).
   (b) Reconstitute the lyophilized powder using 5 mL of sterile water. The solution will contain MSP in 20 mM Tris (pH 7.4), 0.1 M NaCl, 0.5 mM EDTA, and 0.01% NaN_3. Determine the MSP concentration spectrophotometrically using the extinction coefficient indicated on the product sheet from Sigma-Aldrich.

(c) MSP can be stored at 4 °C for several days.

(d) Store at temperatures below −20 °C (remains functional for at least 1 year).

6. 100 mM Sodium cholate [in (PBS)]:

   (a) Cholate is an ideal detergent for solubilizing the phospholipids.

   (b) Filter and store at 4 °C.

7. Amberlite XAD-2 *(Sigma Aldrich) (product #: 10357)*:

   (a) Used for initiating SMPL synaptic membrane formation.

   (b) Store in ddH_2O.

8. Nickel resin elution buffer:

   (a) In ddH_2O: 50 mM Phosphate, 300 mM NaCl, 250 mM imidazole, pH 8.0.

   (b) Store at 4 °C.

9. Dialysis buffer:

   (a) In 4 L ddH_2O: 0.02 M Tris, 0.0825 M NaCl, 0.00065 M EDTA, pH 7.4.

10. Methanol.

11. ddH_2O.

12. Chloroform.

    - Caution! Chloroform liquid and vapors are toxic. Wear a mask, gloves, and chemical safety goggles while handling it. Be aware of the explosive dangers due to peroxide buildup.

13. Sucrose (0.32 M, 2.2 M, 1.1 M, 0.855 M).

14. Protease inhibitors *(Roche)*.

15. N_2 gas.

16. Standard dialysis tubing (MWCO: 3500) *(Spectra/Por Molecularporous Membrane Tubing)*.

AβO-binding protein isolation

17. His-tag-binding Dynabeads *(Invitrogen) (product #: 10–104-D)*.

18. His-tag-binding Dynabeads wash buffer (50 mM sodium phosphate, 300 mM sodium chloride, 0.01% v/v Tween-20).

19. 750 nM Amyloid beta oligomers (AβOs):

    (a) As prepared in Wilcox (2015).

20. Goat anti-mouse IgG Dynabeads (*Thermo Fisher) (product #: 11033)*.

21. His-tag elution buffer (300 mM imidazole, 50 mM sodium phosphate pH 8.0, 300 mM NaCl, 0.01% v/v Tween-20).
22. 0.1% w/v Bovine serum albumin (BSA) in PBS.
23. 0.1 M Citrate (pH 2–3).
24. Coomassie Stain Kit *(Thermo Fisher) (product #: 23236).*
25. Silver stain kit *(Invitrogen) (product #: LC6100).*

3.2 Equipment

1. Vacuum pump.
2. Affinity purification nickel resin and columns.
3. Rotary mixer.
4. FPLC with SEC column (e.g., Superdex 200 10/300 GL).
5. Motor-driven glass/Teflon homogenizer (5000–30,000 rpm, 120 V 50–60 Hz 1.15 A) *(BioSpec Products Inc.) (Model 398).*
6. Beckman Coulter Ultracentrifuge (capable of spinning up to 28,000 rpm, and tube volume/size: 6 × 39 mL, 1 × 3.5 in., 25 × 89 mm).

3.3 Protocol #1: SMPL Synapse Membrane Preparation Encapsulating the Solubilized Synaptic Membrane Proteome into Nanoscale Lipid Bilayers

3.3.1 Day 0

Synaptic Plasma Membrane Isolation. Timing: 6 h

1. Thaw frozen rat cerebral cortex in 8 mL of ice-cold 0.32 M unbuffered sucrose (with protease inhibitors) per gram tissue.
2. Immediately homogenize cortex with 12 up/down strokes at 13,000 rpm on the motor-driven homogenizer with sample immersed in ice.
3. Centrifuge the homogenate at 800 rcf/20 min/4 °C.
4. Separate supernatant from pellet.
5. Centrifuge supernatant at 9000 rcf/20 min/4 °C to obtain crude synaptosomal pellet.
6. Wash synaptosomal pellet once by resuspending in 0.32 M sucrose.
   - Note: The pellet should be completely resuspended during washes.
7. Centrifuge at 9000 rcf/20 min/4 °C.
8. Lyse synaptosomes by resuspending the pellet in ice-cold 20 mM Tris, pH 7.5, and re-homogenizing.
9. Incubate lysate on ice for 30 min.
10. Centrifuge lysate at 25,000 rcf/30 min/4 °C.
11. Resuspend pellet in 4 mL ice-cold ddH_2O and mix with 4 mL of 2.2 M sucrose.

12. Assemble sucrose gradient by layering 8 mL of 0.855 M sucrose and 8 mL of 0.32 M sucrose.
13. Centrifuge at 65,000 rcf/2.5 h/4 °C.
14. Collect the synaptic plasma membrane that accumulated at the interface between the 1.1 and 0.855 M sucrose layers.
15. Wash the synaptic plasma membrane with PBS containing protease inhibitors by resuspending and centrifuging at 9000 rcf/10 min/4 °C.
16. Resuspend in PBS and determine synaptic plasma membrane protein concentration using BCA assay.
17. Store at −80 °C.

3.3.2 Day 1

Synaptic Plasma Membrane Solubilization. Timing: 2 h

1. Thaw synaptic plasma membrane (stored in step 17) on ice and centrifuge at 10,000 rcf/15 min/4 °C.
2. Discard supernatant and suspend pellet containing synaptic plasma membrane in 1% DβM.
   (a) Use a ratio of 1 mL of 1% DβM per 1 mg protein:
      - Note: Ensure that all protein is dissolved.
3. Rotate the resuspended pellet for 25 min/4 °C.
4. Centrifuge at 20,000 rcf/10 min/4 °C.
5. Determine solubilized protein concentration using BCA assay.

POPC Preparation. Timing: 5 h

6. Clean a glass syringe by drawing up and expelling one syringe volume of chloroform three times.
7. Extract the desired amount of POPC using a glass syringe and transfer it to a clean glass test tube.
   (a) Use 9 μmol of POPC per mg solubilized protein (step 22).
   (b) A typical SMPL synaptic membrane preparation contains 40 μg of solubilized protein.
8. Dry POPC under a gentle, indirect stream of N_2 gas (Fig. 1b).
   - Note: Avoid splashing the POPC. Rotate the tube constantly to ensure an even drying pattern.
9. Place the tube in a beaker and cover it loosely with foil.
10. Place the beaker in the vacuum pump and dry for 3 h.

**Critical Step*: Since POPC dissolves in chloroform, the chloroform must be removed completely to ensure the complete drying of the POPC. Dried POPC should have an opaque, white appearance; it should not be glassy.

SMPL Synaptic Membrane Assembly Preparation. Timing: 1–2 h

11. Based on the amount of solubilized protein (step 22) determine the amount of membrane scaffold protein (MSP) required.
    (a) Use 100 nmol MSP per 1 mg solubilized protein.
12. Add 100 mM sodium cholate in PBS to dried POPC to create 50 mM POPC.
13. Vortex the POPC/cholate tube, heat it under hot tap water, and sonicate in an ultrasonic bath until the solution is clear, and no lipid remains on the walls of the tube.

* *Critical Step*: This is the POPC stock to be used for the assembly mixture.

14. Determine the concentration of cholate that will be in the assembly mixture [consisting of protein (step 22), MSP (step 6), and POPC (step 8) volumes] to ensure that it remains over 20 mM.

* *Critical Step*: At least 20 mM cholate in the mixture is required to ensure proper assembly. If the cholate concentration will be above 20 mM, proceed to step 11. If it will be below 20 mM, add cholate.

SMPL Synaptic Membrane Formation. Timing: 16 h

15. Combine:
    (a) Solubilized protein (step 21).
    (b) MSP (step 6).
    (c) POPC (step 2).
    (d) Any additional cholate as determined in step 10 (Fig. 1b).
    - Note: The correct [protein (1 mg):POPC (9 μmol):MSP (100 nmol)] ratio is important for optimal nanodisk assembly.
16. Incubate for 2 h on a rotator at 4 °C.
17. Prewash Amberlite XAD-2 beads 3 × 10 min in methanol followed by 3 × 10 min in ddH_2O. Pellet beads by centrifugation between washes, remove supernatant, and resuspend beads in the appropriate wash buffer.
18. The volume of Amberlite XAD-2 beads used should be half the total assembly mixture volume (step 11).
    (a) Note: The volume of settled beads should equal half the total assembly mixture volume to ensure adequate amount of beads for detergent removal and initiation of SMPL synaptic membrane assembly.

(b) Pour the washed beads into a weigh boat and vacuum to dry using a 10 μL pipette tip.

19. Add the beads into the mixture to initiate assembly (Fig. 1b).
20. Rotate overnight at 4 °C.

* *Critical Step*: Overnight incubation is required to ensure thorough mixing.

3.3.3 Day 2

SMPL Synaptic Membrane Purification. Timing: 17 h

21. Pellet the nickel resin and remove the supernatant by centrifuging the desired volume at room temp.
    (a) Nickel resin volume should be ¼ that of the SMPL synaptic membrane assembly mixture volume (not including the Amberlite XAD-2 beads).
22. Pour resin into column.
23. Wash resin with 2 column volumes of methanol immediately after resin addition, followed by 2 column volumes of ddH_2O.
24. Cap the bottom of column, add 1 column volume of wash buffer (PBS), and cap top. Equilibrate for 30 min/room temp on rotator.
25. Filter out Amberlite XAD-2 beads by using a glass pipette to transfer the SMPL synaptic membrane assembly mixture through an empty column. Collect the SMPL synaptic membrane assembly mixture and keep on ice.
26. Uncap and allow PBS wash buffer to flow through.
27. Re-cap and load the SMPL synaptic membrane assembly mixture into the nickel resin column.
28. Incubate mixture with resin for 10–15 min/4 °C on a rotator (Fig. 1b).
29. Collect the flow-through and reserve if case assembly did not bind beads.
30. Elute using elution buffer: 50 mM Phosphate, 300 mM NaCl, 250 mM imidazole, pH 8.0
    - Note: Collect 1 mL fractions in test tubes as they come out of the column.
31. Identify protein-containing fractions with Coomassie stain.
    (a) Transfer 10 μL of each 1 mL fraction to a new test tube. Add two drops with Pasteur pipet of Coomassie stain to each tube to gauge the protein content of each fraction, indicated by significant color change. Combine the fractions containing significant amounts of protein.

32. Dialyze in 2 L of dialysis buffer at 4 °C for at least 2 h. Exchange 1 L of buffer and dialyze overnight. Exchange another 1 L for at least 2 h after overnight dialysis (Fig. 1b).

3.3.4 Day 3

SMPL Synaptic Membrane Production Analysis.

33. Measure protein concentration via BCA.
34. Confirm protein incorporation into synaptic membrane via SEC using an appropriate column (e.g., Superdex200 10/300 GE Life Sciences) (Fig. 2).
    - Note: SMPL synaptic membranes can be stored at −80°.

4 Identification of an AβO-Binding Protein from SMPL Synaptic Membranes

The second protocol detailed here illustrates the use of SMPL synaptic membranes to identify a toxin receptor for amyloid β oligomers (AβOs). AβOs are widely considered to be pathogenic agents instigating the neural damage leading to Alzheimer's dementia [26–34]. AβOs are found in AD brain and in animal AD models [35–40]. Their elimination rescues memory and preserves brain cell health [41–47], potentially in humans [48, 49] as well as model systems. Exogenous application of AβOs in vivo or in culture instigates virtually all major aspects of AD neuropathology, including tau pathology and loss of synapse plasticity and structure [27–30].

The mechanism of AβO neurotoxicity downstream involves disrupted neural signaling [50], but the first step of the mechanism is less certain. One possibility often suggested is that AβOs can insert into lipid bilayers and act as pores [51, 52]. The pore hypothesis is consistent with Aβ's extensive hydrophobic region and supported by various findings [51–54]. On the other hand, in CNS cultures AβOs show ligand-like properties that are consistent with a mechanism that involves adventitious binding to toxin receptors. Involvement of toxin receptors would explain why AβOs bind and cause damage only to specific subpopulations of neurons. This binding is saturable, has an EC50s in the nanomolar range, and, of particular interest, has been shown to target dendritic spines. The case for the toxin receptor hypothesis and listings of potential receptor candidates have been recently reviewed [31].

The toxin receptor hypothesis has recently been investigated using SMPL synaptic membranes [12], with results providing a good example of how these preparations can be used to isolate difficult membrane proteins. As a starting point, Western blots and SDS-PAGE silver stains verify that SMPL synaptic membranes provide an unbiased collection of synaptic membrane proteins in soluble form. The soluble membrane proteins show enzymatic and ligand-binding activity consistent with conformational integrity, in harmony with the extensive literature describing the use of Nanodisc technology to investigate recombinant membrane proteins [17].

Assays of SMPL synaptic membranes using oligomer-specific ELISAs to measure binding of exogenous AβOs verify the presence of AβO-binding sites. This robust binding is protease sensitive. Protease sensitivity is consistent with findings that bare Nanodiscs, prepared only with synaptic or synthetic lipids, show minimal AβO binding. Binding also shows membrane specificity, saturation, and a high affinity that are similar to those measured using intact cells. Binding depends, moreover, on the same forms of AβOs that show ligand activity in cell culture (i.e., AβOs with mass >50 kDa). Investigations with SMPL synaptic membranes thus show harmony with the hypothesis that AβOs attach to neurons via proteins that act as toxin receptors.

SMPL synaptic membranes are especially well suited for isolation of difficult neural membrane proteins such as those that bind AβOs. Without using Nanodisc technology, when extracting AβO-binding proteins it is difficult to break up large protein-protein complexes without disrupting the sites that bind AβOs. In SMPL synaptic membranes, extracted membrane proteins are repackaged essentially one per Nanodisc, which makes them soluble, and they are surrounded by a lipid bilayer, which maintains conformations capable of AβO binding.

The protocol described below for affinity isolation of AβO-binding proteins yields a greatly reduced preparation. The AβO-dependent band at 110 kDa in silver-stained gels (Fig. 4) has been shown by LC-MS/MS to be the NaK ATPase α3 subunit (NKAα3) [14]. Hippocampal cell culture experiments have verified that AβO-binding sites and NKAα3 co-localize. Co-localization experiments revealed further that relatively short exposures of neurons to AβOs affect the surface distribution of NKAα3, which moves into enlarged clusters. AβO binding to NKAα3 thus could represent the first step in the mechanism of AβO-induced neural damage; redistribution of NKAα3, which serves as a docking station for other membrane proteins, could be the immediate pathogenic consequence. It also appears that AβOs inhibit the NKAα3 ion pump, as found in an independent study showing that AβOs attach to NKAα3 at the ouabain-binding site [14].

It is likely that capturing recombinant NKAα3 in Nanodiscs could provide a platform for high-throughput screens designed to discover receptor antagonists. However, even without purified recombinant NKAα3, it is possible to begin high=throughput screening for toxin receptor antagonists using complete SMPL synaptic membranes. The strategy couples SMPL synaptic membranes to the AlphaScreen HTS platform for a binding assay made feasible by the engineered properties of the MSP [12]. One can thereby begin using SMPL synaptic membranes as one might use an unbiased cell-based assay. The efficiency, however, is much greater for SMPL synaptic membranes. Pilot studies with a small compound library already have identified a high-affinity antagonist that blocks AβO binding to SMPM synaptic membranes, to intact synaptosomes, and to synapses of cultured hippocampal neurons. Remarkably, the same compound, while not useful for therapeutics because of its promiscuity, previously had been shown to rescue behavior in a transgenic mouse AD model [55].

4.1 Protocol #2: Toxin Receptor Isolation Using SMPL Synaptic Membranes to Isolate a Toxin Receptor for Alzheimer's-Related Amyloid Beta Peptide Oligomers (AβOs)

Note—Materials and SMPL synaptic membrane methods are as above.

4.1.1 Day 1

Attachment of Dynabeads® to SMPL synaptic membranes

1. Wash His-tag-binding Dynabeads® in 1 mL wash buffer 3 × 10 min/4 °C on a rotator.
   (a) Use beads at a ratio of 25 μL beads/40 μg SMPL synaptic membrane.
   (b) Typical SMPL synaptic membrane amount is 500 μg per condition.
2. Magnetically immobilize beads to remove supernatant.
3. Add 1 mL new wash buffer.
4. Incubate His-tag-binding Dynabeads® with SMPLs for 2 h/4 °C on a rotator (Fig. 3).
5. Wash bead complex in 1 mL wash buffer 3 × 10 min/4 °C on a rotator.
   - Note: The bead complex contains SMPL synaptic membrane Dynabeads®.

Binding of Aβ oligomers to SMPL synaptic membranes

6. Thaw AβOs on ice immediately before use.
7. Incubate the SMPL synaptic membranes that are attached to Dynabeads® with 750 nM AβOs for 1 h/room temp on a rotator (Fig. 3).
   (a) Prepare a separate control without AβOs.

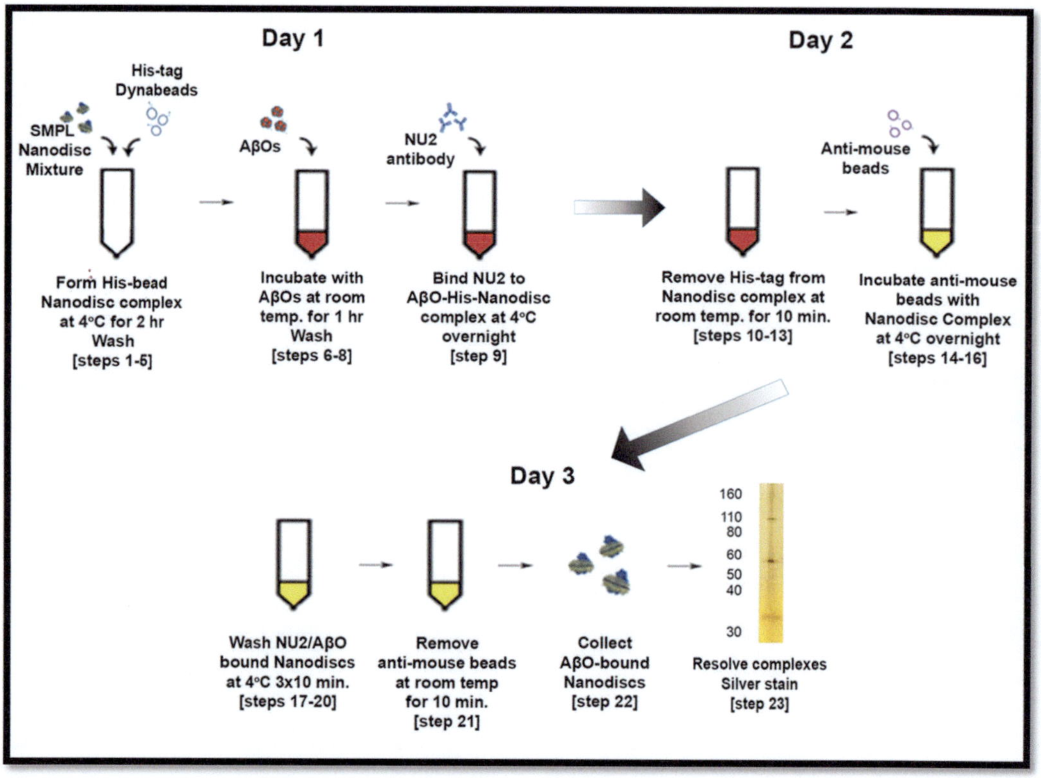

Fig. 3 AβO receptor pull-down

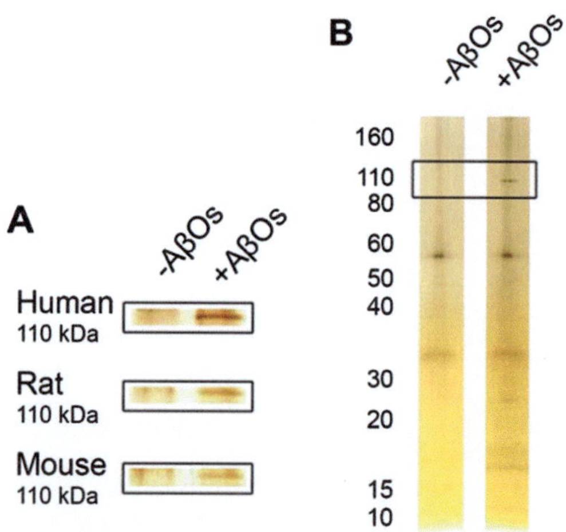

Fig. 4 Identification of AβO-bound receptor

8. Wash bead complex in 1 mL wash buffer 3 × 10 min/4 °C on a rotator.
   - Note: The bead complex contains AβO-SMPL synaptic membrane complex-Dynabeads®.

Binding antibody to oligomers associated with SMPL synaptic membranes

9. Incubate bead complex with an AβO-specific antibody (e.g., NU2; [56]) overnight at 4 °C (Fig. 3).
   - Note: If using NU2 antibody: use 1.5 μg of NU2 per every 25 μL of Dynabeads®.

4.1.2 Day 2

Removal of excess antibody

10. Magnetically immobilize bead complex and discard supernatant.
    - Note: The bead complex contains NU2-AβO-SMPL synaptic membrane-Dynabeads®.
11. Wash bead complex 3 × 10 min in wash buffer.

Eluting SMPLs from His-tag-binding magnetic Dynabeads

12. Elute by incubating with 300 μL His-tag elution buffer for 10 min/room temp on a rotator (Fig. 3).
13. Magnetically immobilize bead complex and transfer supernatant to a new tube.

Attachment of goat anti-mouse IgG magnetic Dynabeads to NU2-AβO-SMPL complex.

14. Wash goat anti-mouse IgG Dynabeads® in 1.5 mL of 0.1 M citrate 4 × 10 min/room temp on a rotator.
15. Resuspend beads in 1.5 mL PBS.
16. Incubate NU2-AβO-SMPLs (step 13) eluate with pre-washed goat anti-mouse IgG Dynabeads® overnight/4 °C on a rotator (Fig. 3).

4.1.3 Day 3

17. Magnetically immobilize and wash bead complex 3 × 10 min in 0.1% BSA/1 mL PBS at 4 °C.
    - Note: Bead complex contains NU2-AβO-SMPL synaptic membrane-goat anti-mouse IgG Dynabeads®.
18. Resuspend bead complex in PBS and transfer to a new tube.
19. Wash bead complex in 1 mL PBS only.
20. Magnetically immobilize bead complex and discard supernatant.

21. Elute the complex off of the beads by incubating with 40 μL of 0.1 M citrate for 10 min/room temp with rotation.
22. Magnetically immobilize beads and transfer eluate to a new tube (Fig. 3).
    - Note: Keep eluate on ice while preparing SDS-PAGE.
    - Note: The elution buffer will disrupt all protein complexes.
23. Resolve complex components via SDS-PAGE and visualize via silver stain according to the manufacturer's instructions (Fig. 4).
    - Note: SDS-PAGE should be run on the same day as elution. If using mass spectrometry analyze soon after elution.

5 SMPL Solutions for the Future

We have described here the preparation of SMPL synaptic membranes. To illustrate their value for the CNS membrane protein toolkit, we have shown how these libraries have been used to pursue Alzheimer's mechanisms and drug discovery. Focusing on AβO binding with SMPL synapse membranes can be seen as a first step, and future applications are readily envisioned:

SMPL synaptic membranes could be used to discover other physiological or pathological receptors such as those for exosomes or for prion-like forms of tau protein or synuclein. They also could be exploited in searches for new physiological ligands, e.g., for endogenous ouabain-like molecules that compete with AβOs for binding to NKAα3. (The possibility also exists that AβOs themselves are endogenous ouabain-like regulators under non-pathological conditions.)

Nanodisc technology makes it straightforward to assemble a soluble membrane protein library, opening the door to creation of libraries of various intracellular organelles (e.g., SMPL synaptic vesicles, SMPL endosomes, SMPL mitochondria, or SMPL ER). For some purposes, more complex libraries could be assembled (SMPL synaptosomes, e.g., also yield NKAα3 as a prominent AβO-binding protein, and they also can be used for HTS). Coupled to appropriate platforms, these SMPL solutions could be used for HTS of targets localized intracellularly.

Physiological investigations of membrane protein-protein interactions are amenable to molecular level studies by adjusting the MSP belt size of the Nanodiscs or by combining different SMPL preparations. With appropriate baits or experimental design, one could investigate interactions involving, e.g., GPCRs and G proteins and adenylate cyclase; oligomerization of GPCRs; endocytosis; exocytosis; synaptic adhesion codes; and assembly of signaling scaffolds analogous to past work on constitutes of focal contacts [18, 20, 57, 58].

To close, we want to say that at first glance, the application of nanotechnology to investigations of CNS membrane proteins might appear daunting. We expect you will find, however, that the benefits outweigh the difficulties. The methods themselves in fact are straightforward and quickly become routine.

References

1. Whittaker VP (1993) Thirty years of synaptosome research. J Neurocytol 22(9):735–742
2. McIlwain H (1975) The Second Thudichum Lecture. Cerebral isolates and neurochemical discovery. Biochem Soc Trans 3(5):579–590
3. Abbott MA, Wells DG, Fallon JR (1999) The insulin receptor tyrosine kinase substrate p58/53 and the insulin receptor are components of CNS synapses. J Neurosci 19(17):7300–7308
4. Fein JA, Sokolow S, Miller CA, Vinters HV, Yang F, Cole GM, Gylys KH (2008) Co-localization of amyloid beta and tau pathology in Alzheimer's disease synaptosomes. Am J Pathol 172(6):1683–1692. https://doi.org/10.2353/ajpath.2008.070829
5. Henkins KM, Sokolow S, Miller CA, Vinters HV, Poon WW, Cornwell LB, Saing T, Gylys KH (2012) Extensive p-tau pathology and SDS-stable p-tau oligomers in Alzheimer's cortical synapses. Brain Pathol 22(6):826–833. https://doi.org/10.1111/j.1750-3639.2012.00598.x
6. Sokolow S, Henkins KM, Williams IA, Vinters HV, Schmid I, Cole GM, Gylys KH (2012) Isolation of synaptic terminals from Alzheimer's disease cortex. Cytometry A 81(3):248–254. https://doi.org/10.1002/cyto.a.22009
7. Kahle PJ, Neumann M, Ozmen L, Muller V, Jacobsen H, Schindzielorz A, Okochi M, Leimer U, van Der Putten H, Probst A, Kremmer E, Kretzschmar HA, Haass C (2000) Subcellular localization of wild-type and Parkinson's disease-associated mutant alpha -synuclein in human and transgenic mouse brain. J Neurosci 20(17):6365–6373
8. Chen RH, Wislet-Gendebien S, Samuel F, Visanji NP, Zhang G, Marsilio D, Langman T, Fraser PE, Tandon A (2013) Alpha-Synuclein membrane association is regulated by the Rab3a recycling machinery and presynaptic activity. J Biol Chem 288(11):7438–7449. https://doi.org/10.1074/jbc.M112.439497
9. Betzer C, Movius AJ, Shi M, Gai WP, Zhang J, Jensen PH (2015) Identification of synaptosomal proteins binding to monomeric and oligomeric alpha-synuclein. PLoS One 10(2):e0116473. https://doi.org/10.1371/journal.pone.0116473
10. Denisov IG, Sligar SG (2017) Nanodiscs in membrane biochemistry and biophysics. Chem Rev 117(6):4669–4713. https://doi.org/10.1021/acs.chemrev.6b00690
11. Bayburt TH, Grinkova YV, Sligar SG (2002) Self-assembly of discoidal phospholipid bilayer nanoparticles with membrane scaffold proteins. Nano Lett 2(8):853–856. https://doi.org/10.1021/nl025623k
12. Wilcox KC, Marunde MR, Das A, Velasco PT, Kuhns BD, Marty MT, Jiang H, Luan CH, Sligar SG, Klein WL (2015) Nanoscale synaptic membrane mimetic allows unbiased high throughput screen that targets binding sites for Alzheimer's-Associated Abeta Oligomers. PLoS One 10(4):e0125263. https://doi.org/10.1371/journal.pone.0125263
13. Marty MT, Wilcox KC, Klein WL, Sligar SG (2013) Nanodisc-solubilized membrane protein library reflects the membrane proteome. Anal Bioanal Chem 405(12):4009–4016. https://doi.org/10.1007/s00216-013-6790-8
14. DiChiara T, DiNunno N, Clark J, Bu RL, Cline EN, Rollins MG, Gong Y, Brody DL, Sligar SG, Velasco PT, Viola KL, Klein WL (2017) Alzheimer's toxic amyloid beta oligomers: unwelcome visitors to the Na/K ATPase alpha3 Docking Station. Yale J Biol Med 90(1):45–61
15. Ritchie TK, Grinkova YV, Bayburt TH, Denisov IG, Zolnerciks JK, Atkins WM, Sligar SG (2009) Chapter 11 – reconstitution of membrane proteins in phospholipid bilayer nanodiscs. Methods Enzymol 464:211–231. https://doi.org/10.1016/S0076-6879(09)64011-8
16. Luthra A, Gregory M, Grinkova YV, Denisov IG, Sligar SG (2013) Nanodiscs in the studies of membrane-bound cytochrome P450 enzymes. Methods Mol Biol 987:115–127. https://doi.org/10.1007/978-1-62703-321-3_10
17. Denisov IG, Sligar SG (2016) Nanodiscs for structural and functional studies of membrane

proteins. Nat Struct Mol Biol 23(6):481–486. https://doi.org/10.1038/nsmb.3195

18. Bayburt TH, Leitz AJ, Xie G, Oprian DD, Sligar SG (2007) Transducin activation by nanoscale lipid bilayers containing one and two rhodopsins. J Biol Chem 282(20):14875–14881. https://doi.org/10.1074/jbc.M701433200
19. Bayburt TH, Sligar SG (2010) Membrane protein assembly into nanodiscs. FEBS Lett 584(9):1721–1727. https://doi.org/10.1016/j.febslet.2009.10.024
20. Bayburt TH, Vishnivetskiy SA, McLean MA, Morizumi T, Huang CC, Tesmer JJ, Ernst OP, Sligar SG, Gurevich VV (2011) Monomeric rhodopsin is sufficient for normal rhodopsin kinase (GRK1) phosphorylation and arrestin-1 binding. J Biol Chem 286(2):1420–1428. https://doi.org/10.1074/jbc.M110.151043
21. Civjan NR, Bayburt TH, Schuler MA, Sligar SG (2003) Direct solubilization of heterologously expressed membrane proteins by incorporation into nanoscale lipid bilayers. BioTechniques 35(3):556–560 562–553
22. Marty MT, Zhang H, Cui W, Blankenship RE, Gross ML, Sligar SG (2012) Native mass spectrometry characterization of intact nanodisc lipoprotein complexes. Anal Chem 84(21):8957–8960. https://doi.org/10.1021/ac302663f
23. Marty MT, Das A, Sligar SG (2012) Ultra-thin layer MALDI mass spectrometry of membrane proteins in nanodiscs. Anal Bioanal Chem 402(2):721–729. https://doi.org/10.1007/s00216-011-5512-3
24. Marty MT, Hoi KK, Robinson CV (2016) Interfacing membrane mimetics with mass spectrometry. Acc Chem Res 49(11):2459–2467. https://doi.org/10.1021/acs.accounts.6b00379
25. Hoi KK, Robinson CV, Marty MT (2016) Unraveling the composition and behavior of heterogeneous lipid nanodiscs by mass spectrometry. Anal Chem 88(12):6199–6204. https://doi.org/10.1021/acs.analchem.6b00851
26. Lambert MP, Barlow AK, Chromy BA, Edwards C, Freed R, Liosatos M, Morgan TE, Rozovsky I, Trommer B, Viola KL, Wals P, Zhang C, Finch CE, Krafft GA, Klein WL (1998) Diffusible, nonfibrillar ligands derived from Abeta1-42 are potent central nervous system neurotoxins. Proc Natl Acad Sci U S A 95(11):6448–6453
27. De Felice FG, Velasco PT, Lambert MP, Viola K, Fernandez SJ, Ferreira ST, Klein WL (2007) Abeta oligomers induce neuronal oxidative stress through an N-methyl-D-aspartate receptor-dependent mechanism that is blocked by the Alzheimer drug memantine. J Biol Chem 282(15):11590–11601. https://doi.org/10.1074/jbc.M607483200
28. Lacor PN, Buniel MC, Furlow PW, Clemente AS, Velasco PT, Wood M, Viola KL, Klein WL (2007) Abeta oligomer-induced aberrations in synapse composition, shape, and density provide a molecular basis for loss of connectivity in Alzheimer's disease. J Neurosci 27(4):796–807. https://doi.org/10.1523/jneurosci.3501-06.2007
29. De Felice FG, Wu D, Lambert MP, Fernandez SJ, Velasco PT, Lacor PN, Bigio EH, Jerecic J, Acton PJ, Shughrue PJ, Chen-Dodson E, Kinney GG, Klein WL (2008) Alzheimer's disease-type neuronal tau hyperphosphorylation induced by A beta oligomers. Neurobiol Aging 29(9):1334–1347. https://doi.org/10.1016/j.neurobiolaging.2007.02.029
30. De Felice FG, Vieira MN, Bomfim TR, Decker H, Velasco PT, Lambert MP, Viola KL, Zhao WQ, Ferreira ST, Klein WL (2009) Protection of synapses against Alzheimer's-linked toxins: insulin signaling prevents the pathogenic binding of Abeta oligomers. Proc Natl Acad Sci U S A 106(6):1971–1976. https://doi.org/10.1073/pnas.0809158106
31. Viola KL, Klein WL (2015) Amyloid beta oligomers in Alzheimer's disease pathogenesis, treatment, and diagnosis. Acta Neuropathol 129(2):183–206. https://doi.org/10.1007/s00401-015-1386-3
32. Selkoe DJ (2008) Soluble oligomers of the amyloid beta-protein impair synaptic plasticity and behavior. Behav Brain Res 192(1):106–113. https://doi.org/10.1016/j.bbr.2008.02.016
33. Mucke L, Selkoe DJ (2012) Neurotoxicity of amyloid beta-protein: synaptic and network dysfunction. Cold Spring Harb Perspect Med 2(7):a006338. https://doi.org/10.1101/cshperspect.a006338
34. Selkoe DJ, Hardy J (2016) The amyloid hypothesis of Alzheimer's disease at 25 years. EMBO Mol Med 8(6):595–608. https://doi.org/10.15252/emmm.201606210
35. Gong Y, Chang L, Viola KL, Lacor PN, Lambert MP, Finch CE, Krafft GA, Klein WL (2003) Alzheimer's disease-affected brain: presence of oligomeric A beta ligands (ADDLs) suggests a molecular basis for reversible memory loss. Proc Natl Acad Sci U S A 100(18):10417–10422. https://doi.org/10.1073/pnas.1834302100
36. Lacor PN, Buniel MC, Chang L, Fernandez SJ, Gong Y, Viola KL, Lambert MP, Velasco PT, Bigio EH, Finch CE, Krafft GA, Klein WL (2004) Synaptic targeting by Alzheimer's-

related amyloid beta oligomers. J Neurosci 24(45):10191–10200. https://doi.org/10.1523/jneurosci.3432-04.2004

37. Oddo S, Caccamo A, Tran L, Lambert MP, Glabe CG, Klein WL, LaFerla FM (2006) Temporal profile of amyloid-beta (Abeta) oligomerization in an in vivo model of Alzheimer disease. A link between Abeta and tau pathology. J Biol Chem 281(3):1599–1604. https://doi.org/10.1074/jbc.M507892200
38. Tomiyama T, Matsuyama S, Iso H, Umeda T, Takuma H, Ohnishi K, Ishibashi K, Teraoka R, Sakama N, Yamashita T, Nishitsuji K, Ito K, Shimada H, Lambert MP, Klein WL, Mori H (2010) A mouse model of amyloid beta oligomers: their contribution to synaptic alteration, abnormal tau phosphorylation, glial activation, and neuronal loss in vivo. J Neurosci 30(14):4845–4856. https://doi.org/10.1523/jneurosci.5825-09.2010
39. Ferretti MT, Bruno MA, Ducatenzeiler A, Klein WL, Cuello AC (2012) Intracellular Abeta-oligomers and early inflammation in a model of Alzheimer's disease. Neurobiol Aging 33(7):1329–1342. https://doi.org/10.1016/j.neurobiolaging.2011.01.007
40. Price KA, Varghese M, Sowa A, Yuk F, Brautigam H, Ehrlich ME, Dickstein DL (2014) Altered synaptic structure in the hippocampus in a mouse model of Alzheimer's disease with soluble amyloid-beta oligomers and no plaque pathology. Mol Neurodegener 9:41. https://doi.org/10.1186/1750-1326-9-41
41. Xiao C, Davis FJ, Chauhan BC, Viola KL, Lacor PN, Velasco PT, Klein WL, Chauhan NB (2013) Brain transit and ameliorative effects of intranasally delivered anti-amyloid-beta oligomer antibody in 5XFAD mice. J Alzheimers Dis 35(4):777–788. https://doi.org/10.3233/jad-122419
42. Dorostkar MM, Burgold S, Filser S, Barghorn S, Schmidt B, Anumala UR, Hillen H, Klein C, Herms J (2014) Immunotherapy alleviates amyloid-associated synaptic pathology in an Alzheimer's disease mouse model. Brain 137(Pt 12):3319–3326. https://doi.org/10.1093/brain/awu280
43. Knight EM, Kim SH, Kottwitz JC, Hatami A, Albay R, Suzuki A, Lublin A, Alberini CM, Klein WL, Szabo P, Relkin NR, Ehrlich M, Glabe CG, Gandy S, Steele JW (2016) Effective anti-Alzheimer Abeta therapy involves depletion of specific Abeta oligomer subtypes. Neurol Neuroimmunol Neuroinflamm 3(3):e237. https://doi.org/10.1212/NXI.0000000000000237
44. Dodart JC, Bales KR, Gannon KS, Greene SJ, DeMattos RB, Mathis C, DeLong CA, Wu S, Wu X, Holtzman DM, Paul SM (2002) Immunization reverses memory deficits without reducing brain Abeta burden in Alzheimer's disease model. Nat Neurosci 5(5):452–457. https://doi.org/10.1038/nn842
45. Nitsch RM, Hock C (2008) Targeting beta-amyloid pathology in Alzheimer's disease with Abeta immunotherapy. Neurotherapeutics 5(3):415–420. https://doi.org/10.1016/j.nurt.2008.05.013
46. Klyubin I, Betts V, Welzel AT, Blennow K, Zetterberg H, Wallin A, Lemere CA, Cullen WK, Peng Y, Wisniewski T, Selkoe DJ, Anwyl R, Walsh DM, Rowan MJ (2008) Amyloid beta protein dimer-containing human CSF disrupts synaptic plasticity: prevention by systemic passive immunization. J Neurosci 28(16):4231–4237. https://doi.org/10.1523/jneurosci.5161-07.2008
47. Gimenez-Llort L, Rivera-Hernandez G, Marin-Argany M, Sanchez-Quesada JL, Villegas S (2013) Early intervention in the 3xTg-AD mice with an amyloid beta-antibody fragment ameliorates first hallmarks of Alzheimer disease. MAbs 5(5):665–677. https://doi.org/10.4161/mabs.25424
48. Tucker S, Moller C, Tegerstedt K, Lord A, Laudon H, Sjodahl J, Soderberg L, Spens E, Sahlin C, Waara ER, Satlin A, Gellerfors P, Osswald G, Lannfelt L (2015) The murine version of BAN2401 (mAb158) selectively reduces amyloid-beta protofibrils in brain and cerebrospinal fluid of tg-ArcSwe mice. J Alzheimers Dis 43(2):575–588. https://doi.org/10.3233/jad-140741
49. Sevigny J, Chiao P, Bussiere T, Weinreb PH, Williams L, Maier M, Dunstan R, Salloway S, Chen T, Ling Y, O'Gorman J, Qian F, Arastu M, Li M, Chollate S, Brennan MS, Quintero-Monzon O, Scannevin RH, Arnold HM, Engber T, Rhodes K, Ferrero J, Hang Y, Mikulskis A, Grimm J, Hock C, Nitsch RM, Sandrock A (2016) The antibody aducanumab reduces Abeta plaques in Alzheimer's disease. Nature 537(7618):50–56. https://doi.org/10.1038/nature19323
50. Paula-Lima AC, Brito-Moreira J, Ferreira ST (2013) Deregulation of excitatory neurotransmission underlying synapse failure in Alzheimer's disease. J Neurochem 126(2):191–202. https://doi.org/10.1111/jnc.12304
51. Serra-Batiste M, Ninot-Pedrosa M, Bayoumi M, Gairi M, Maglia G, Carulla N (2016) Abeta42 assembles into specific beta-barrel

pore-forming oligomers in membrane-mimicking environments. Proc Natl Acad Sci U S A 113(39):10866–10871. https://doi.org/10.1073/pnas.1605104113

52. Bode DC, Baker MD, Viles JH (2017) Ion channel formation by amyloid-beta42 oligomers but not amyloid-beta40 in cellular membranes. J Biol Chem 292(4):1404–1413. https://doi.org/10.1074/jbc.M116.762526
53. Di Scala C, Troadec JD, Lelievre C, Garmy N, Fantini J, Chahinian H (2014) Mechanism of cholesterol-assisted oligomeric channel formation by a short Alzheimer beta-amyloid peptide. J Neurochem 128(1):186–195. https://doi.org/10.1111/jnc.12390
54. Di Scala C, Yahi N, Boutemeur S, Flores A, Rodriguez L, Chahinian H, Fantini J (2016) Common molecular mechanism of amyloid pore formation by Alzheimer's beta-amyloid peptide and alpha-synuclein. Sci Rep 6:28781. https://doi.org/10.1038/srep28781
55. Lee M, Guo JP, Schwab C, McGeer EG, McGeer PL (2012) Selective inhibition of the membrane attack complex of complement by low molecular weight components of the aurin tricarboxylic acid synthetic complex. Neurobiol Aging 33(10):2237–2246. https://doi.org/10.1016/j.neurobiolaging.2011.12.005
56. Lambert MP, Velasco PT, Chang L, Viola KL, Fernandez S, Lacor PN, Khuon D, Gong Y, Bigio EH, Shaw P, De Felice FG, Krafft GA, Klein WL (2007) Monoclonal antibodies that target pathological assemblies of Abeta. J Neurochem 100(1):23–35. https://doi.org/10.1111/j.1471-4159.2006.04157.x
57. Bayburt TH, Sligar SG (2003) Self-assembly of single integral membrane proteins into soluble nanoscale phospholipid bilayers. Protein Sci 12(11):2476–2481. https://doi.org/10.1110/ps.03267503
58. Gregory MC, McLean MA, Sligar SG (2017) Interaction of KRas4b with anionic membranes: a special role for PIP2. Biochem Biophys Res Commun 487(2):351–355. https://doi.org/10.1016/j.bbrc.2017.04.063

Part VI

Studying Development and Disease Using Synaptosomes

Chapter 14

Profiling of MicroRNA and Protein from Purified Synaptoneurosomes in a Neurodegenerative Disease Model

Amrit S. Boese, Anna Majer, and Stephanie A. Booth

Abstract

The synaptic connections between neurons become lost at early stages of prion disease, often long before clinical signs of disease are detectable. In addition, microRNAs play an important role in modulating the localized protein synthesis at the synapse by binding to and regulating mRNA transcripts, of which some microRNAs have been attributed to dendritic remodeling. In the method described here, we simultaneously extracted protein and total RNA from purified synaptoneurosomes that were isolated from disease-affected forebrain and hippocampus tissue from mice infected with the RML strain of scrapie, a model of neurodegeneration. The synaptic protein samples contained enough material to perform Western blot analysis for detection of individual proteins of interest and to perform a more global proteome screen using mass spectrometry. For the isolated RNA, we describe how to screen hundreds of unique microRNAs using a RT-qPCR array. The concurrent detection of changes in protein and RNA at synapses during neurodegenerative disease, particularly at early stages where synaptic loss is reversible, is useful to illuminate the synaptic landscape of not only prion diseases but also other neurodegenerative conditions.

Key words MicroRNA, Prion disease, Neurodegeneration, Synapses, Neurons, Synaptoneurosome, Brain

1 Introduction

Disruption followed by progressive loss of synapses is the earliest pathology identified in neurodegenerative diseases such as Alzheimer's [1] reviewed in [2] and prion disease [3–10]. In fact, this pathology is a correlate of cognitive impairment [11–13] and behavioral changes that commence well before adverse clinical symptoms are detected [9, 12]. In animal models of prion disease, synaptic density is reduced by up to 25% before the onset of symptoms, reaching up to 55% reduction at terminal disease [1, 14–17]. Despite this damage, the brain is capable of self-repair as demonstrated in animal models of prion disease. Specifically, genetic or pharmacological modulation of a cellular pathway linked to

Kathryn M. Murphy (ed.), *Synaptosomes*, Neuromethods, vol. 141,
https://doi.org/10.1007/978-1-4939-8739-9_14,

neuronal degeneration reversed early synaptic damage, cognitive dysfunction, and prion-induced pathology in the brain, especially when treatment was initiated at presymptomatic stages of disease [18–20]. However, the molecular mechanisms involved in the progressive damage and loss of synapses during neurodegenerative diseases remain poorly understood.

One method used to study molecular changes that occur at synapses during neurodegenerative disease is the isolation of synaptoneurosomes from brain tissue. The synaptoneurosome preparations consist largely of synapses and pinched-off dendritic spines that belong to presynaptic and postsynaptic compartments [21,22]. The presynaptic compartment includes neurotransmitter-containing vesicles while the pre- and postsynaptic compartments contain protein translational machinery including polyribosomes, tRNAs, translational initiation and elongation factors, co-translational protein sorting organelles [23–25], microRNAs (miRNAs), and RNA-inducing silencing complex (RISC) [26, 27]. Detection of these molecular components suggests that modulation of protein abundance and functional activities at both posttranscriptional and posttranslational levels occurs locally at each individual synapse. This allows each synapse to modulate its own protein complement by changing abundance and/or protein activity in response to environmental cues (reviewed in [28]). This is an important factor to consider in neurodegenerative diseases because these structures are the first to be damaged in disease. Furthermore, accumulating disease-related protein aggregates that associate with synapses in Alzheimer's [29–32] and prion [14, 33] diseases may be responsible for toxicity at the synapse. As such, disease-affected synapses may modulate local protein expression/activity in response to this toxicity, prior to the involvement of de novo transcription at the nucleus. Although more research is required to determine if this is the case, cellular assembly and organization were the most disrupted functions identified at the synapse in Alzheimer's disease patients by mass spectrometry [34]. In a similar manner, numerous genes were deregulated in synaptoneurosomes from incipient Alzheimer's disease patients [35]. These experiments highlight the local changes taking place at synapses during neurodegenerative diseases.

A major player in regulating neuronal structure and function is small, noncoding RNA species called miRNAs. These RNA species are important posttranscriptional regulators of numerous biological processes including neuronal development, plasticity, and function (reviewed in [36]). Not surprising, deregulation of miRNA expression was documented in various neurological and neurodegenerative diseases including Alzheimer's and prion diseases [37–41]. Recent studies have begun to unravel the functional role that certain miRNAs play at the synapse under healthy conditions. Indeed, several miRNAs including miR-132, miR-134, miR-138, among others have been found to regulate synaptic plasticity and dendrite formation (reviewed in [42, 43]). However, identification of key miRNAs that are detected at the synapse in neurodegenerative diseases remains understudied.

Herein we describe the protocol used for isolating synaptoneurosomes from specific brain regions of prion-infected animals (Fig. 1) [44]. We followed the protocol described by Verity [45] to isolate synaptoneurosomes at two time points post-infection (presymptomatic and symptomatic stages) from two brain regions known to be affected in disease (hippocampus and cortex). From the purified synaptoneurosomes, we extracted total RNA and proteins that were subsequently used for miRNA profiling and western blot analysis, respectively. This method produced purified synapto-

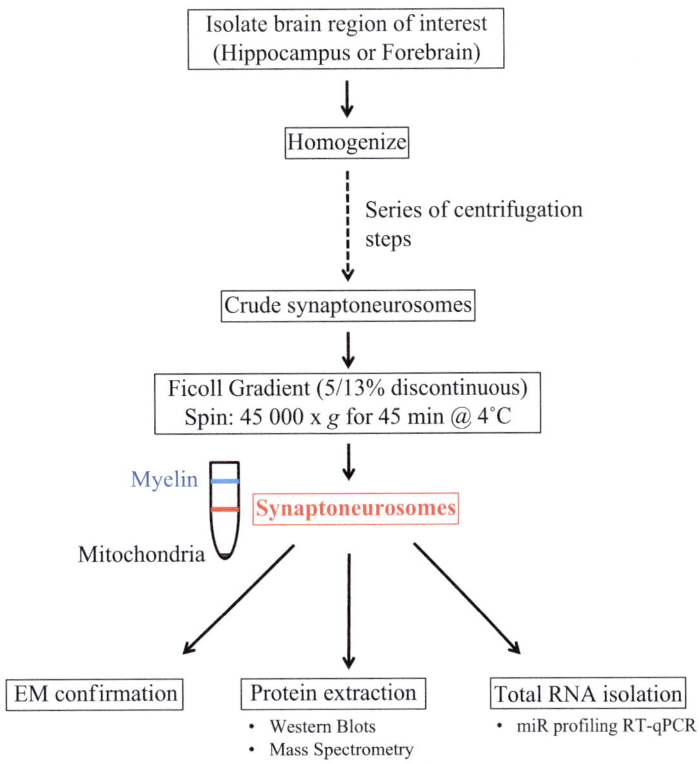

Fig. 1 The experimental design for obtaining total RNA and protein from the same purified synaptoneurosome sample. The brain region of interest, forebrain and hippocampus in our experiment, is macro-dissected and homogenized. This is followed by a number of centrifugation steps in order to remove cell debris and nuclei. The resulting crude synaptoneurosome is laid onto a discontinuous 5/13% Ficoll gradient in order to fractionate the subcellular layers of myelin and mitochondria, resulting in a purified synaptoneurosome layer. The collected synaptoneurosome sample can be further processed to visualize purity of the sample by electron microscopy, and extract total RNA and proteins for subsequent profiling experiments

neurosome preparations that contained both synaptic vesicles and postsynaptic densities as detected by electron microscopy (EM). We also performed western blot analysis to detect both pre- and postsynaptic proteins by using specific antibodies for synapsin and postsynaptic density 95, respectively. One consideration when analyzing synaptoneurosomes isolated from neurodegenerating tissue is the temporal accumulation of molecules related to disease that may not originate from neurons but that can be detected in the synaptoneurosome fraction. A major hallmark of prion disease is astrocytosis that is marked by the presence of glial fibrillary acidic protein (GFAP). This is important when interpreting the nature of miRNA function because of the cell-type-specific roles of miRNAs that are well documented (reviewed in [46]). We specifically looked for GFAP in our synaptoneurosome preparations by western blot and found that the amount of protein detected did not vary between mock- and prion-infected samples, suggesting that our preparation was relatively devoid of astrocyte material. The protocol described here can be followed to profile miRNA changes at synapses concurrently with protein expression from whole brain, or specific brain regions, from any neurodegenerative disease or neurological condition in order to help illuminate the structural/functional changes that occur at the synapse during disease.

2 Materials

2.1 Synaptoneurosome Isolation

1. Whole-brain tissue from mouse model.
2. Homogenization buffer (0.1 mM ethylenediaminetetraacetic acid (EDTA, catalog # 324503100GM, Fisher Scientific), 0.25 mM dithiothreitol (DTT, catalog #43816, Sigma), 2 mM 2-[4-(2-hydroxyethyl)piperazin-1-yl] ethanesulfonic acid (HEPES, catalog # H0887, Sigma), 0.35 M sucrose (catalog #S5–500, Fisher Scientific) [pH 7.4]).
3. Ficoll 400 (catalog #F2637, Sigma).
4. Sucrose 0.35 M solution.
5. Deoxyribonuclease 1 from bovine pancreas (catalog # DN25-10MG, Sigma).
6. Dumont #5/45 Forceps (catalog #11251–35, Fine Science Tools).
7. Razor blade (catalog #55411–050, VWR).
8. Spring scissors (catalog #15005–08, Fine Science Tools).
9. Dounce tissue grinder set (7 mL) (catalog #D9063-1SET SIGMA).
10. Sterile Petri dishes (catalog #FB0875713A, Fisher Scientific).

11. 2.0 mL RNase-free microfuge tubes (catalog #AM12425, Thermo Fisher Scientific).
12. 1.5 mL RNase-free microfuge tubes (catalog #AM12400 Thermo Fisher Scientific).

2.2 Protein Extraction and Western Blot

1. Cell disruption buffer from mirVana Paris kit (catalog #AM1556, Thermo Fisher).
2. BCA Protein Assay Kit (catalog #23227, Thermo Fisher Scientific).
3. Acetone (catalog #A18-1, Fisher Scientific).
4. NuPAGE® LDS Sample Buffer (catalog #NP0007, Thermo Fisher Scientific).
5. NuPAGE® MOPS SDS Running Buffer (catalog #NP0001, Thermo Fisher Scientific).
6. NuPAGE™ 10% Bis-Tris Protein Gels, 1.0 mm, 12-well (catalog #NP0302BOX, Thermo Fisher Scientific).
7. MagicMark™ XP Western Protein Standard (catalog #LC5603, Thermo Fisher Scientific).
8. Precision Plus Protein™ Kaleidoscope™ Prestained Protein Standards (catalog #1610375, Bio-Rad).
9. iBlot® Transfer Stack, PVDF, mini (catalog #IB401032, Thermo Fisher Scientific).
10. Blocking solution [5% skim milk powder (catalog #1153630500, EMD Millipore) in 1× Tris-buffered saline, TBS (catalog #T5912, Sigma) with 0.1% Tween-20, T (catalog #P9416, Sigma)].
11. Antibody diluent (dilute blocking solution 1:10 with 1X TBS-T to achieve 0.5% skim milk).
12. Polyclonal anti-synapsin antibody raised in rabbit (catalog #ab64581, Abcam).
13. Polyclonal anti-postsynaptic density 95 raised in rabbit (catalog #ab18258, Abcam).
14. Monoclonal anti-alpha-tubulin raised in mouse (catalog #T9026, Sigma).
15. Polyclonal anti-GFAP raised in rabbit (catalog #ab7260, Abcam).
16. Wash solution (1X Tris-buffered saline with 0.1% Tween-20).
17. Polyclonal goat anti-rabbit secondary antibody conjugated to horseradish peroxidase (catalog #P0448, Dako).
18. Polyclonal goat anti-mouse secondary antibody conjugated to horseradish peroxidase (catalog #P0447, Dako).

19. SuperSignal™ West Pico Chemiluminescent Substrate (catalog #34077, Thermo Fisher Scientific).
20. GE Healthcare Amersham™ Hyperfilm™ ECL (catalog #45001507, Fisher Scientific).

2.3 Mass Spectrometry

1. Water, Optima™ LC/MS Grade (catalog #W6500, Fisher scientific).
2. DTT (catalog #43816, Sigma).
3. Nanosep 10 K omega cartridge (catalog #OD010C34, Pall Life Sciences).
4. Urea (catalog #45-000-234, Fisher Scientific).
5. Urea exchange buffer (UEB) [add 4.8 g urea to 1 mL of 0.5 M HEPES and 6.5 mL Optima LC/MS water].
6. Iodoacetamide (IAA) solution [add 9.26 mg IAA (catalog #I1149, Sigma) to 1 mL UEB].
7. HEPES solution (catalog #H0887, Sigma).
8. Benzonase solution [4 μL of 25 mM $MgCl_2$ (Roche), 1000 U of Benzonase (Roche), and 42 μL of 50 mM HEPES].
9. Trypsin (catalog #T8658, Sigma).

2.4 RNA Extraction and qPCR Analysis

1. RNase Zap (catalog #AM9780, Thermo Fisher Scientific).
2. mirVana miRNA Isolation kit (catalog #AM1561, Thermo Fisher Scientific).
3. Acid-phenol:chloroform (catalog #AM9720, Thermo Fisher Scientific).
4. RNA 6000 Pico Kit (catalog #5067–1513, Agilent).
5. Universal cDNA Synthesis Kit II (catalog #203301, Exiqon).
6. ExiLENT SYBR® Green master mix (catalog #203421, Exiqon).
7. microRNA Ready-to-Use PCR, Mouse & Rat panel I (catalog #203715, Exiqon).
8. GenEx Software (MultiD analyses AB, Göteborg, Sweden).

3 Methods

3.1 Isolation of Brain Tissue

We used freshly dissected tissues for this protocol but others have used frozen postmortem human brain tissue with success for the crude synaptoneurosome pellet [35]. Ensure that ethics are in place with your institution before commencing any animal experiments. All procedures involving handling of infectious prion material were performed in a biosafety cabinet within a biosafety containment level (BSL) 3 laboratory. Materials from mock-infected samples were processed in a biosafety cabinet when possible in BSL-2 labs. Anesthetize animals with 5% isoflurane before euthanasia through cervical decapitation. Dissect the brain out of the skull cap using surgical scissors and forceps, and remove meninges if present. Place brains carefully into sterile 50 mL Falcon tube containing 10 mL homogenization buffer. For forebrain isolation, place brain in a sterile Petri dish and obtain the two cerebral cortices by removing the olfactory bulbs, midbrain, and hindbrain with a sterilized razor blade. Remove and keep the hippocampus in a 1.5 mL tube with homogenization buffer using fine spring scissors. Then continue with the forebrain preparation by mincing the two cerebral cortices using a razor blade. Using the razor blade, place the minced forebrain into the Dounce tissue grinder unit and add 5 mL of homogenization buffer. Homogenize the tissue using a maximum of eight strokes with the small pestle (*see* **Note 1**). For hippocampal isolation, remove excess fornix tissue surrounding the hippocampus collected using a back-and-forth cutting motion with two sterile 18 gauge needles, using one to hold the hippocampus and the other to remove the fornix. Then place the cleaned-up hippocampal tissues into the Dounce tissue grinder without mincing; these tissues will homogenize readily. For our studies, we pooled the hippocampi of eight mice to obtain enough material for a detectable synaptoneurosome layer in the subsequent centrifugation.

3.2 Synaptoneurosome Preparation

Pipette homogenate into 2.0 mL RNase-free tubes and spin down in a tabletop centrifuge at 2000 × *g* for 1 min to pellet out the cell debris and nuclei, using as many tubes as required to accommodate the volume from each suspension. If desired, the pellet can be washed with homogenization buffer and spun down again to obtain more supernatant. Place supernatant from the last step into a new 2.0 mL DNase/RNase-free tube and spin down at 14,000 × *g* for 4 min to pellet out the crude synaptoneurosomes. Resuspend the synaptoneurosome pellet from the last step in 1 mL of homogenization buffer, using aliquots if multiple pellets were obtained. For example, if three pellets were derived from one mouse forebrain, use 333 μL of homogenization buffer per pellet and combine. Add the crude synaptoneurosome suspension to the prechilled 5/13% discontinuous Ficoll gradient using 1 mL of suspension per 4 mL of total Ficoll gradient (*see* **Note 2**). Ficoll gradient was prepared with 0.35 M sucrose which contains 10 μg/mL DNase in the ultracentrifuge tube. Following preparation of all the ultracentrifuge tubes, use a balance to ensure that they have the same weight. If weights differ, add homogenization buffer to equilibrate. Carefully load the ultracentrifuge tubes into the tube holders on the rotor, place rotor into the ultracentrifuge, and spin at 45,000 × *g* for 45 min (*see* **Note 3**) at 4 °C. Carefully remove the ultracentrifuge rotor from the centrifuge and remove the tubes slowly so as not to disturb the gradient. There are three subcellular components in the gradient; the bottom pellet contains mitochondria, the middle band contains purified synaptoneurosomes, and the top layer contains myelin sheaths. Using 1000 μL tips, remove the top myelin layer and place into a 1.5 mL RNase-free tube for later analysis if desired. Isolate the middle synaptoneurosome layer and place in a 1.5 mL RNase-free tube. In our experience, we obtain ~1200 μL of synaptoneurosomes from one mouse forebrain. The amount that will be used for either RNA isolation, protein extraction, or obtaining images with the electron microscope (EM) should be considered prior to commencing the experiment. In our case, we performed EM on synaptoneurosomes from wild-type animals to ensure that our protocol was robust as the containment restrictions for the RML scrapie prion agent prevented us from performing EM on the infected samples at our facility. For protein isolation, mix 600 μL of the synaptoneurosome layer with 600 μL of homogenization buffer and pellet at 14,000 × *g*. Remove the supernatant and add an appropriate amount of protein isolation buffer to the pellet. We used cell disruption buffer from a mirVana PARIS kit which is compatible with downstream applications such as BCA and western blotting, but not mass spectrometry (*see* **Note 4**). For RNA isolation, place half of the synaptoneurosomes (~600 μL) with 600 μL of homogenization buffer and pellet at 14,000 × *g*. Remove the supernatant and add lysis/binding buffer from the mirVana RNA isolation kit, vortex, and either immediately freeze the lysate at −80 °C or perform complete RNA extraction (*see* **Note 5**). A fraction of neat synaptoneurosomes can be fixed for later EM imaging (*see* **Note 6**).

3.3 Protein Detection Using Western Blot

Establish protein concentration using the BCA kit, ensuring that BSA protein standard is diluted with N-PER rather than PBS. We obtained approximately 360 μg per half of the synaptoneurosome layer for the eight pooled hippocampi. Combine 10 μg of protein with 2.5 μL NuPAGE LDS sample buffer lysis buffer, 1 μL of NuPAGE reducing agent, and deionized water for a final volume of 10 μL (*see* **Note 7**). Denature at 70 °C for 10 min, cool on ice, then add samples to a 10% Bis-Tris gel, and run with prepared 1× MOPS buffer for 50 min at 200 V constant. Transfer gel to a PVDF membrane using the iBLOT transfer system with the 7-min setting. Block the membrane in blocking solution for 30 min at room temperature with mixing. Incubate with primary antibody diluted 1:1000 in antibody diluent overnight at 4 °C with mixing. Wash the membrane three times with wash solution for 5 min shaking per wash. Incubate with secondary antibody diluted in 0.5% skim milk in TBS-T, depending on the animal the primary antibody was raised in; we used either goat anti-rabbit at 1:2000 or goat anti-mouse at 1:2000 for 1 h at room temperature. Wash the membrane three times with wash solution for 5 min with shaking.

Incubate the blot with SuperSignal West Pico Chemiluminescent substrate for 5 min at room temperature to detect the horseradish peroxidase. Blot off excess West Pico substrate and detect signal using X-ray film or imaging system.

3.4 Protein Detection Using Mass Spectrometry

To obtain a global proteome profile, prepare 100 μg of protein for unlabeled mass spectrometry. If a noncompatible buffer was used (cell disruption buffer), acetone precipitate as per **Note 4**. Reconstitute the air-dried protein pellet in 70 μL MS water, 10 μL 0.5 M HEPES buffer, and 20 μL 20% SDS and heat to 95 °C for 5 min. After cooling to room temperature, add 10 μL of 1 M DTT and incubate at 95 °C for 5 min. Bring samples to room temperature before adding 490 μL UEB, vortex, and incubate at room temperature for 10 min. Equilibrate a Nanosep 10 K omega cartridge by adding 200 μL MS water, centrifuge for 2 min at 10,000 × *g*, discard supernatant, and then repeat step with 200 μL UEB. After discarding the flow through, load 300 μL of protein sample onto the cartridge and centrifuge for 10 min at 10,000 × *g* until sample passes through but ensuring that the filter column does not dry out. Then wash three times with 250 μL UEB by spinning at 10,000 × *g*, and discarding the flow through after each wash. Add 100 μL iodoacetamide and place in the thermomixer to shake at 600 rpm for 1 min, followed by a 20-min incubation at room temperature in the dark. Then spin at 10,000 × *g* for 10 min and discard the flow through. Perform three washes with 100 μL UEB per wash, spin at 10,000 × *g* for 10 min, and discard flow through. Then wash two times with 150 μL of 50 mM HEPES with 150 μL per wash and a 10-min spin at 10,000 × *g*. Add 50 μL of Benzonase solution to each cartridge and mix at 600 rpm for 2 min, followed by a 30-min incubation, all at room temperature. Wash a total of three times with 100 μL of 50 mM HEPES, centrifuge at 10,000 × *g* for 10 min, and discard flow through. Transfer the filter cartridge to a new 1.5 mL tube, add 1 ug trypsin, and mix at 600 rpm for 1 min at room temperature. Place the cartridge assembly into a moisture chamber, 50 mL Falcon tube with 1 mL of MS water, and incubate at 37 °C overnight. The next morning, perform peptide elution by cutting off the top rim of the Nanosep cartridge with a razor blade. Then invert the filter into a new collection tube, spin at 10,000 × *g* for 3 min, and *keep* the flow through containing the peptides. Turn the cartridge right side up, add 50 μL of 50 mM HEPES buffer, and shake the assembly at 600 rpm for 2 min at room temperature. Then invert the filter cartridge and spin at 10,000 × *g* for 3 min. Perform the HEPES elution two more times for a total of three elutions, and process accordingly for mass spectrometry.

3.5 RNA Extraction and Profiling Using Real-Time PCR

Prior to performing RNA work, clean all surfaces including pipettes, tube holders, and gloves with RNaseZap to prevent RNA degradation. Perform RNA extraction using the mirVana miRNA Isolation kit (*see* **Note 8**) as per the manufacturer's instructions. We used a 1:1 ratio of synaptoneurosome to lysis/binding buffer, starting with 600 μL synaptoneurosome layer. The isolated RNA can be either used immediately or stored at −80 °C until use. To determine the quantity and quality of the obtained total RNA we used the RNA 6000 Pico kit and the Bioanalyzer 2100 following the manufacturer's recommendations. The resulting chromatograms show the amount of 18S and 28S ribosomal mammalian RNA in each sample and use the ratio of 28S/18S with an algorithm to give the RNA integrity number (RIN). RINs for our samples ranged between 5.5 and 6.9 (Fig. 2). The RNA yield was approximately 335 and 410 ng from the eight pooled hippocampi. The starting material was half of the total synaptoneurosome layer obtained. For miRNA detection, we used the Rodent miRNome panels in which a universal primer reverse transcription produces a pool of cDNA which is then probed with specific locked nucleic acid primers in the qPCR step. We performed RT-qPCR over microarray for profiling as this method had good sensitivity and specificity in comparison to other chemistries [47]. Perform cDNA step using 20 ng of RNA following the manufacturer's protocol for one panel including 1 μL of the UniSp6 spike-in RNA (*see* **Note 9**) to later compare the equivalency of reverse transcription rates. The cDNA can be stored at −20 °C until qPCR can be performed. Dilute cDNA with 1980 μL nuclease-free water, add 2 mL of ExiLENT SYBR Green Master mix, and mix well by pipetting up and down. It is best to make this up in a sterile trough to prevent volume loss. The passive reference dye ROX is not included in the master mix, and was not used in our assays. Using a Repeater pipette, place 10 μL of the mixture into each well of the 384-well panels. Seal the plate with optical film, and spin down at 1500 rpm for 2 min. Load plate onto ViiA7 thermal cycler and run the following program: 95 °C for 10 min followed by 40 cycles of 95 °C for 10 s with 60 °C for 1 min, and standard ramp rate (1.6 °C/s) followed by melting curve analysis, ensuring that no passive reference is selected as ROX is not incorporated. The melt curves should be examined to ensure that only one product is within each PCR reaction and in general the curves should be examined to ensure that false Cts were not called in error during Ct analysis. The miRNome panels are a 384-well format which contain 376 unique miRNAs, alongside controls consisting of 1 blank, 3 UniSp3 interplate calibrators, 1 spike-in RNA primer, and 3 small RNAs U6, RNU5G, and RNU1A. The GenEx analysis of miRNome panels is a streamlined method to incorporate the various controls into the data analysis. Load the data into the GenEx software and analyze as per Exiqon's recommendations for the miR-

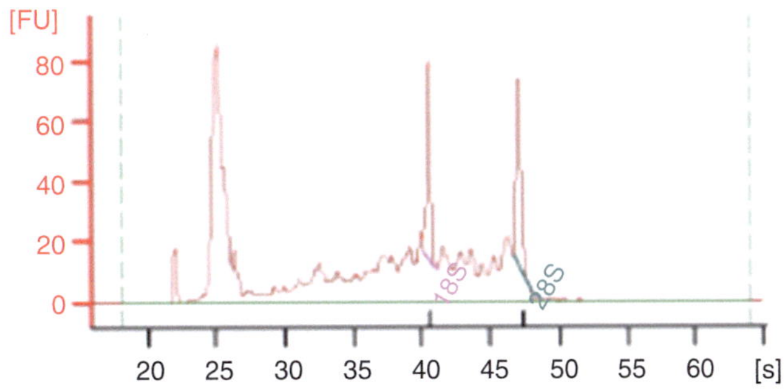

Overall Results for sample 1 : **syn ctrl h 1**

RNA Area:	989.2
RNA Concentration:	3,412 pg/µl
rRNA Ratio [28s / 18s]:	1,2
RNA Integrity Number (RIN):	5,5 (B.02.08)
Result Flagging Color:	
Result Flagging Label:	RIN: 5.50

Fragment table for sample 1 : **syn ctrl h 1**

Name	Start Time [s]	End Time [s]	Area	% of total Area
18S	40.14	41.02	51.7	5.2
28S	46.65	47.95	61.4	6.2

Fig. 2 A sample readout from the Bioanalyzer 2100 run with the Pico 6000 chip with 1 µL of purified synaptoneurosome RNA from the hippocampus. The area under the signal for 18S and 28S ribosomal RNA on the electropherogram is subject to an algorithm to obtain the RNA integrity number (RIN) of 5.5

Nome panels. Remove any non-detected miRNAs or those with Cts above desired threshold; ours was Ct >35. Use the Interplate calibrator option to calibrate data between plates. Use the global mean of detected miRNAs to normalize Cts before calculating fold changes in miR abundance. After obtaining fold changes, go back to ensure that the miRNAs identified had proper qPCR curves and only one product in the melting curve analysis.

4 Notes

1. If possible use a clean Dounce-Potter apparatus for each sample; if not possible, then maintain at least two, one for control and the other for infected; clean between samples by first spraying with 70% ethanol and then rinsing out with nuclease-free water, noting that this does not inactivate the prion agent. If mass spectrometry is to be performed, and the brain material is designated as safe for BSL-2, ensure that the work is done in BSL-2.

2. Equilibrate the discontinuous Ficoll gradient, which contains DNase, at 4 °C for 1 h, prior to loading synaptoneurosome suspension.
3. We have successfully isolated synaptoneurosomes using 26,000 × *g* for 45 min without significant reduction in synaptoneurosome purity, and suggest using this speed if a high-speed ultracentrifuge is unavailable.
4. Some samples may require acetone precipitation if 10 μg of protein is in too large a volume to fit in the wells of the gel or for mass spectrometry, as the cell disruption buffer contains detergents that interfere with mass spectrometry protein detection. It is also possible to substitute SDS as a protein lysis buffer, which should be compatible with mass spectrometry, but we have not tested this. To perform acetone precipitation, add 5× volume of −20 °C acetone, vortex, incubate in −20 °C freezer for 30 min or overnight, spin down at 10,000 × *g* for 10 min, decant supernatant, and air-dry pellet before resuspending in PBS or for mass spectrometry prepare as indicated in the procedure.
5. We have not experienced RNA degradation after freezing the lysate prior to the RNA isolation.
6. As electron microscopy is a very specialized technique we suggest collaborating with a lab proficient in performing this task.
7. This step can be delayed by storing overnight at −20 °C as the trypsin digestion step can take all day. The sample should then be thawed before proceeding to the urea exchange step.
8. The RNA extraction can be performed with any suitable kit that preserves both small and large RNAs so that miRs can be detected from the same sample, and so that mRNA is preserved for later examination of genes if desired.
9. Calculations for 20 ng of RNA should be performed ahead of time, with dilutions as necessary so that pipetting of small volumes is avoided. Ideally, a master mix of RNA with 2–3× the volume required should be made.

5 Conclusions

The mechanism of synaptic loss and degeneration in neurodegenerative diseases remains poorly understood. Here we have described a method to isolate pure synaptoneurosomes from mouse brain tissue infected with prions in order to isolate and detect changes in the profiles of miRNAs and proteins during neurodegeneration. The procedure itself is relatively straightforward with the main limitation being to generate enough brain tissue to obtain sufficient quantities of purified synaptoneurosomes. However, we showed that it is possible to examine molecular changes even in discrete brain regions such as the hippocampus. This method is easily adaptable for use in other neurodegenerative diseases modeled in rodents such as Alzheimer's, Parkinson's, and Huntington's diseases, to name a few. Although we were restricted to western blot analysis due to the infectivity of prion-infected samples, mass spectrometry analysis of disease tissue in other neurodegenerative diseases is feasible. Promising advancements in synaptoneurosome isolation techniques suggest that frozen tissues can also be used as a source material for synaptoneurosome preparations. Furthermore, high-throughput miRNA profiling strategies such as next-generation sequencing could also be performed from similar quantities of tissue. Clearly, understanding how synapses are degraded and lost in neurodegenerative diseases is an ongoing research topic. This procedure provides a means to concurrently identify the changes in miRNAs and proteins that occur to better understand the molecular mechanism(s) by which synapses are disrupted and lost in these diseases.

References

1. Scheff SW, Price DA (2006) Alzheimer's disease-related alterations in synaptic density: neocortex and hippocampus. J Alzheimers Dis 9:101–115
2. Koffie RM, Hyman BT, Spires-Jones TL (2011) Alzheimer's disease: synapses gone cold. Mol Neurodegener 6:63. https://doi.org/10.1186/1750-1326-6-63
3. Ferrer I, Costa F, Grau Veciana JM (1981) Creutzfeldt-Jacob disease: a golgi study. Neuropathol Appl Neurobiol 7:237–242
4. Landis DM, Williams RS, Masters CL (1981) Golgi and electronmicroscopic studies of spongiform encephalopathy. Neurology 31:538–549
5. Hogan RN, Baringer JR, Prusiner SB (1987) Scrapie infection diminishes spines and increases varicosities of dendrites in hamsters: a quantitative Golgi analysis. J Neuropathol Exp Neurol 46:461–473
6. Johnston AR, Black C, Fraser J, MacLeod N (1997) Scrapie infection alters the membrane and synaptic properties of mouse hippocampal CA1 pyramidal neurones. J Physiol 500(Pt 1):1–15
7. Belichenko PV, Brown D, Jeffrey M, Fraser JR (2000) Dendritic and synaptic alterations of hippocampal pyramidal neurones in scrapie-infected mice. Neuropathol Appl Neurobiol 26:143–149 doi:nan232 [pii]
8. Jeffrey M, Halliday WG, Bell J, Johnston AR, MacLeod NK, Ingham C, Sayers AR, Brown DA, Fraser JR (2000) Synapse loss associated with abnormal PrP precedes neuronal degeneration in the scrapie-infected murine hippocampus. Neuropathol Appl Neurobiol 26:41–54 doi:nan216 [pii]
9. Fuhrmann M, Mitteregger G, Kretzschmar H, Herms J (2007) Dendritic pathology in prion disease starts at the synaptic spine. J Neurosci 27:6224–6233 doi:27/23/6224 [pii]
10. Campeau JL, Wu G, Bell JR, Rasmussen J, Sim VL (2013) Early increase and late decrease of purkinje cell dendritic spine density in prion-infected organotypic mouse cerebellar cultures. PLoS One 8:e81776. https://doi.org/10.1371/journal.pone.0081776
11. Terry RD, Masliah E, Salmon DP, Butters N, DeTeresa R, Hill R, Hansen LA, Katzman R (1991) Physical basis of cognitive alterations in Alzheimer's disease: synapse loss is the major correlate of cognitive impairment. Ann Neurol 30:572–580. https://doi.org/10.1002/ana.410300410
12. Cunningham C, Deacon R, Wells H, Boche D, Waters S, Diniz CP, Scott H, Rawlins JN, Perry VH (2003) Synaptic changes characterize early behavioural signs in the ME7 model of murine prion disease. Eur J Neurosci 17:2147–2155 doi:2662 [pii]
13. Hilton KJ, Cunningham C, Reynolds RA, Perry VH (2013) Early hippocampal synaptic

loss precedes neuronal loss and associates with early behavioural deficits in three distinct strains of prion disease. PLoS One 8:e68062. https://doi.org/10.1371/journal.pone.0068062

14. Kitamoto T, Shin RW, Doh-ura K, Tomokane N, Miyazono M, Muramoto T, Tateishi J (1992) Abnormal isoform of prion proteins accumulates in the synaptic structures of the central nervous system in patients with Creutzfeldt-Jakob disease. Am J Pathol 140:1285–1294
15. Clinton J, Forsyth C, Royston MC, Roberts GW (1993) Synaptic degeneration is the primary neuropathological feature in prion disease: a preliminary study. Neuroreport 4:65–68
16. Scheff SW, Price DA, Schmitt FA, DeKosky ST, Mufson EJ (2007) Synaptic alterations in CA1 in mild Alzheimer disease and mild cognitive impairment. Neurology 68:1501–1508 doi:68/18/1501 [pii]
17. Siskova Z, Page A, O'Connor V, Perry VH (2009) Degenerating synaptic boutons in prion disease: microglia activation without synaptic stripping. Am J Pathol 175:1610–1621. https://doi.org/10.2353/ajpath.2009.090372
18. Mallucci GR, White MD, Farmer M, Dickinson A, Khatun H, Powell AD, Brandner S, Jefferys JG, Collinge J (2007) Targeting cellular prion protein reverses early cognitive deficits and neurophysiological dysfunction in prion-infected mice. Neuron 53:325–335 doi:S0896-6273(07)00008-6 [pii]
19. Moreno JA, Halliday M, Molloy C, Radford H, Verity N, Axten JM, Ortori CA, Willis AE, Fischer PM, Barrett DA, Mallucci GR (2013) Oral treatment targeting the unfolded protein response prevents neurodegeneration and clinical disease in prion-infected mice. Sci Transl Med 5:206ra138. https://doi.org/10.1126/scitranslmed.3006767
20. Halliday M, Radford H, Sekine Y, Moreno J, Verity N, le Quesne J, Ortori CA, Barrett DA, Fromont C, Fischer PM, Harding HP, Ron D, Mallucci GR (2015) Partial restoration of protein synthesis rates by the small molecule ISRIB prevents neurodegeneration without pancreatic toxicity. Cell Death Dis 6:e1672. https://doi.org/10.1038/cddis.2015.49
21. Hollingsworth EB, McNeal ET, Burton JL, Williams RJ, Daly JW, Creveling CR (1985) Biochemical characterization of a filtered synaptoneurosome preparation from Guinea pig cerebral cortex: cyclic adenosine 3′:5′-monophosphate-generating systems, receptors, and enzymes. J Neurosci 5:2240–2253
22. Quinlan EM, Olstein DH, Bear MF (1999) Bidirectional, experience-dependent regulation of N-methyl-D-aspartate receptor subunit composition in the rat visual cortex during postnatal development. Proc Natl Acad Sci U S A 96:12876–12880
23. Steward O, Falk PM, Torre ER (1996) Ultrastructural basis for gene expression at the synapse: synapse-associated polyribosome complexes. J Neurocytol 25:717–734
24. Steward O, Wallace CS, Lyford GL, Worley PF (1998) Synaptic activation causes the mRNA for the IEG Arc to localize selectively near activated postsynaptic sites on dendrites. Neuron 21:741–751 doi:S0896-6273(00)80591-7 [pii]
25. Gardiol A, Racca C, Triller A (1999) Dendritic and postsynaptic protein synthetic machinery. J Neurosci 19:168–179
26. Schratt G (2009) microRNAs at the synapse. Nat Rev Neurosci 10:842–849. https://doi.org/10.1038/nrn2763
27. Lugli G, Larson J, Demars MP, Smalheiser NR (2012) Primary microRNA precursor transcripts are localized at post-synaptic densities in adult mouse forebrain. J Neurochem 123:459–466. https://doi.org/10.1111/j.1471-4159.2012.07921.x
28. Hanus C, Schuman EM (2013) Proteostasis in complex dendrites. Nat Rev Neurosci 14:638–648. https://doi.org/10.1038/nrn3546
29. Snyder EM, Nong Y, Almeida CG, Paul S, Moran T, Choi EY, Nairn AC, Salter MW, Lombroso PJ, Gouras GK, Greengard P (2005) Regulation of NMDA receptor trafficking by amyloid-beta. Nat Neurosci 8:1051–1058 doi:nn1503 [pii]
30. Kelly BL, Ferreira A (2006) Beta-amyloid-induced dynamin 1 degradation is mediated by N-methyl-D-aspartate receptors in hippocampal neurons. J Biol Chem 281:28079–28089 doi: M605081200 [pii]
31. Lacor PN, Buniel MC, Furlow PW, Clemente AS, Velasco PT, Wood M, Viola KL, Klein WL (2007) Abeta oligomer-induced aberrations in synapse composition, shape, and density provide a molecular basis for loss of connectivity in Alzheimer's disease. J Neurosci 27:796–807 doi:27/4/796 [pii]
32. Viola KL, Velasco PT, Klein WL (2008) Why Alzheimer's is a disease of memory: the attack on synapses by A beta oligomers (ADDLs). J Nutr Health Aging 12:51S–57S
33. Chiesa R, Piccardo P, Biasini E, Ghetti B, Harris DA (2008) Aggregated, wild-type prion protein causes neurological dysfunction and synaptic abnormalities. J Neurosci 28:13258–

13267. https://doi.org/10.1523/JNEUROSCI.3109-08.2008

34. Chang RY, Etheridge N, Nouwens AS, Dodd PR (2015) SWATH analysis of the synaptic proteome in Alzheimer's disease. Neurochem Int 87:1–12. https://doi.org/10.1016/j.neuint.2015.04.004
35. Williams C, Mehrian Shai R, Wu Y, Hsu YH, Sitzer T, Spann B, McCleary C, Mo Y, Miller CA (2009) Transcriptome analysis of synaptoneurosomes identifies neuroplasticity genes overexpressed in incipient Alzheimer's disease. PLoS One 4:e4936. https://doi.org/10.1371/journal.pone.0004936
36. McNeill E, Van Vactor D (2012) MicroRNAs shape the neuronal landscape. Neuron 75:363–379. https://doi.org/10.1016/j.neuron.2012.07.005
37. Saba R, Goodman CD, Huzarewich RL, Robertson C, Booth SA (2008) A miRNA signature of prion induced neurodegeneration. PLoS One 3:e3652. https://doi.org/10.1371/journal.pone.0003652
38. Majer A, Medina SJ, Niu Y, Abrenica B, Manguiat KJ, Frost KL, Philipson CS, Sorensen DL, Booth SA (2012) Early mechanisms of pathobiology are revealed by transcriptional temporal dynamics in hippocampal CA1 neurons of prion infected mice. PLoS Pathog 8:e1003002. https://doi.org/10.1371/journal.ppat.1003002
39. Montag J, Hitt R, Opitz L, Schulz-Schaeffer WJ, Hunsmann G, Motzkus D (2009) Upregulation of miRNA hsa-miR-342-3p in experimental and idiopathic prion disease. Mol Neurodegener 4:36. https://doi.org/10.1186/1750-1326-4-36
40. Karnati HK, Panigrahi MK, Gutti RK, Greig NH, Tamargo IA (2015) miRNAs: key players in neurodegenerative disorders and epilepsy. J Alzheimers Dis 48:563–580. https://doi.org/10.3233/JAD-150395
41. Lukiw WJ (2007) Micro-RNA speciation in fetal, adult and Alzheimer's disease hippocampus. Neuroreport 18:297–300. https://doi.org/10.1097/WNR.0b013e3280148e8b
42. Bicker S, Lackinger M, Weiss K, Schratt G (2014) MicroRNA-132, −134, and −138: a microRNA troika rules in neuronal dendrites. Cell Mol Life Sci 71:3987–4005. https://doi.org/10.1007/s00018-014-1671-7
43. Ye Y, Xu H, Su X, He X (2016) Role of MicroRNA in governing synaptic plasticity. Neural Plast 2016:4959523. https://doi.org/10.1155/2016/4959523
44. Boese AS, Saba R, Campbell K, Majer A, Medina S, Burton L, Booth TF, Chong P, Westmacott G, Dutta SM, Saba JA, Booth SA (2015) MicroRNA abundance is altered in synaptoneurosomes during prion disease. Mol Cell Neurosci 71:13–24 doi:S1044-7431(15)30041-5 [pii]
45. Verity MA (1972) Cation modulation of synaptosomal respiration. J Neurochem 19:1305–1317
46. Wanet A, Tacheny A, Arnould T, Renard P (2012) miR-212/132 expression and functions: within and beyond the neuronal compartment. Nucleic Acids Res 40:4742–4753. https://doi.org/10.1093/nar/gks151
47. Mestdagh P, Hartmann N, Baeriswyl L, Andreasen D, Bernard N, Chen C, Cheo D, D'Andrade P, DeMayo M, Dennis L, Derveaux S, Feng Y, Fulmer-Smentek S, Gerstmayer B, Gouffon J, Grimley C, Lader E, Lee KY, Luo S, Mouritzen P, Narayanan A, Patel S, Peiffer S, Ruberg S, Schroth G, Schuster D, Shaffer JM, Shelton EJ, Silveria S, Ulmanella U, Veeramachaneni V, Staedtler F, Peters T, Guettouche T, Wong L, Vandesompele J (2014) Evaluation of quantitative miRNA expression platforms in the microRNA quality control (miRQC) study. Nat Methods 11:809–815. https://doi.org/10.1038/nmeth.3014

Chapter 15

Use of Synaptoneurosome Samples to Study Development and Plasticity of Human Cortex

Caitlin R. Siu, Simon P. Beshara, Justin L. Balsor, Steven J. Mancini, and Kathryn M. Murphy

Abstract

Translation from animal models of visual system development and plasticity to human studies is difficult due to many obstacles in comparing results. Animal models provide important data about the neurobiological mechanisms that support cortical function and behavior, but identifying the same mechanisms in human cortex can be challenging. Many neurobiological techniques used in animal models cannot be used in humans, hindering our understanding of visual system development in the human brain. Western blotting using synaptoneurosomes prepared from postmortem human tissue, however, is a simple and reliable way to study synaptic protein expression in both animal and human brains. Synaptic proteins are linked with specific aspects of visual system development and plasticity necessary to establish functional neural circuitry. Our lab has implemented a filtered synaptoneurosome preparation using human cortical tissue to study the development of human visual cortex. This approach provides human researchers with much-needed information about neurobiological development and potential targets for treatments or therapies of visual disorders that have been previously tested in animal models. The protocol detailed in this chapter provides the step-by-step information needed for making synaptoneurosomes from human postmortem brain tissue, testing and equating antibodies for Western blotting using human brain tissue, and studying the expression of synaptic proteins. We provide strengths and limitations for using synaptoneurosomes to link structure and function in the human brain. This chapter highlights Western blotting of human synaptoneurosomes as an effective tool for studying the human brain and helping to narrow the translation gap.

Key words Human, Visual cortex, Postmortem, Western blotting, Synaptic proteins, Synaptoneurosome, Translation, Synaptic plasticity, Development

1 Introduction

A fundamental problem in neuroscience is the difficulty translating discoveries from animal models into clinical applications. One way to address this challenge is by using similar techniques across species; yet, few techniques are amenable to this goal. In animal models, especially rodents, there are many sophisticated neuroanatomical, neurophysiological, molecular, and imaging techniques that allow precise experimental control and generate rich datasets. Most of

Kathryn M. Murphy (ed.), *Synaptosomes*, Neuromethods, vol. 141,
https://doi.org/10.1007/978-1-4939-8739-9_15,

those techniques, however, are either too invasive or impractical to use in human studies. One technique, however, quantifying neural proteins using synaptosomes or synaptoneurosomes, has emerged as a reliable approach for studying neurodevelopment and plasticity in both animal and human brains.

1.1 Translation Challenges

In vivo single-cell neurophysiology provides unparalleled resolution for measuring the properties of individual neurons and their contributions to perception and behavior. The critical period for ocular dominance plasticity and orientation columns in primary visual cortex (V1) were discovered using single-cell physiology [1, 2]. In humans, in vivo physiology has been used to study mirror neurons [3], but is restricted to patients who are undergoing operations for pathological conditions, such as epilepsy, because it is too invasive for common use and subject to significant sampling bias. Pharmacological manipulation is another tool commonly used to study the role of receptors and cell types in the developing brain. For example, application of diazepam, a $GABA_A$ receptor agonist, revealed that $GABA_A$ receptors play a role in controlling the timing of the critical period [4, 5]. Pharmaceuticals have also been used in studies of human visual perception to link specific receptors with visual behavior [6] but research in humans is restricted by dose, delivery method, and eligible participants, all of which adds to the translation challenge.

The emergence of precise genetic manipulations with in vivo control of neurons is revealing how the function of neural circuits gives rise to behavior [e.g., 5, 7]. Furthermore, optogenetic manipulations allow for specific control of the membrane potential in a brain region, or single cell type or subcellular location. All of that can be done with millisecond precision leading to new insights into the role of different neural populations in V1 for visual perception [8–10]. This remarkable spatial and temporal specificity is accelerating our understanding of the neural basis of plasticity and function in animal models but it is not currently applicable for human studies. Thus, while genetic and optogenetic manipulations provide precise experimentation in animals, their findings are not readily translated to human cortical function.

Several imaging and neurophysiological recording techniques, like fMRI and EEG, have been designed and optimized for studying in vivo human cortical activity. Those tools have been used to make great strides linking structure [11, 12] with the function of human V1 [13–15]. There are translation challenges, however, because MRI and EEG techniques have poor spatial resolution, are indirect measures of the underlying neurobiology, and are not commonly used in animal studies [16]. Those reasons have made it hard to link findings from human brain imaging with the modern high-resolution imaging techniques (e.g., multiphoton) commonly used for animal studies.

Synaptic proteins link structure and function in animal and human models.

Studying synaptic proteins has been a successful approach for linking development and plasticity in V1 of animal models with function [17–23]. Synaptic proteins are highly conserved across species making them ideal candidates for studying the human brain and comparing its development with animal models. This approach has been applied to different cohorts, including cases with neuro-psychiatric disease [24–26], or cases representing different stages of the life span [27–29]. Synaptic proteins, however, are of low abundance and can be difficult to quantify in a whole-homogenate sample preparation and hence the need to use a preparation such as filtered synaptoneurosomes that enriches pre- and postsynaptic membranes threefold or more over homogenate samples [30, 31]. Finally, Western blotting using synaptoneurosome samples is a rapid, reliable, and relatively inexpensive technique to directly measure and compare synaptic protein expression across species.

2 Materials

2.1 Laboratory Equipment for Human Synaptoneurosome Preparation and Western Blotting

1. −80 °C Freezer for storing human tissue samples.
2. Benchtop homogenizer (Fast Prep-24 Tissue and Cell Homogenizer, MP Biomedicals) located inside a biosafety cabinet certified for using human tissue (*see* **Note 1**).
3. Scale.
4. Coarse filters (100 μm mesh pores) cut to fit holder and Swinnex Filter Holder (EMD Millipore, Billerica, MA).
5. 3 mL Syringes and 18 gauge needle tips (at least two per tissue sample), labeled with sample code.
6. Pipette (0.5–10 μL) and pipette tips (>25 μL, gel loading) (Diamed Lab Services Inc., Mississauga, ON, Canada).
7. Centrifugal filter units (fine filters) (Ultrafree-CL Durapore PVDF 5.0 μm pore, EMD Millipore, Billerica, MA).
8. Scalpel, surgical blades, forceps, scoopula, 2 mL centrifuge tubes, and weigh boats, all placed in cooler with dry ice for use with frozen tissue.
9. Centrifuge, cooled to 4 °C (e.g., Sorvall Legend Micro 21R).
10. Personal protective equipment for working with human tissue, including gloves, face masks, and lab coats.
11. 96-Well microplate.
12. iMark Microplate Absorbance Reader (Bio-Rad Laboratories, Hercules, CA, USA).
13. Incubator set to 45 °C.

14. Gel tank and gel holder (Thermo Fisher Scientific, Carlsbad, CA, USA) and power source (PowerPac 200).
15. Immobilon-FL PVDF Blot Membrane, wool transfer pads, and transfer filter paper, all cut to approximately 6x8cm rectangles (Millipore Corporation, Billerica, MA, USA).
16. Plastic square petri dishes for incubation of membranes in antibodies, PBS washing, antibody blocking, and antibody stripping solutions.
17. Aluminum foil to protect membranes from light.
18. Orbital shaker (Thermolyne Rotomix, Type 50,800, Marshall Scientific, Hampton, NH, USA).
19. Transfer cassette and small glass vial to roll and flatten transfer layers.
20. Large styrofoam box to house transfer tanks, filled with wet ice.
21. Ice pack to place in transfer containers.
22. Fluorescent flatbed scanner (Odyssey Infrared Scanner, Li-Cor, Lincoln, NE, USA).

2.2 Chemicals for Human Synaptoneurosome Prep, Tissue Homogenization, and Western Blotting

1. Dry and wet ice.
2. Homogenization buffer (*see* **Note 2**) chilled to 4 °C.
3. 10% Sodium dodecyl sulfate (SDS) at room temperature, 1% SDS boiling (100 °C).
4. Sample loading buffer (M260 NextGel® Sample Loading Buffer 4×, Amresco LLC, Solon, OH, USA).
5. Laemmli buffer (Cayman Chemical Company, Ann Arbor, MI, USA).
6. BCA Protein Assay Reagents A & B (Pierce, Thermo Fisher Scientific, Rockford, IL, USA).
7. Bovine serum albumin (BSA) protein standards, diluted for 4 concentrations: 2 mg/mL, 1 mg/mL, 0.5 mg/mL, 0.25 mg/mL (Pierce Bovine Serum Albumin Standard, 2 mg/mL, Bio-Rad Laboratories, Hercules, CA, USA).
8. Novex WedgeWell 4–20% Tris-Glycine Mini Gels, 15-well (Thermo Fisher Scientific, Waltham, MA, USA) (*see* **Note 3**).
9. Selected primary and secondary antibodies.
10. 10× Tris/glycine/SDS running buffer, 10× Tris/glycine transfer buffer (Bio-Rad Laboratories, Hercules, CA, USA).
11. Distilled water.
12. 99% Methanol.
13. ProSieve QuadColor Protein Marker, 4.6–300 kDa (Lonza Group, Basel, Switzerland).

14. PBS (PBS tablets) and PBS-Tween.
15. Blocking buffer (Li-Cor, Lincoln, NE, USA).
16. Blot Restore Membrane Rejuvenation Kit, 10× (EDM Millipore, Temecula, CA, USA).

3 Methods

To use human synaptoneurosome samples for effective and efficient quantification of synaptic proteins, we have implemented a three-stage approach to measure and equate samples before probing with antibodies. Here we outline our approach for (1) obtaining, (2) preparing, and (3) analyzing protein expression in human synaptoneurosomes for use in (4) Western blotting.

3.1 Obtaining High-Quality Human Tissue Samples

To ensure high-quality, reliable protein data from human synaptoneurosomes, it is important to obtain human cortical samples from a brain bank with rigorous protocols for tissue quality and handling. We obtained our human brain samples from the University of Maryland Brain and Tissue Bank, an organization that has a strong reputation for providing high-quality human brain samples for some of the most important human developmental brain studies over the last 20 years [32–34].

At the brain and tissue bank, the brain and section of brainstem are carefully removed, typically within 24 h postmortem, and cooled on wet ice for 30 min to increase ease of sectioning. The brain is then cut into right and left hemispheres and sectioned coronally in 1 cm intervals. The right hemisphere is preserved in 10% formalin, and the left is flash frozen by freezing in isopentane/dry ice at −30 to −40 °C and then stored in a freezer at −80 °C. Only the fresh frozen human tissue is suitable to be used for semiquantitative SDS-PAGE Western blotting, since formalin fixation causes cross-linking of proteins, and this prevents proteins from being separated on gels based on linear molecular weight. This protocol is only modified if serious injury or disease prevents typical sectioning. This standard protocol ensures that human postmortem tissue has undergone minimal structural or cellular disruptions so that specific areas of interest can be easily examined, sectioned, and prepared according to sulcal and gyral landmarks.

There are many considerations for how to obtaining a suitable tissue piece for synaptoneurosome preparation from the larger cortical sample. When the frozen tissue sample is removed from the −80 °C freezer it must be kept dry. We transfer the sample in a cooler with dry ice and then place it on a block of dry ice covered with aluminum foil, located inside a biosafety cabinet, to cut a small sample for the synaptoneurosome preparation. Begin by orienting the coronal section to identify sulcal and gyral landmarks before cutting a small sample from the area of interest. Use a new scalpel blade to cut/break off a small piece of frozen tissue (~50–100 mg). This piece must be kept frozen until homogenization to avoid denaturing of proteins. To do this, freeze all tools before touching the tissue piece, including the weigh boat and Lysing Matrix D tube to weigh and homogenize the sample, respectively.

3.1.1 Human Tissue Collection Protocol

1. All experiments with human tissues must be approved by the institute's guidelines set forth by the Standard Operating Protocols and Research Ethics Board.
2. Label all vials with respective sample ID for each human case and each type of tissue prepared (Homogenate, Synaptoneurosome, Supernatant), and freeze all surgical tools, Lysing Matrix D tubes, centrifuge tubes, and weigh boats in preparation for working with frozen tissue (*see* **Note 4**).
3. Prepare inside of biosafety cabinet with a slab of dry ice covered in aluminum foil, frozen tools, tubes, weigh boats kept on dry ice, and scale (*see* **Note 5**).
4. Tare scale with empty frozen weigh boat to zero.
5. Retrieve frozen coronal sections stored in sealed plastic bags from −80 °C freezer and keep on crushed dry ice until ready to use.
6. Place whole coronal section on aluminum foil on top of the dry ice block. Determine the orientation of the coronal section, and identify gyral and sulcal landmarks. Draw a sketched map of the tissue, including gyral and sulcal landmarks and shape, and identify the area needed for tissue collection (i.e., V1 primary visual cortex).
7. Identify the area of interest, and using surgical blade, scalpel, and forceps carefully cut a 1 mm × 1 mm × 1 mm tissue piece with frozen blade. Quickly place sectioned tissue piece in frozen weigh boat.
8. Weigh frozen sample in weigh boat on the tared scale and record weight. If the sample is too small (<50 mg) or too large (>100 mg), make note, and if necessary cut another small piece or section tissue further (*see* **Note 6**).
9. Using frozen forceps, place tissue piece into frozen Lysing Matrix D tube or frozen centrifuge tube, and keep on crushed dry ice (*see* **Note 7**).
10. Return original coronal section of human tissue to a sealed plastic bag on dry ice, to be returned to the −80 °C freezer.
11. Repeat collection steps for each coronal sample, using a different frozen, labeled tube for each region of interest and case.
12. If doing the synaptoneurosome preparation at a later time, keep tissues frozen by storing each tissue piece in −80 °C freezer (*see* **Note 8**).

3.2 Preparing Human Synaptoneurosome Samples

Human brain tissue is not perfused, and requires an extra filtration step before starting the synaptoneurosome preparation described in Chap. 1. Filtering the homogenized tissue sample through a coarse mesh pore filter (100 μm), fit into a Swinnex Filter Holder, captures large cellular debris like blood cells that remains in tissue samples that are not perfused. The subsequent steps in the synaptoneurosome preparation, including the subsequent use of 5 μm mesh pore fine filtering to filter non-synaptic components, are the same for human and animal brain samples.

3.2.1 Synaptoneurosome Preparation for Human Cortical Tissue

1. Prepare homogenization buffer according to recipe, enough for about 3 mL per sample and keep chilled on ice (for buffer recipe *see* **Note 2**).
2. Transfer 1 mm × 1 mm × 1 mm frozen tissue pieces into labeled Lysing Matrix D tubes.
3. Add the appropriate amount of homogenization buffer to each frozen sample (*see* **Note 6**).
4. Homogenize human tissue in the Fast Prep-24 benchtop homogenizer using Lysing Matrix D tubes for 40 s at 6 m/s. If the tissue is not visibly homogenized, then run Fast Prep-24 again.
5. Insert 100 μm coarse filter into Swinnex filter holder.
6. Add 10 μL of 10% SDS to each of the centrifuge tubes labeled "Homogenate" and "Supernatant."
7. Pipette 90 μL of homogenized tissue into centrifuge tubes labeled "Homogenate" and keep chilled on ice.
8. Take up the remaining amount of homogenized tissue in 3 mL syringe from the Lysing Matrix D tube, and replace needle tip with prepared Swinnex filter holder.
9. Using constant pressure, filter the tissue through the coarse filter into the centrifuge tube labeled "Supernatant" (*see* **Note 9**).
10. Once human homogenate tissue is filtered, follow the remainder of the synaptoneurosome preparation exactly as listed in Chap. 1 beginning with the fine filtration step using fine filter Ultrafree Spin tubes (Chap. 1, Sect. 3.2.3).
11. Continue to protein assay step or store synaptoneurosome samples in −80 °C freezer until ready to use (*see* **Note 10**).

3.3 Synaptoneurosome Protein Assay

The reliable detection of synaptic proteins by Western blotting is improved when a synaptosome or synaptoneurosome preparation is used because those protocols enrich synaptic proteins by three- to fivefold over homogenate preparations (Fig. 1). The measurement and equating of protein concentration are key steps for ensuring high quality and reliable. A loading control is typically used to aid

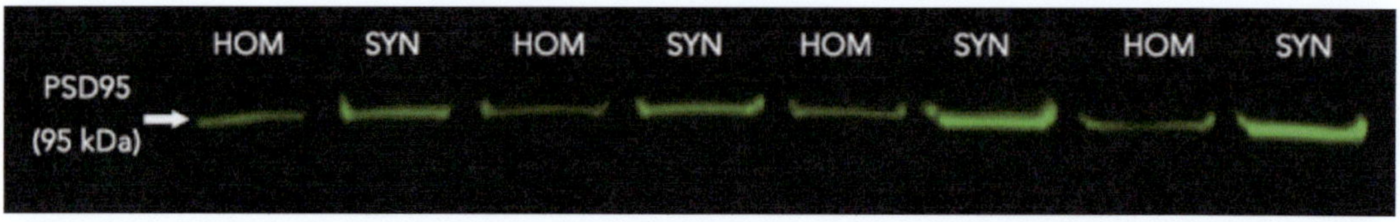

Fig. 1 An example blot showing the relative expression for a synaptic protein (PSD-95) in homogenate (HOM) versus synaptoneurosome (SYN) samples. Visual inspection shows that the synaptoneurosome bands are brighter than the homogenate bands. When the bands were quantified there was a three- to fivefold enrichment for synaptic proteins in synaptoneurosome samples prepared following the protocol described in this chapter

in these steps but the housekeeping proteins used as loading controls (e.g., as GAPDH or β-tubulin) are reduced about tenfold (Balsor, Chap. 1) by the synaptoneurosome preparation. The loss of housekeeping proteins occurs because the small volume of sealed pre- and postsynaptic membranes in a synaptoneurosome are dominated by synaptic proteins. As a result, we do not advise using one of those loading controls because the reduced amount of loading control proteins can have the unintended effect of inflating the apparent expression of synaptic proteins (*see* Sect. 3.5 for more details on loading controls in synaptoneurosomes). Instead, we developed a stringent three-stage protocol to measure the protein concentrations and equate the sample amounts loaded into the gel.

First, we use a bicinchoninic acid (BCA) assay to measure and compare the protein concentrations in the synaptoneurosomes against known albumin protein standards. The colorimetric change is quantified for each sample and for the protein standards. Second, the absorbance results for the protein standards are plotted relative to their known concentrations, and fit with a linear function. The fit for the protein standards must be $R^2 > 0.99$, and if that is not achieved then the assay is rerun (*see* **Note 11**). The absorbance values for the synaptoneurosome samples are also measured and the function for the linear fit to the standards is used to the determine the amount of Laemmli buffer and sample buffer needed to equate the samples to have a concentration of 1 μg/μL. Third, to achieve loading of equivalent sample volumes into the gel we use a high-quality pipette, do monthly in-lab calibrations, and have annual servicing of the pipette (*see* **Note 12**). These steps can take the place of a traditional loading control for Western blotting with human synaptoneurosome tissue.

3.3.1 Protein Assay Protocol

1. Collect diluted BSA protein standards (2.0, 1.0, 0.5, 0.25 μg/mL) and synaptoneurosome protein samples out of the freezer to thaw (*see* **Notes 10** and **13**).

2. Organize an assay map so that each sample, including the protein standards and synaptoneurosomes, is loaded three separate times. Also, include a full column of blank wells for control (*see* **Note 14**).
3. Prepare BCA assay solution for about 300 μL/sample loaded + blank lanes (*see* **Note 15**).
4. Vortex and load 3 μL of each protein standard and synaptoneurosome sample into well-plate according to the assay map organization.
5. Add 300 μL of BCA solution to each of the wells that have samples and standards, and also to the designated "blank" wells (*see* **Note 16**).
6. Incubate plate for 45 min in an incubator set to 45 °C (*see* **Note 17**).
7. After incubation, insert plate into Microplate Absorbance Reader and obtain absorbance measurement values for each well.
8. Record absorbance values for protein standards, calculate the average of the three runs for each standard, and calculate the correlation of observed absorbance values against expected.
9. Obtain the correlation (R^2) value for a standard curve fit to the data points. The R^2 value of the standard curve must be equal or greater than 0.99 to proceed with quantification of assay results. If the R^2 value is not 0.99 then the plate with protein standards must be run again (*see* **Note 11**).
10. Calculate the average absorbance of each synaptoneurosome sample loaded into the three wells, and using the slope and y-intercept of the protein standards to calculate the protein concentration with the following equation: [(average sample absorbance/slope) + y-intercept] (*see* **Note 18**).
11. Calculate and add the appropriate amounts of 4× sample buffer and Laemmli buffer to each sample to equal 1 μg/μL protein concentration (*see* **Note 19**).
12. Continue to Western blotting protocol, or store prepared samples in −20 °C freezer (*see* **Note 8**).

3.4 Postmortem Interval Analysis

The postmortem interval for collecting of human brain tissue is typically counted in hours and that delay can affect the quality of the tissue, including synaptic protein integrity, cytoskeletal levels, and receptor binding stability [29, 35, 36]. To test if any variations in synaptic protein expression among samples are caused by the postmortem interval, the total protein concentration for each sample can be tested against postmortem interval in hours. In our studies, we have not found an effect of postmortem interval on expression of individual synaptic protein or total protein [27, 37].

3.5 Western Blotting Using Human Synaptoneurosomes

Western blotting provides robust quantitative data on protein expression from synaptoneurosome samples when care is taken to equate the amount of protein loaded, generate an appropriate control sample, and test the antibodies to be used.

Equating the amount of sample protein is typically done by probing the blot for a ubiquitous cellular housekeeping or cytoskeletal protein and then using that to normalize expression of the synaptic protein. This is called the loading control and it can be used to equate the amount of sample loaded into each well, and to ensure good transfer to the membrane. The idea is that if the same amount of sample is loaded into each well then expression of the loading control will not vary among samples or lanes, because the housekeeping and cytoskeletal proteins are assumed to be stable under various conditions as well as ages. Loading controls, however, often display high variability [38]. Furthermore, those proteins are significantly diminished in a synaptoneurosome preparation because of the small volume of the pre- and postsynaptic membranes and adjacent synaptic compartments. In addition, the size and shape of dendritic spines and synapses change across development [39, 40] and aging, [41] which can affect the amount of those loading control proteins [42]. These issues led us to develop a stringent protocol to measure and equate protein concentrations before running gels rather than using an unreliable loading control after running the gel.

It is necessary to run a common "control" sample on each gel so that samples can be normalized to the control and compared across gels. For our studies of human visual cortex, the control sample is made by mixing a small amount from each human synaptoneurosome sample and then running that control sample on every gel. Expression of protein from each sample on the blot is divided by the expression of the control sample to allow for comparison samples run on different blots.

Finally, the expression of a protein on a blot is visualized using an antibody to visualize the band and then quantify it by densitometry to infer the amount of protein expression. The specificity of an antibody is typically validated by using it to probe for protein expression in a knockout mouse, but for human samples antibody validation is done by other means, including testing the antibodies on a multispecies blot to ensure consistency among species for molecular weight and expression (*see* **Note 20**). In addition, many of the antibodies used in our studies of human visual cortex have been verified in the Human Protein Atlas ([43] www.proteinatlas.org). Accurate protein quantification also depends on titrating the concentration of primary and secondary antibody to identify the linear range where an increase in antibody concentration causes a linear increase in the optical density of the band (*see* **Note 21**). Finally, we use secondary antibodies that fluoresce in the near-infrared range because there is less background and a sensitive scanner with a high dynamic range to accurately image even small difference in protein amounts (Odyssey Infrared Scanner, Li-Cor, Lincoln, NE, USA).

3.5.1 Western Blotting Protocol for Human Synaptoneurosomes

1. Organize gel maps, to use as a guide for loading each gel (i.e., Protein ladder, Control Sample, Sample 1, Sample 2 …).
2. Prepare transfer, running, and blocking buffers for the desired number of blots.
3. Load gels into tank; fill with running buffer, inside and outside of the gels; and open the wells by aspirating some running buffer into each well, ensuring that there are no bubbles.
4. Pipette 1.5 μL of protein ladder into the first well.
5. Pipette 25–30 μg of control sample into the second well.
6. Pipette the same amount (titrated between 20 and 30 μg) of each sample into the following wells, using a gel-loading-tipped pipette with a slow-speed setting, according to the gel map. The last lane is typically not loaded to ensure equal transfer among all lanes.
7. Cover the gel box with the lid and plug leads into the power source set to 150 V for about 1 h, watching to ensure that the bands do not run off the gel; add more time if the bands have not run far enough or if you are interested in heavier proteins.
8. Cut PVDF (Immobilon-FL) membranes into rectangular squares large enough to fit a gel (typically about 6 × 8 cm), soak in methanol for 10 s, and put in transfer buffer for 10 min under agitation.
9. Soak wool pads, transfer paper in transfer buffer, and prepare transfer sandwiches for gels in the following order: wool pad, transfer paper, gel, PVDF membrane, transfer paper, and wool pad. This step helps eliminate bubbles between layers (*see* **Note 22**).
10. Close the transfer sandwich in the cassette and fit into the transfer tank according to tank specifications. Fill with transfer buffer and an ice pack, and surround completely by ice, as the tank and contents will become hot while it is running.
11. Connect the transfer box to a power source and start the transfer at 40 V for 2 h.
12. Once transfer is over, remove the blot membrane from the cassette, and place in methanol for 10 s, and then PBS for 10 min to prepare for the blocking process.

13. Add 20 mL of blocking buffer and place blot on orbital shaker for 1 h. Take primary antibody out of −20 °C freezer to thaw at this time. After 1 h, pour the blocking buffer back into the original container as it can be reused for approximately 1 month.
14. Add ~20 mL of primary antibody diluted to a working concentration (*see* **Note 21**), place on orbital shaker plate in the 4 °C fridge, and incubate for 14–18 h or overnight.
15. Take off primary antibody, rinse, and wash blots with PBS-Tween 3 × 10 min, on an orbital shaker. Primary antibody can be reused typically about 4–6 times.
16. Rinse with PBS, and add 20 mL of secondary antibody diluted to a working concentration (*see* **Note 21**). Cover petri dish with aluminum foil to shield from the light (*see* **Note 23**), and place on an orbital shaker for 1 h at room temperature.
17. Take off secondary antibody; it can be reused typically about 6–10 times. Rinse and wash blots with PBS-Tween 3 × 10 min on an orbital shaker, under aluminum foil to keep blots in the dark.
18. Rinse with PBS and then bring covered blots to scanner.
19. Scan blots according to specifications of the scanner, and using the corresponding wavelength of laser appropriate for the secondary antibody. Identify the molecular weight of the band(s) against the ladder loaded in lane 1, making sure that this band represents the protein of interest. Quantify the integrated intensity value of each band, divided by the width of the lane. Normalize the expression value of each sample to the value of the control band on the same blot.
20. Blots can be stripped using a membrane blot, restore stripping solution, and re-probe with additional antibodies typically 4–6 times without any decrease in the quality of the blot.

4 Notes

1. Biosafety cabinet must be certified by the establishments' biosafety office for use with human brain tissue; the type of cabinet may vary depending on the kind of experiment and must be part of the standard operating procedures for using human tissue in the lab.
2. Homogenization buffer recipe, makes 50 mL (see more details in Chap. 1):
   (a) 5 mL, 10 mM HEPES
   (b) 5 mL, 2 mM EDTA
   (c) 5 mL, 2 mM EGTA

(d) 125 μL, 0.5 mM DTT
(e) 500 μL, 10 mg/L Leupeptin
(f) 500 μL, 50 mg/L Soybean trypsin inhibitor
(g) 50 μL, 100 nM Microcystin 24
(h) 33.825 mL Water

3. Novex WedgeWell 4–20% Tris-Glycine Mini Gels can easily accommodate large volumes of tissue, which may be necessary when loading synaptoneurosome tissue. A single gel contains 15 wells and each well can hold up to 35 μL of sample, which minimizes the amount (and hence cost) of chemical reagents and saves experimenter time.
4. Labeling the vials is an important part of planning the experiment. Before any tissue is disturbed from the −80 freezer, organize all tools and equipment, so tissue is not lost due to unnecessary thaw. Before beginning tissue collection, ensure that all experimenters are prepared with lab notebooks to record notes and that checkpoints are in place in case it is necessary to stop the procedure and start again at another time. Tissue collection (Sect. 3.1) can be done on a separate day from tissue preparation (Sect. 3.2) and protein assay (Sect. 3.3), and Western blotting (Sect. 3.5) can be done later. We recommend, however, that tissue preparation (Sect. 3.2) and protein assay (Sect. 3.3) be done on the same day to avoid an additional freeze-thaw cycle that might damage the proteins.
5. The tools used to dissect a piece of human tissue from the larger block should be cooled using dry ice to avoid the scalpel blade or forceps from thawing the tissue when cutting or picking up the small tissue samples. Keeping the tissue frozen preserves the integrity of the proteins to be analyzed by Western blotting.
6. It is important to record the weight of the sample to determine how much homogenization buffer to add. The amount of homogenization buffer added to the sample depends on the weight of the excised piece, for example, approximately 1 mL buffer for 50 mg, 1.25 mL buffer for 75 mg, and 1.5 mL buffer for 100 mg.
7. If continuing the homogenization step, then frozen tissue pieces can be put into frozen Lysing Matrix D tubes. Otherwise, put samples into 2 mL centrifuge tubes and store in the −80 °C freezer.
8. Human brain tissue samples must be stored in a −80 °C freezer before they are prepared with sample, loading, and Laemmli buffers. Once the samples have been prepared with the correct amounts of buffers, they may be stored in the −20 °C freezer for up to 1 month, and in the −80 °C freezer for longer periods.

9. Filtering non-perfused homogenate tissue may require replacing the coarse filter with a new one as a filter may become clogged with debris. As the filter becomes clogged the force needed to push the syringe plunger will increase indicating that it is time to change the filter. The filter may have to be changed multiple times even for one sample.
10. It is best to limit the number of freeze/thaw cycles for each synaptoneurosome sample. Therefore, if possible it is ideal to plan to homogenize, prepare the tissue, and run on the same day.
11. Calculate a standard curve by plotting the average absorbance of each protein standard on a scatter plot, with expected protein concentration (0.25, 0.5, 1.0, 2.0) on the *x*-axis and observed concentration on the *y*-axis. Fit a linear trend line to the points and use this to identify the slope, *y*-axis, and R^2 of the fit. The correlation value R^2 must be at least 0.99 to be used as a protein standard for the assay. If R^2 is not >0.99, the entire assay must be run again. This value can be difficult to achieve and is usually dependent on the experience of the experimenter in pipetting and titration of samples.
12. To ensure that the experimenter can dispense a consistent volume from the pipette, it is useful to practice pipetting and calibrating the experimenter. This is done by pipetting small volumes of dH2O into a weight-boat sitting on an analytic balance, then recording the weight of each amount pipetted, and analyzing the variability to ensure that it is within the error specifications from the pipette manufacturer. The pipette should also be sent out annually for servicing that includes certification. Our lab has tried various pipettes and concluded that the Picus Electronic Pipette (Sartorius, Goettingen, Germany) is optimal for this application.
13. Albumin protein standards are bovine serum albumin (BSA) protein solutions that provide reference standards for use in most protein assays. This standard is typically supplied with a concentration of 2 mg/mL (Pierce Bovine Serum Albumin Standard Ampules, Thermo Fisher Scientific) and must be diluted with homogenization buffer to make the following concentrations: 1.0, 0.5, and 0.25 mg/mL to use in the assay. These standards can be stored in −20 °C freezer for future use.
14. Assay map must show a diagram of a 96-well plate with each sample, including protein standards, labeled in three different wells, and must also contain one blank column of wells.
15. The BCA solution provides selective colorimetric detection that produces an intense purple-colored reaction dependent on the amount of protein concentration in the sample; for

more details refer to https://www.thermofisher.com/order/catalog/product/23225.

16. Blank wells loaded with BCA solution only and no protein will provide a lane of controls that establish a baseline for zero protein concentration. If the blank wells show some amount of protein after the assay is complete, the BCA solution must be remade, and the assay must be completed again.
17. The length of time and temperature of incubation determine the rate of color development of the BCA solution with the proteins in the sample. Ensure that the spectrophotometer in the microplate reader is set to a wavelength of 562 nm, or the appropriate wavelength for your solution.
18. Once the average protein concentration is obtained for each sample, calculate the amount of sample buffers needed to dilute the sample to 1 μg/μL (*see* **Note 19**). Calculate how many runs (of 25 μg) are possible for each human synaptoneurosome sample to be used in Western blotting.
19. Sample loading buffer (M260 NextGel® Sample Loading Buffer 4×, Amresco LLC, Solon, OH, USA) and Laemmli buffer (Cayman Chemical Company, Ann Arbor, MI, USA) are added sample buffers for protein denaturation, as well as adding a dye and density to the sample to make them easier to run on SDS-PAGE. Sample loading buffer must be diluted 1 part buffer:3 parts protein sample. After that, enough Laemmli buffer is added to dilute the sample to 1 μg/μL protein concentration. Do not dilute the samples to less than 1 μg/μL.
20. One way to verify antibodies used on human synaptoneurosome tissue is to probe on a multispecies blot (*see* example blot, Fig. 2). To do this, prepare a single blot with at least three different species run (twice each). For example, we run a multispecies blot with control samples of cortical tissue from rat, cat, monkey, and human. This blot is run with multiple different antibodies at different concentrations to test that a band(s) of the same molecular weights are found across species. If the band is found in the human tissue at the same molecular weight and expression as the other species, it is considered a reliable antibody to use to quantify accurate data.
21. Different primary and secondary antibodies can work within their linear dynamic range at many different concentrations depending on the type of tissue, protein concentration, age of antibody, or even amount of freeze/thaw cycles. It is important to test the dilution factor of each antibody before use by performing a titration starting with the suggested working dilution. It is possible and helpful to mix primary antibodies

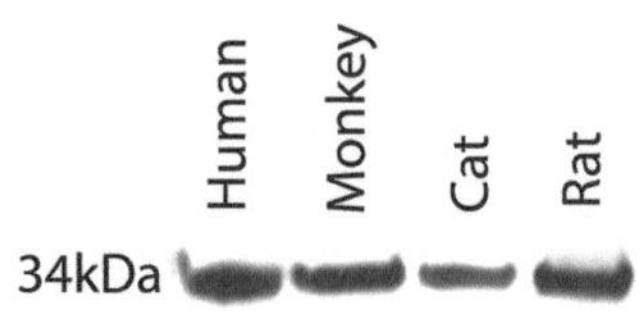

Fig. 2 An example image of a multispecies test blot using synaptoneurosome samples (25 μg) from human, monkey, rat, and cat V1 that were loaded into adjacent wells in a gel. The blot was probed with an antibody for the presynaptic protein synaptophysin and the image shows the similar appearance of the bands among the four species

that probe for proteins of differing molecular weights or correspond to different secondary antibodies to create a "cocktail." This is most useful when testing a new antibody to mix with a known antibody that shows dependable antigenicity so that any error in the new antibody will not be mistaken for sample error, as the known antibody will show consistent expression.

22. When making the transfer sandwiches, use a small glass vial to pour small amounts of transfer buffer between the layers; this ensures that there are no bubbles. Between each layer, except for the gel, roll the glass tube horizontally across the layer with equal pressure to facilitate a smooth transfer surface.

23. Keep secondary antibody in the dark at all times. The aluminum foil is used as a cover to protect the blots from the light of the room; an option is to also turn the lights off or down in the lab to further protect the fluorescence of the secondary antibody from light exposure.

References

1. Wiesel TN, Hubel DH (1963) Single-cell responses in striate cortex of kittens deprived of vision in one eye. J Neurophysiol 26:1003–1017
2. Hubel DH, Wiesel TN (1968) Receptive fields and functional architecture of monkey striate cortex. J Physiol 195:215–243
3. Mukamel R, Ekstrom AD, Kaplan J et al (2010) Single-neuron responses in humans during execution and observation of actions. Curr Biol 20:750–756. https://doi.org/10.1016/j.cub.2010.02.045
4. Fagiolini M, Hensch TK (2000) Inhibitory threshold for critical-period activation in primary visual cortex. Nature 404:183–186. https://doi.org/10.1038/35004582
5. Hensch TK, Fagiolini M (2005) Excitatory-inhibitory balance and critical period plasticity in developing visual cortex. Prog Brain Res 147:115–124. https://doi.org/10.1016/S0079-6123(04)47009-5
6. Meuwese JDI, van Loon AM, Scholte HS et al (2013) NMDA receptor antagonist ketamine impairs feature integration in visual perception. PLoS One 8:e79326. https://doi.org/10.1371/journal.pone.0079326
7. Levelt CN, Hübener M (2012) Critical-period plasticity in the visual cortex. Annu Rev Neurosci 35:309–330. https://doi.org/10.1146/annurev-neuro-061010-113813
8. Atallah BV, Bruns W, Carandini M, Scanziani M (2012) Parvalbumin-expressing interneu-

rons linearly transform cortical responses to visual stimuli. Neuron 73:159–170. https://doi.org/10.1016/j.neuron.2011.12.013

9. Wilson NR, Runyan CA, Wang FL, Sur M (2012) Division and subtraction by distinct cortical inhibitory networks in vivo. Nat Publ Group 488:343–348. https://doi.org/10.1038/nature11347
10. Lee S-H, Kwan AC, Zhang S et al (2012) Activation of specific interneurons improves V1 feature selectivity and visual perception. Nat Publ Group 488:379–383. https://doi.org/10.1038/nature11312
11. Engel SA (1994) fMRI of human visual-cortex. Nature 369(6481):525
12. Engel S, Zhang XM, Wandell B (1997) Colour tuning in human visual cortex measured with functional magnetic resonance imaging. Nature 388:68–71. https://doi.org/10.1038/40398
13. Polonsky A, Blake R, Braun T, Heeger DJ (2000) Neuronal activity in human primary visual cortex correlates with perception during binocular rivalry. Nat Neurosci 3:1153–1159. https://doi.org/10.1038/80676
14. Haynes J-D, Rees G (2005) Predicting the orientation of invisible stimuli from activity in human primary visual cortex. Nat Neurosci 8:686–691. https://doi.org/10.1038/nn1445
15. Lunghi C, Berchicci M, Morrone MC, Di Russo F (2015) Short-term monocular deprivation alters early components of visual evoked potentials. J Physiol 593:4361–4372. https://doi.org/10.1113/JP270950
16. Mirsattari SM, Bihari F, Leung LS et al (2005) Physiological monitoring of small animals during magnetic resonance imaging. J Neurosci Methods 144:207–213. https://doi.org/10.1016/j.jneumeth.2004.11.019
17. Fagiolini M, Katagiri H, Miyamoto H et al (2003) Separable features of visual cortical plasticity revealed by N-methyl-D-aspartate receptor 2A signaling. Proc Natl Acad Sci 100:2854–2859. https://doi.org/10.1073/pnas.0536089100
18. Ramoa AS, Mower AF, Liao D, Jafri S (2001) Suppression of cortical NMDA receptor function prevents development of orientation selectivity in the primary visual cortex. J Neurosci 21:4299–4309
19. Rivadulla C, Sharma J, Sur M (2001) Specific roles of NMDA and AMPA receptors in direction-selective and spatial phase-selective responses in visual cortex. J Neurosci 21:1710–1719
20. Hensch TK, Fagiolini M, Mataga N et al (1998) Local GABA circuit control of experience-dependent plasticity in developing visual cortex. Science 282:1504–1508. https://doi.org/10.1126/science.282.5393.1504
21. Philpot BD, Sekhar AK, Shouval HZ, Bear MF (2001) Visual experience and deprivation bidirectionally modify the composition and function of NMDA receptors in visual cortex. Neuron 29:157–169
22. Larsen RS, Smith IT, Miriyala J et al (2014) Synapse-specific control of experience-dependent plasticity by presynaptic NMDA receptors. Neuron 83:879–893. https://doi.org/10.1016/j.neuron.2014.07.039
23. Gainey MA, Hurvitz-Wolff JR, Lambo ME, Turrigiano GG (2009) Synaptic scaling requires the GluR2 subunit of the AMPA receptor. J Neurosci 29:6479–6489. https://doi.org/10.1523/JNEUROSCI.3753-08.2009
24. Curley AA, Arion D, Volk DW et al (2011) Cortical deficits of glutamic acid decarboxylase 67 expression in schizophrenia: clinical, protein, and cell type-specific features. Am J Psychiatry 168:921–929. https://doi.org/10.1176/appi.ajp.2011.11010052
25. Barksdale KA, Lahti AC, Roberts RC (2014) Synaptic proteins in the post-mortem anterior cingulate cortex in schizophrenia: relationship to treatment and treatment response. Neuropsychopharmacology 39:2095–2103. https://doi.org/10.1038/npp.2014.57
26. Glantz LA, Gilmore JH, Overstreet DH et al (2010) Pro-apoptotic Par-4 and dopamine D2 receptor in temporal cortex in schizophrenia, bipolar disorder and major depression. Schizophr Res 118:292–299. https://doi.org/10.1016/j.schres.2009.12.027
27. Pinto JGA, Hornby KR, Jones DG, Murphy KM (2010) Developmental changes in GABAergic mechanisms in human visual cortex across the lifespan. Front Cell Neurosci 4:16. https://doi.org/10.3389/fncel.2010.00016
28. Murphy KM, Beston BR, Boley PM, Jones DG (2005) Development of human visual cortex: a balance between excitatory and inhibitory plasticity mechanisms. Dev Psychobiol 46:209–221. https://doi.org/10.1002/dev.20053
29. Glantz LA, Gilmore JH, Hamer RM et al (2007) Synaptophysin and postsynaptic density protein 95 in the human prefrontal cortex from mid-gestation into early adulthood. Neuroscience 149:582–591. https://doi.org/10.1016/j.neuroscience.2007.06.036
30. Quinlan EM, Olstein DH, Bear MF (1999) Bidirectional, experience-dependent regulation of N-methyl-D-aspartate receptor subunit composition in the rat visual cortex during

postnatal development. Proc Natl Acad Sci 96:12876–12880

31. Murphy KM, Balsor J, Beshara S et al (2014) A high-throughput semi-automated preparation for filtered synaptoneurosomes. J Neurosci Methods 235:35–40. https://doi.org/10.1016/j.jneumeth.2014.05.036
32. Huttenlocher PR, Dabholkar AS (1997) Regional differences in synaptogenesis in human cerebral cortex. J Comp Neurol 387:167–178
33. Lu T, Pan Y, Kao SY et al (2004) Gene regulation and DNA damage in the ageing human brain. Nature 429:883–891. https://doi.org/10.1038/nature02661
34. Ernst A, Alkass K, Bernard S et al (2014) Neurogenesis in the striatum of the adult human brain. Cell 156:1072–1083. https://doi.org/10.1016/j.cell.2014.01.044
35. Blair JA, Wang C, Hernandez D et al (2016) Individual case analysis of post-mortem interval time on brain tissue preservation. PLoS One 11:e0151615. https://doi.org/10.1371/journal.pone.0151615
36. Kontur PJ, al-Tikriti M, Innis RB, Roth RH (1994) Post-mortem stability of monoamines, their metabolites, and receptor binding in rat brain regions. J Neurochem 62:282–290
37. Siu CR, Beshara SP, Jones DG, Murphy KM (2017) Development of glutamatergic proteins in human visual cortex across the lifespan. J Neurosci 37:6031. https://doi.org/10.1523/JNEUROSCI.2304-16.2017
38. Lee H-G, Jo J, Hong H-H et al (2016) State-of-the-art housekeeping proteins for quantitative western blotting: revisiting the first draft of the human proteome. Proteomics 16:1863–1867. https://doi.org/10.1002/pmic.201500344
39. Michel AE, Garey LJ (1984) The development of dendritic spines in the human visual cortex. Hum Neurobiol 3:223–227
40. Petanjek Z, Judaš M, Šimic G et al (2011) Extraordinary neoteny of synaptic spines in the human prefrontal cortex. Proc Natl Acad Sci U S A 108:13281–13286. https://doi.org/10.1073/pnas.1105108108
41. Dickstein DL, Weaver CM, Luebke JI, Hof PR (2013) Dendritic spine changes associated with normal aging. Neuroscience 251:21–32. https://doi.org/10.1016/j.neuroscience.2012.09.077
42. Pinto JGA, Jones DG, Murphy KM (2013) Comparing development of synaptic proteins in rat visual, somatosensory, and frontal cortex. Front Neural Circuits 7:97. https://doi.org/10.3389/fncir.2013.00097
43. Uhlen M (2007) Mapping the human proteome using antibodies. Mol Cell Proteomics 6:1455–1456

Chapter 16

The Use of Synaptosomes in Studying Autism Spectrum Disorder and Other Neurodevelopmental Disorders

Nadeem Murtaza, Vickie Kwan, Biren Dave, and Karun K. Singh

Abstract

Autism spectrum disorder (ASD) is one of the most common neurodevelopmental disorders in children. Atypical neural connectivity and signalling are major contributors behind the pathology of ASD and genome sequencing studies of individuals affected by ASD have revealed the association of synaptic proteins to the disorder. By studying synaptosomes isolated from human or mouse tissue, it is possible to study the role of proteins on synaptic function and localization. We have shown that Dixdc1 and Mark1 localization in the synapse is important for its function in synapse formation through actin polymerization. However, further study of *Dixdc1* and other ASD-associated genes requires the generation of newer mouse models that recapitulate the genetic disruptions seen in humans or by studying the genes in a human context and more broad methods of studying the synaptic protein networks involved. The CRISPR/Cas9 genome editing system offers an efficient and easier method of creating ASD mouse models or human IPSC-derived neuron models to study the effect on the synaptic compartment. The generation of brain organoids further advances the capability of studying synaptic formation and function in a human context at an early developmental time point. Lastly, the use of BioID allows tagging of even transient protein interactions that can be used to identify disruptions in synaptic protein networks through proteomic analysis. Overall, the use of synaptosomes is widely important in studying the mechanisms involved in the development of ASD, which is linked with atypical synaptic formation and function.

Key words ASD, Neurodevelopmental disorders, Synapses, Synaptosomes, Brain organoids, CRISPR/Cas9, BirA

1 Introduction

Neurodevelopmental disorders are one of the most common developmental disabilities that affect children; these disorders include attention-deficit hyperactivity disorder (ADHD), obsessive compulsive disorder (OCD), intellectual disabilities (ID), and autism spectrum disorder (ASD) [1]. In the most recent study conducted by the Center for Disease Control and Prevention (CDC), 1 in 68 children were diagnosed with ASD, reflecting the high prevalence in younger individuals [2]. ASD encompasses a broad range of more complex neurodevelopmental disorders,

Kathryn M. Murphy (ed.), *Synaptosomes*, Neuromethods, vol. 141,
https://doi.org/10.1007/978-1-4939-8739-9_16,

characterized by two core symptoms: impaired social communication and reciprocal interaction and repetitive behaviors and restricted interests. Many affected children may also display many secondary symptoms including epilepsy, anxiety, hyperactivity, atypical sensory responsiveness, irritability, aggression, and sleep disruptions [3]. Although treatments do exist for children affected by ASD, they are centered towards treating the secondary symptoms and do not treat the core symptoms themselves [3]. For example, risperidone is an antipsychotic drug that has been repurposed to treat hyperactivity and irritability [4]. Alternatively, behavioral therapies have had high success rates in treating ASD phenotypes in children; however these therapies are limited in accessibility, and therefore cannot be administered to every child affected by ASD. The lack of robust and easy-to-access treatments is directly related to the lack of knowledge in the pathophysiology of ASD and requires further research into the biological pathology.

A large component of idiopathic ASD is attributed to genetics and has led to many large-scale sequencing studies to identify ASD-associated genes [5, 6]. Multiple genome sequencing studies have identified genetic variants, including loss-of-function (LoF) and de novo mutations in genes encoding synaptic proteins [7–10]. Two major hypotheses behind the pathophysiology of ASD are atypical connectivity and imbalances in excitatory and inhibitory signals within the brain and the majority of these connectivity and signaling imbalances can be attributed to disruptions in synaptic proteins [11]. Many studies of ASD-associated genes have focused on studying the effect on synapse formation and function using mouse or human neuronal models of synaptic proteins including cell adhesions proteins (NLGN, NRXN) [12, 13], ion channels (CACNA1C, SCN2A) [14, 15], actin cytoskeleton regulators (DIXDC1, SHANK2/3, SYNGAP1) [16–18], and translational proteins (JAKMIP1, FMRP, CYFIP1, UBE3A) [19–23]. The cellular deficits in certain genetic models of ASD involving synaptic proteins have also been pharmacologically rescued, suggesting that protein networks related to synaptic function and formation are highly relevant to the development of ASD [24–28]. Further study can be used to identify novel therapies for ASD; the study of synaptic proteins will greatly advance the understanding of pathways involved in the atypical connectivity and signalling associated with ASD.

Here we describe how our lab is studying synaptic protein function in regard to ASD pathology using synaptosomes and new techniques that will aid in studying neurodevelopment in a human context.

Synaptosomes are an important and significant way to study changes in synaptic protein levels, activity, and localization within the neuronal compartments. Synaptosomes can be isolated from

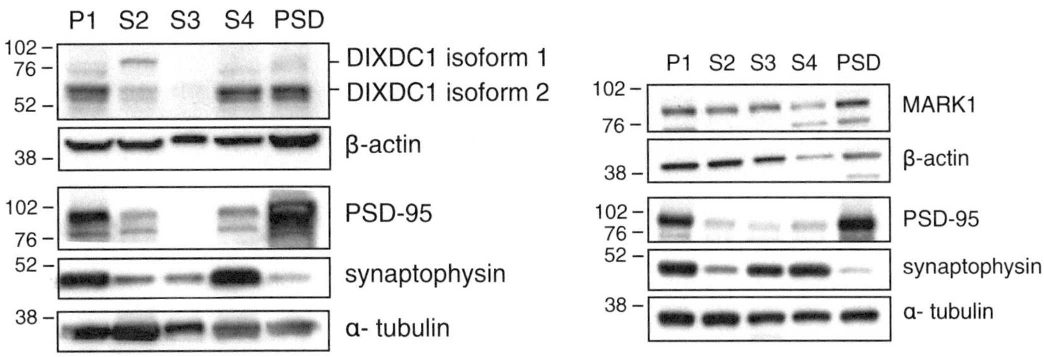

Fig. 1 *Left:* Postsynaptic densities (PSD) fractionated from P28 mouse brains were probed for DIXDC1, actin, PSD95, synaptophysin, and tubulin. *Right:* Postsynaptic densities (PSDs) fractionated from 1-month-old CD-1 mouse brains probed for MARK1, actin, PSD95, synaptophysin, and tubulin. P1, nuclear fraction; S2, crude cytoplasm; S3, crude synaptic vesicle; S4, crude synaptosomal membrane fraction; PSD, postsynaptic density

neuronal cultures or brain tissue and the fractionation process results in multiple fractions, including a nuclear fraction, crude synaptosome, synaptic vesicles, presynaptic fraction, and postsynaptic density (PSD). The separation of these fractions can be used to study both differences in protein localization and expression levels. Other neurological disease models, such as schizophrenia and Alzheimer's disease, have used synaptosomes for multiple purposes, including the study of protein levels and RNA expression specifically in synapses [29–31], protein activity assays [32], and as models for drug screening [33]. Many studies of ASD models have looked at synaptosomes to study changes in protein levels in the synapse or to show enrichment of ASD-associated proteins within the synaptic compartment. Our lab has recently published about the enriched localization of Dixdc1 and Mark1 in the postsynaptic density of dendritic spines in postnatal day 30 mouse brains (Fig. 1) [16]. The localization of Dixdc1 and Mark1 in the PSD is important for actin polymerization through Rac1 and Arp2/3 complex in dendritic spines. Future experiments regarding Dixdc1 will involve identification of synaptic pathways altered by disruption of *Dixdc1* through proteomic analysis. Altogether, synaptosome isolation provides a robust method of studying protein expression, activity, and interaction within the synapse.

2 PSD Preparations via Differential Centrifugation

2.1 Crude Synaptosome Preparation

1. Dissect brains on ice and collect sufficient whole cortical and hippocampal tissues using forceps:
   (a) If necessary tissue can be rapidly frozen on dry ice and stored at −80 °C.

2. Homogenize brain tissue in 3.5 mL ice-cold *homogenization buffer* and transfer to a 15 mL Falcon tube.
3. Centrifuge homogenized brain tissue at 1000 × *g* for 10 min at 4 °C to yield the *nuclear fraction (P1)*.
4. Transfer supernatant *(S1)* to 3.5 mL thick-walled ultracentrifuge tube (Thermo Scientific, 46,102).
5. Centrifuge supernatant *(S1)* at 10,000 × *g* for 15 min to yield the crude *synaptosomal pellet (P2)*.
6. Collect *S2 (crude cytoplasm)* and wash the pellet in 5 mL HEPES-buffered sucrose:
   (a) Do not dislodge the pellet, as this will cause improper lysis.
7. Lyse *P2* fraction by hypo-osmotic shock in 3 mL ice-cold *Buffer A*.
8. Triturate with a 1000 μL pipette to homogenize pellet.
9. Rotate homogenized pellet for 30 min at 4 °C.
10. Centrifuge the lysate at 25,000 × *g* for 20 min to yield *crude synaptic vesicle fraction (supernatant—S3)* and lysed synaptosomal membrane fraction *(pellet—P3)*.

2.2 PSD Fraction Preparation

1. Resuspend *P3* in 3 mL of ice-cold *Buffer B*.
2. Rotate for 15 min at 4 °C.
3. Centrifuge at 32,000 × *g* for 20 min to obtain PSD pellet.
4. Collect crude *synaptosomal membrane fraction (S4)*.
5. Resuspend pellet *(P5)* in 500 μL ice-cold *resuspension buffer* to yield the *PSD fraction (S5)*.

2.3 Notes

- All steps should be performed at 4 °C and on ice.
- If synaptosome fractions will be used for proteomic analysis experiments, substitute HEPES with 20 mM $NaHCO_3$.

2.4 Apparatus

- Dumont #5 Forceps (Fine Science Tools).
- Ultracentrifuge (10,000–32,000 × *g*).
- Thick-walled ultracentrifuge tubes (Thermo Scientific, 46,102).
- Tube rotator/revolver.

2.5 Solutions

2.5.1 Homogenization Buffer

- 0.32 mM Sucrose.
  2 mM EDTA.
  4 mM HEPES.
  Protease inhibitor cocktail (1×).

2.5.2 Buffer A

4 mM HEPES.

Protease inhibitor cocktail (1×).

2.5.3 Buffer B

50 mM HEPES.

2 mM EDTA.

Protease inhibitor cocktail (1×).

0.5% Triton X-100.

2.5.4 Resuspension Buffer

50 mM HEPES.

2 mM EDTA.

Protease inhibitor cocktail (1×).

To further study changes in synaptic activity and function, our lab is utilizing newer techniques that complement traditional synaptosome isolation techniques. Traditionally, ASD studies in the human context have relied heavily on postmortem samples [34–38]. However, these samples may not show any relevant differences and usually do not reflect the time point of interest (early childhood). A seminal advancement in the study of ASD was the derivation of induced pluripotent stem cells (iPSCs) from patient fibroblasts or peripheral blood and the differentiation of these iPSCs into various subtypes of human neurons, specifically cortical and hippocampal neurons [39–42]. This breakthrough has allowed for the modeling of many neurodevelopmental disorders in the human context [43–47]. Although human neuronal models of ASD have been created and studied, only the synaptic protein SHANK3, a well-established ASD-risk gene, has been studied thoroughly in human iPSC-derived neurons [48–51]. A possible reason for the lack of studies in synaptic proteins in the human context is the difficulty involved in obtaining mature human neurons with functional synapses. This process often requires long culturing times (months) and co-culturing with glial cells. However, recent advancements in the generation of human brain organoids can be used to better recapitulate the synaptic compartments in a human context at early developmental stages. Currently, many protocols exist to create various region-specific brain organoids, including the forebrain and midbrain [52–55]. Notably, neurodevelopmental differences have been seen when using ASD-derived organoids and organoids derived from iPSCs with knockout of an ASD-associated gene (e.g., CHD8) using the CRISPR/Cas9 system [56, 57]. Furthermore, a recent protocol was able to show functional connectivity within the organoid, suggesting that functional synapses exist in these 3D structures. By isolating synaptosomes from these organoids, it may be possible to study differences in synaptic proteins and RNA at an early human developmental time point. Although brain organoids require at least 1–2 months for proper synapse function, they recapitulate the organization within the brain to a higher degree than in neuron cultures, and therefore would be a better representation of synaptic connectivity and activity.

The use of synaptosomes to study neurodevelopmental disorders has been limited by the ability to create knockout or knock-in mouse models and by the limitations of using mice to model a complex human disorder at an early developmental time point. The discovery and development of the CRISPR/Cas9 genome editing system have allowed the generation of mouse models with ease that were once difficult to produce [12, 16–18, 58]. Additionally, the CRISPR/Cas9 system can be used efficiently for the manipulation of human iPSCs to study the role of ASD-associated genes, particularly synaptic genes, and thereby expanding the use of studying synaptosomes beyond the murine model. Furthermore, this genome editing system allows for the study of neurological disorders in a more patient-relevant context. Patient-derived iPSCs can be genetically modified to introduce or correct mutations that are enriched in ASD cohorts to investigate their role in the pathophysiology of ASD. In these ways, CRISPR/Cas9 has greatly increased the number of ASD-related human and mouse models that can be generated and studied. The isolation and study of synaptosomes from these models will significantly increase our understanding behind the synaptic pathology in ASD.

In addition to editing of genes, the CRISPR/Cas9 editing system can be used to create endogenously expressed fusions between genes that encode synaptic proteins with fluorescent proteins and labelling proteins. The fusion proteins, when coupled with synaptosome isolation, can be used for identifying protein complexes and protein networks. One major technique that has become a popular method for identifying novel protein interactions and binding partners is the BioID system [59]. The BioID system uses the BirA protein fused to a protein of interest that then biotinylates any proteins within a 10 nm radius of the BirA fusion protein. BirA is a biotin ligase isolated from *E. coli* that was originally used to biotinylate the 15-amino-acid Avi tag. However with recent bioengineering, a smaller and more promiscuous mutant BirA (BirA*) has been generated that can be fused to proteins of interest, where the promiscuous nature allows tagging of all neighboring proteins rather than only the Avi tag [60, 61]. Traditionally, studies that wished to identify protein interactions required the introduction of a recombinant tagged protein, such as the Avi tag, and then multiple purification steps that are labor intensive. These steps may result in many off-target interactions that make it difficult to identify novel interacting partners [62, 63]. However, with the use of the CRISPR/Cas9 system, BirA* can be fused to virtually any gene of interest endogenously. Biotinylated proteins can then be purified using streptavidin bead columns, and proteins can be identified using mass spectrometry or western blotting. BioID coupled with mass spectrometry has already been accomplished in vivo to study protein complexes within inhibitory and excitatory synapses using BirA-fused gephyrin and PSD-95, respectively [64]. The use of synaptosome fractionation will make possible, the identification of specific interacting partners in both the presynaptic and postsynaptic compartments, allowing the study of location-specific interactions. BioID has the advantage of identifying transient interactions within the synapse, where many interactions and changes in activity occur on the scale of a millisecond. BioID can be used in both in vitro cultures of human neurons derived from human iPSCs or in vivo through generation of a mouse model expressing BirA fused to a gene of interest. However, the combination of using CRISPR/Cas9 and BioID in human iPSCs could allow creation of brain organoids that can be used to identify changes in protein complexes and proteomic networks that better mimic a human context through isolation of their synaptosomes.

3 Conclusion

A major contributor of ASD pathophysiology is atypical neural connectivity and signalling, which can be attributed to disruptions in the function of synaptic proteins. The isolation of synaptosomes is an important method to study changes in protein and RNA expression specifically in the synapse. We have used synaptosome isolation to show specific localization and enrichment of Dixdc1 and Mark1 in the synapses of mouse brains, and that this localization is important for its function in spine formation through actin polymerization. Future work will aim to study the role of ASD-associated genes in synaptic connectivity using a human model, which has not been thoroughly done using current models of neurodevelopmental disorders. By isolating synaptosomes from brain organoids or from human iPSC-derived neurons that express the BirA-tagged protein, we can study the impact these genes have on proper synapse function and the affected synaptic protein networks in various ASD models. Additionally, synaptosome isolation from brain organoids that are expressing recombinant BirA may be a possible avenue of exploration that will allow for the identification of human disease-relevant networks that may be disrupted in ASD at early age developmental time points. The identification of these networks through mouse and human models will be imperative in creating treatments for individuals affected by ASD.

References

1. Boyle CA et al (2011) Trends in the prevalence of developmental disabilities in US children, 1997–2008. Pediatrics 127:1034–1042
2. Jon Baio C, Report MW, Jon Baio C (2014) Prevalence of autism spectrum disorder among children aged 8 years – autism and developmental disabilities monitoring network, 11 sites, United States, 2010. MMWR Surveill Summ 63:1–21
3. Doyle CA, McDougle CJ (2012) Pharmacologic treatments for the behavioral symptoms associated with autism spectrum disorders across the lifespan. Dialogues Clin Neurosci 14:263–279
4. Ghanizadeh A, Sahraeizadeh A, Berk M (2014) A head-to-head comparison of aripiprazole and risperidone for safety and treating autistic disorders, a randomized double blind clinical trial. Child Psychiatry Hum Dev 45:185–192
5. Colvert E et al (2015) Heritability of autism Spectrum disorder in a UK population-based twin sample. JAMA Psychiat 72:415
6. Gaugler T et al (2014) Most genetic risk for autism resides with common variation. Nat Genet 46:881–885
7. O'Roak BJ et al (2012) Sporadic autism exomes reveal a highly interconnected protein network of de novo mutations. Nature 485:246–250
8. Jiang Y-H et al (2013) Detection of clinically relevant genetic variants in autism Spectrum disorder by whole-genome sequencing. Am J Hum Genet 93:249–263
9. Stessman HAF et al (2017) Targeted sequencing identifies 91 neurodevelopmental-disorder risk genes with autism and developmental-disability biases. Nat Genet 49:515. https://doi.org/10.1038/ng.3792
10. Chang J, Gilman SR, Chiang AH, Sanders SJ, Vitkup D (2015) Genotype to phenotype relationships in autism spectrum disorders. Nat Neurosci 18:191–198
11. Chen JA, Peñagarikano O, Belgard TG, Swarup V, Geschwind DH (2015) The emerging picture of autism Spectrum disorder: genetics and pathology. Annu Rev Pathol 10:111–144
12. Etherton MR, Blaiss C a, Powell CM, Südhof TC (2009) Mouse neurexin-1alpha deletion causes correlated electrophysiological and behavioral changes consistent with cognitive impairments. Proc Natl Acad Sci U S A 106:17998–18003
13. Jamain S et al (2003) Mutations of the X-linked genes encoding neuroligins NLGN3 and NLGN4 are associated with autism. Nat Genet 34:27–29
14. Bader PL et al (2011) Mouse model of Timothy syndrome recapitulates triad of autistic traits. Proc Natl Acad Sci U S A 108:15432–15437
15. C Yuen RK et al (2017) Whole genome sequencing resource identifies 18 new candidate genes for autism spectrum disorder. Nat Neurosci 20:13–17
16. Kwan V et al (2016) DIXDC1 phosphorylation and control of dendritic morphology are impaired by rare genetic variants. Cell Rep 17:1892–1904
17. Schmeisser MJ et al (2012) Autistic-like behaviours and hyperactivity in mice lacking ProSAP1/Shank2. Nature 486:256–260
18. Peça J et al (2011) Shank3 mutant mice display autistic-like behaviours and striatal dysfunction. Nature 472:437–442
19. Berg JM et al (2015) JAKMIP1, a novel regulator of neuronal translation, modulates synaptic function and autistic-like behaviors in mouse. Neuron 88:1173–1192
20. Ascano M Jr et al (2012) FMRP targets distinct mRNA sequence elements to regulate protein expression. Nature 492:382–386
21. Napoli I et al (2008) The fragile X syndrome protein represses activity-dependent translation through CYFIP1 , a New 4E-BP. Cell 134:1042–1054. https://doi.org/10.1016/j.cell.2008.07.031
22. Oguro-Ando A et al (2014) Increased CYFIP1 dosage alters cellular and dendritic morphology and dysregulates mTOR. Mol Psychiatry 20:1–10
23. Smith SEP et al (2011) Increased gene dosage of Ube3a results in autism traits and decreased glutamate synaptic transmission in mice. Sci Transl Med 3:103ra97
24. Bidinosti M et al (2016) CLK2 inhibition ameliorates autistic features associated with SHANK3 deficiency. Science 351:1199–1203
25. Marchetto MC et al (2016) Altered proliferation and networks in neural cells derived from idiopathic autistic individuals. Mol Psychiatry 22:820. https://doi.org/10.1038/mp.2016.95
26. Gkogkas CG, Khoutorsky A, Cao R, Jafarnejad SM, Prager-khoutorsky M (2014) Pharmacogenetic inhibition of eIF4E-dependent Mmp9 mRNA translation reverses fragile X syndrome-like phenotypes. Cell Rep 9:1742–1755
27. Aguilar-Valles A et al (2015) Inhibition of group I metabotropic glutamate receptors reverses autistic-like phenotypes caused by deficiency of

the translation repressor eIF4E binding protein 2. J Neurosci 35:11125–11132
28. Harony-Nicolas H et al (2017) Oxytocin improves behavioral and electrophysiological deficits in a novel Shank3-deficient rat. elife 6:1–2
29. Smalheiser NR et al (2014) Expression of micrornas and other small RNAs in prefrontal cortex in schizophrenia, bipolar disorder and depressed subjects. PLoS One 9:1–12
30. Talbot K et al (2011) Synaptic dysbindin-1 reductions in schizophrenia occur in an isoform-specific manner indicating their subsynaptic location. PLoS One 6:1–14
31. Han MHJ et al (2014) Dysbindin-associated proteome in the P2 synaptosome fraction of mouse brain. J Proteome Res 13:4567–4580
32. Pacheco-Quinto J, Eckman CB, Eckman EA (2016) Major amyloid-β–degrading enzymes, endothelin-converting enzyme-2 and neprilysin, are expressed by distinct populations of GABAergic interneurons in hippocampus and neocortex. Neurobiol Aging 48:83–92
33. Prieto GA et al (2017) Pharmacological rescue of long-term potentiation in Alzheimer diseased synapses. J Neurosci 37:1197–1212
34. Hutsler JJ, Zhang H (2010) Increased dendritic spine densities on cortical projection neurons in autism spectrum disorders. Brain Res 1309:83–94
35. Voineagu I et al (2011) Transcriptomic analysis of autistic brain reveals convergent molecular pathology. Nature 474:380–384
36. Tang G et al (2014) Article loss of mTOR-dependent macroautophagy causes autistic-like synaptic pruning deficits. Neuron 1:13. https://doi.org/10.1016/j.neuron.2014.07.040
37. Stockman JA (2013) Neuron number and size in prefrontal cortex of children with autism. Yearb Pediatr 2013:395–397
38. Chow ML et al (2012) Age-dependent brain gene expression and copy number anomalies in autism suggest distinct pathological processes at young versus mature ages. PLoS Genet 8:e1002592
39. Takahashi K et al (2007) Induction of pluripotent stem cells from adult human fibroblasts by defined factors. Cell 131:861–872
40. Staerk J et al (2010) Reprogramming of peripheral blood cells to induced pluripotent stem cells. Cell Stem Cell 7:20–24
41. Yu DX et al (2014) Modeling hippocampal neurogenesis using human pluripotent stem cells. Stem Cell Reports 2:295–310
42. Ho S et al (2015) Rapid Ngn2 -induction of excitatory neurons from hiPSC-derived neural progenitor cells. Methods 101:1–12
43. Brennand KJ, Gage FH (2011) Concise review: the promise of human induced pluripotent stem cell-based studies of schizophrenia. Stem Cells 29:1915–1922
44. Lee G et al (2009) Modelling pathogenesis and treatment of familial dysautonomia using patient-specific iPSCs. Nature 461:402–406
45. Xu J et al (2016) Inhibition of STEP61 ameliorates deficits in mouse and hiPSC-based schizophrenia models. Mol Psychiatry 1:11. https://doi.org/10.1038/mp.2016.163
46. Maroof AM et al (2013) Directed differentiation and functional maturation of cortical interneurons from human embryonic stem cells. Cell Stem Cell 12:559–572
47. Chailangkarn T et al (2016) A human neurodevelopmental model for Williams syndrome. Nature 536(7616):338–343
48. Griesi-Oliveira K et al (2014) Modeling non-syndromic autism and the impact of TRPC6 disruption in human neurons. Mol Psychiatry 20:1–16
49. Yi F et al (2016) Autism-associated SHANK3 haploinsufficiency causes Ih channelopathy in human neurons. Science 352:aaf2669
50. Shcheglovitov A et al (2013) SHANK3 and IGF1 restore synaptic deficits in neurons from 22q13 deletion syndrome patients. Nature 503:267–271
51. Germain ND et al (2014) Gene expression analysis of human induced pluripotent stem cell-derived neurons carrying copy number variants of chromosome 15q11-q13.1. Mol Autism 5:44
52. Birey F et al (2017) Assembly of functionally integrated human forebrain spheroids. Nature 545:54. https://doi.org/10.1038/nature22330
53. Lancaster MA, Knoblich JA (2014) Generation of cerebral organoids from human pluripotent stem cells. Nat Protoc 9:2329–2340
54. Jo J et al (2016) Midbrain-like organoids from human pluripotent stem cells contain functional dopaminergic and Neuromelanin-producing neurons. Cell Stem Cell 19:248–257
55. Monzel AS et al (2017) Derivation of human midbrain-specific organoids from neuroepithelial stem cells. Stem Cell Reports 8:1–11
56. Mariani J et al (2015) FOXG1-dependent dysregulation of GABA/glutamate neuron differentiation in autism Spectrum disorders. Cell 162:375–390
57. Wang P et al (2015) CRISPR/Cas9-mediated heterozygous knockout of the autism gene CHD8 and characterization of its transcriptional networks in neurodevelopment. Mol Autism 6:55

58. Katayama Y et al (2016) CHD8 haploinsufficiency results in autistic-like phenotypes in mice. Nature 1:19. https://doi.org/10.1038/nature19357
59. Gautier A, Hinner MJ (2015) Site-specific protein labeling: methods and protocols. Springer, New York, pp 1–267. https://doi.org/10.1007/978-1-4939-2272-7
60. Fallis A (2013) An improved smaller biotin ligase for BioID proximity labeling. J Chem Inf Model 53:1689–1699
61. Choi-Rhee E, Schulman H, Cronan JE (2004) Promiscuous protein biotinylation by Escherichia coli biotin protein ligase. Protein Sci 13:3043–3050
62. Gregan J et al (2007) Tandem affinity purification of functional TAP-tagged proteins from human cells. Nat Protoc 2:1145–1151
63. Fernández E et al (2009) Targeted tandem affinity purification of PSD-95 recovers core postsynaptic complexes and schizophrenia susceptibility proteins. Mol Syst Biol 5:269
64. Buneman OP, et al. (2016) Identification of an elaborate complex mediating postsynaptic inhibition. 353:960–962

Chapter 17

Flow Cytometric Evaluation of Crude Synaptosome Preparation as a Way to Study Synaptic Alteration in Neurodegenerative Diseases

Nadia O. Postupna, Caitlin S. Latimer, C. Dirk Keene, Kathleen S. Montine, Thomas J. Montine, and Martin Darvas

Abstract

Neurodegenerative diseases, the most common among them Alzheimer's disease (AD) and Lewy body disease (LBD), are a group of progressive incurable illnesses. In both AD and LBD, abundant evidence points to the synapse as the critical and early focus of pathological changes. Here we present a method for the isolation and flow cytometric analysis of synaptosomes prepared from postmortem human brain tissue, which we also applied to animal models, including mice and nonhuman primates. The use of flow cytometry for analysis allows for relatively fast and efficient examination of thousands of synaptosome particles in a matter of minutes, and also makes it possible to use crude, rather than purified, synaptosomal preparation, thus conserving tissue resources. We have applied this method to study synaptic alteration in several brain regions in human research participants and animal models.

Key words Flow cytometry, Human, Neurodegeneration, Postmortem, Alzheimer's, Amyloid beta

1 Introduction

1.1 Neurodegenerative Diseases

Neurodegenerative diseases are a heterogeneous group of debilitating illnesses that result in progressive degeneration of nerve cells. They affect millions of people worldwide, and no effective cure is available [1].

The most prevalent neurodegenerative disease is Alzheimer's disease (AD), which is also the most common cause of dementia in the elderly, affecting more than 40 million people worldwide (Alzheimer's Association, 2015). Pathological features of AD include neuronal loss, and two characteristic lesions: neuritic plaques, which predominantly accumulate fibrillary forms of amyloid beta (Aβ) peptides, and neurofibrillary tangles (NFTs), which are largely composed of a hyperphosphorylated form of a

Kathryn M. Murphy (ed.), *Synaptosomes*, Neuromethods, vol. 141,
https://doi.org/10.1007/978-1-4939-8739-9_17,

microtubule-associated protein τ (paired-helical filament-τ, or PHF-τ) [2].

Another widespread neurodegenerative disease is Lewy body disease (LBD), the second most common cause of dementia in the elderly, which includes dementia with Lewy bodies (DLB) and Parkinson's disease (PD). Pathological features of LBD include neuronal loss, and the formation of the Lewy bodies and Lewy neurites, intracellular protein inclusions, that can be seen in a portion of the surviving cells [3]. A principal component of these inclusions is α-synuclein, a protein that is ubiquitously expressed in neurons, but principally localized to the synaptic region [4, 5].

1.2 Synaptic Dysfunction Is Key to Pathophysiology of AD and LBD

Although neuronal loss and characteristic lesions are the most pronounced features of both AD and LBD, mounting evidence suggests that synaptic dysfunction and synapse loss are early events driving the pathologic process. In 2003, Li et al. [6] introduced the term "synaptopathy" and suggested that synaptic dysfunction may be the critical first step in the pathophysiology of several neurodegenerative diseases that have been traditionally explained by loss of neurons. Li and colleagues used this term to describe changes in the brain of patients with Huntington's disease, but the notion has been since extended to other diseases, such as AD and LBD.

There is evidence to suggest that in AD, synaptic dysfunction may preclude neuronal loss by many years or even decades [7, 8]. Both AD and mild cognitive impairment (MCI), which often precede AD, are characterized by significant synapse loss in frontal and temporal cortex, and also the dentate gyrus of hippocampus [9–13]. Moreover, there is a significant correlation between synapse loss and cognition in both AD and MCI [10, 14], with cortical synaptic density showing better correlation with psychometric indices than plaques and tangles [13].

Despite mounting evidence that synaptic dysfunction is integral to AD pathophysiology, and that it occurs early in the disease process, the exact mechanism leading to impaired synaptic function in AD is not known. In general, synaptic dysfunction can be caused by molecular alterations that begin inside the pre- or postsynaptic neuron or that result from pathological changes in the environment surrounding the synapse. It has been shown that loss of synapses in AD is most pronounced in the proximity of Aβ-containing plaques [15], and that many neural processes around plaques are dystrophic and contain aggregates of PHF-τ [16–21]. It has also been demonstrated in vitro and in animal models that Aβ, both in fibrillar and oligomeric form, is toxic to neurons and can induce synaptic dysfunction [22], such as loss of dendritic spines, alteration of electrophysiological properties [23–25], and impaired memory formation [26, 27]. Oligomeric forms of Aβ have been shown to be significantly more toxic to synaptic function than fibrils [28, 29]. It is currently unclear whether Aβ's effect on the synapse is mediated by its interaction with a specific receptor [7]. Aβ is known to bind to many receptors, including NMDA, metabotropic glutamate receptors, and nicotinic acetylcholine receptors [30], but the relative importance of these interactions in the synaptic effects of Aβ has not been established.

The relationship between PHF-τ-containing neurofibrillary tangles and impairment in synaptic function is less clear. It has been demonstrated that in the brain of AD patients, surviving neurons containing neurofibrillary tangles receive fewer synapses onto their somata and express less synaptic proteins than neighboring neurons [31–33]. The most likely mechanism of PHF-τ toxicity is disruption in microtubule-associated cellular transport, which impairs transport of mitochondria and other essential cargo to pre- and postsynaptic terminals [34, 35]. In addition, recent evidence suggests that the effects of Aβ and PHF-τ at the synapse are most likely synergistic [36–40].

In PD and other disorders with Lewy body pathology, accumulating pathological forms of α-synuclein appear to be central to the synaptic dysfunction. The normal function of α-synuclein is not completely understood, but it has been demonstrated to be involved in regulating neurotransmitter release at the presynaptic terminal [4, 41]. α-Synuclein can undergo conformational changes to form oligomers, protofibrils, and fibrils, all of which can accumulate in cells [42, 43] and induce synaptic dysregulation [5, 44]. Results from in vitro and in vivo studies in rodents suggest that pathological forms of α-synuclein, such as oligomers or protofibrils, affect protein degradation, impair vesicle trafficking at the synapse, and compromise long-term potentiation [45–47]. In dopaminergic neurons, fibrillar α-synuclein can bind to and permeabilize the membrane of dopamine-containing synaptic vesicles, spilling dopamine into the cytosol [48, 49], which in turn can lead to oxidative stress and neurotoxicity [50–52]. In addition to α-synuclein, other proteins implicated in the pathogenesis of PD, such as parkin and LRRK2, have also recently been shown to impair synaptic vesicle recycling and autophagy, providing additional evidence for synaptic dysfunction as an early event in the disease process [53].

1.3 Challenges to Studies of Human Synapses

It appears that in both AD and LBD, the pathological process either begins at the synapse, or, at the very least, synaptic structure and function become affected very early in the course of the disease. Therefore, developing a better understanding of the pathological process at the synaptic level could provide important insights into the development of early therapeutic interventions. Our current understanding of mechanisms of synaptic dysregulation in AD and LBD came from in vitro studies and animal models, which can never fully re-create human disease process. But studying synapses in human brain tissue can be challenging. Traditional histochemical techniques were perfected for studying whole cells, and are not well suited for visualization of single-nerve terminals. High-resolution microscopy technique allows detailed examination of fine synaptic architecture, and can show association of pathological elements with synaptic structures, but is labor intensive and not practical for studying multiple brain regions in large-scale research cohorts [54]. Methods that involve quantifying proteins in tissue homogenates, such as Western blotting, allow for precise quantification of synaptic proteins, physiological and pathological, but don't distinguish between proteins that are present at the synapse and the ones that are ultimately destined for the synapse, but are not there yet. In short, all of these methods have their limitations, with the general trade-off between quantification, precise localization of the pathological changes, and scalability.

1.4 Using Flow Cytometry to Study Synaptosomes

Preparing synaptosomes from brain tissue homogenates is a well-established technique, which has been extensively used in animal studies of synaptic physiology [55, 56]. Synaptosomes obtained from postmortem human tissue are not generally suitable for functional studies, because of their compromised electrochemical integrity, but can be a great resource for testing hypotheses about molecular alterations in human synapses.

A recent approach, pioneered by Karen Gylys's group [57], combines synaptosome preparations with flow cytometry methodologies in order to study synaptic alterations and associations of synapse subtypes with Aβ and tau species in AD in human subjects and animal models [57–61].

To better understand synaptic changes that occur in cerebral cortex and striatum in neurodegeneration, especially AD and LBD, we have adapted the method developed by Gylys and colleagues to probe the molecular composition of synaptic particles in multiple regions of human brain in several disease groups [62, 63]. Because the previous studies by both groups used autopsy material obtained from research cohorts, they may present a somewhat skewed view of the development and type of synaptic injury that occurs with aging. In addition, the effect of postmortem interval (PMI), unavoidable in human studies, may have systematically affected the changes that were observed. To address these issues, we examined the molecular composition of synaptosomal preparations obtained from several brain regions from aged nonhuman primates with no PMI (sacrificed as part of a separate study). We have also effectively used this method to study synaptic changes in transgenic mouse models, such as a model of dopaminergic degeneration [62] and mice with human *ApoE ε3* or *ApoE ε4* alleles [63].

2 Materials and Methods

Our protocol for analyzing synaptosome preparation flow cytometry [62] was adapted from Gylys et al. [57] and includes the following steps:

1. Tissue collection (and storage if needed)
2. Preparation of the crude synaptosome fraction (and storage if needed)
3. Immunolabeling of the crude synaptosome fraction
4. Flow cytometry analysis of the labeled synaptosome fraction

2.1 Tissue Collection

Our protocol has been established for human tissue samples that are obtained at autopsy or during surgery, and animal tissue samples that are collected immediately after sacrifice. At autopsy, we aim for collecting approximately 0.5 g of tissue from each structure of interest, which provides enough material to be later divided into 20 experimental samples. We only collect tissue from regions that are free of macroscopic lesions (hemorrhage, etc.). For human cerebral cortex samples we focus on collecting gray matter. To achieve that, we dissect out cortical gray matter with about 1 mm of underlying white matter from a 5 mm thick coronal section taken from the region of interest. For human hippocampal samples, a coronal section containing the hippocampus at the level of lateral geniculate nucleus is taken, and the whole hippocampus (dentate gyrus and cornu ammonis) is dissected out in one piece. Basal ganglia structures (putamen and head of the caudate nucleus) are collected from coronal slices at the level of anterior commissure. We carefully remove white matter surrounding the nuclei, to the extent that it can be done without compromising the structure of interest. For the caudate nucleus and putamen, it is usually necessary to use two adjacent coronal sections to provide enough material, because caudate nucleus samples are divided into equal dorsolateral and ventromedial subregions.

Animal tissue samples are obtained immediately after sacrifice. At collection, mouse cortex is split into anterior and posterior parts, striatum is collected as one piece, and hippocampus is also collected as one piece.

Upon collection, all tissue samples are placed in Petri dishes, and minced into smaller (approximately 2 mm × 2 mm × 2 mm) pieces using two razor blades. The samples are then transferred to cryotube vials and frozen in 1 mL of 0.32 M sucrose with 10% DMSO using a "Mr. Frosty™" freezing container (Sigma-Aldrich, MO), which is designed to achieve an optimal cooling rate of ~−1 °C/min, and later stored at −80 °C. We only collect human tissue from cases with a PMI of 8 h or less. To test if a PMI of this length resulted in significant degradation of synaptic material, we prepared synaptosomes from human cortex resected during surgery for intractable temporal lobe epilepsy, and labeled them with the antibody against glutamate transporter VGLUT1 [62]. The percentage of VGLUT1-positive particles in the surgically resected samples were not significantly different from those obtained at autopsy. In addition, we recently [63] compared relative levels of synaptosomes positive for several different markers—SNAP-25 (a component of the SNARE complex), VGLUT1 and VGLUT2 (markers for glutamatergic synapses), VGAT (vesicular GABA transporter, a marker for GABAergic synapses), SERT (serotonin transporter, a marker for serotoninergic synapses), $A\beta_{42}$, and PHF-τ—between autopsy samples (PMI ≤8 h) of human frontal cortex and striatum and samples from aged nonhuman primates (euthanized for a different study) immediately after sacrifice. The levels of SNAP-25, VGLUT1, VGLUT2, SERT, Aβ42, and PHF-τ were not decreased in the human samples, although levels of VGAT were lower in human frontal cortex (but not striatum), allowing us to conclude that a PMI of 8 h or less is unlikely to elicit enough degradation of synaptic material to significantly affect our results. We also concluded that agonal events, unavoidable in human cases, but eliminated in the monkey samples, did not have a significant effect on preparation. Moreover, longer PMIs may be acceptable under certain circumstances. In wild-type mice, we tested the effect of longer (12, 18, and 24 h) PMIs on VGLUT1, VGAT, and DAT level in cortex and striatum and found that levels of these proteins changed significantly only in the tissue collected 24 h after euthanasia.

2.2 Preparation of the Crude Synaptosome Fraction

Our protocol produces a crude synaptosome fraction—obtained from minced tissue samples by homogenization and a series of centrifugations, but not gradient purification. Although gradient purification is undoubtedly beneficial for removing non-synaptosome material, it significantly reduces the overall yield. Instead, we use a strategy of gating on particle size during flow cytometry to restrict our analysis to particles of the correct size (Fig. 1).

To obtain this crude fraction, the samples are thawed by shaking in a 37 °C water bath for approximately 2 min. The contents of the tube are then added to 10 volumes of 0.32 M sucrose in 10 mM Tris buffer (pH 7.4) with cOmplete™ protease inhibitor cocktail (Roche, IN) and phosphatase inhibitor cocktails 2 and 3 (Sigma-Aldrich, MO). Tissue is then homogenized using a glass/Teflon homogenizer (clearance 0.1–0.15 mm) with 6 to 8 up-and-down strokes. The homogenate is transferred to a 15 mL conical tube and centrifuged at 1000 × *g* at 4 °C for 10 min. The supernatant is then collected, taking extra care to avoid collecting any pellet material, transferred into 1.75 mL centrifuge tubes (20 tubes for one 0.5 × *g* sample), and centrifuged again at 10,000 × *g* at 4 °C for 20 min. The resulting pellets are resuspended in 200 μL

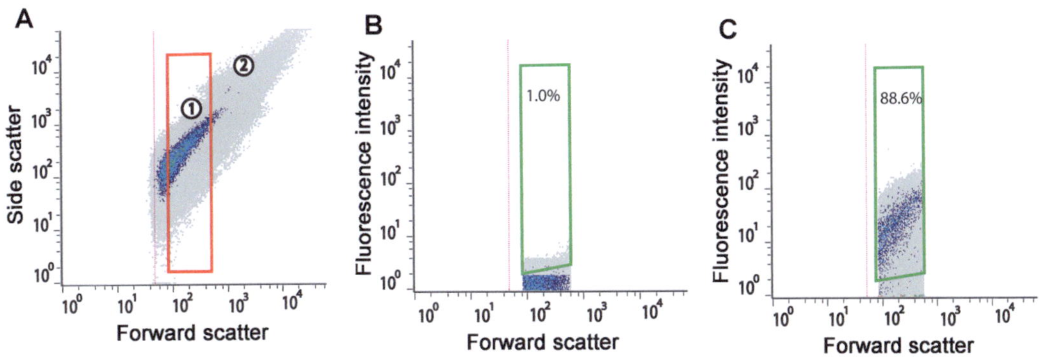

Fig. 1 Flow cytometry of a synaptosomal preparation of human frontal cortex. (**a**) Size gating of particles using silica bead standards (numbered circles): 1–0.6 μm beads, 2–1.2 μm beads. The size gate for synaptosomes is drawn in red. (**b**) Isotype control antibody labeling of preparations. The gate (drawn in green) is set to include 1 ± 0.1% of all size-gated positive particles in the isotype control sample, and then applied to the sample labeled with SNAP-25 antibody. (**c**) SNAP-25 antibody labeling of samples. Indicated is the percentage of positive particles within the gate (green)

of a sucrose/Tris solution (to facilitate resuspension after thawing) and stored at −80 °C for future use.

2.3 Immunolabeling of the Crude Synaptosome Fraction

The synaptosome pellets are thawed by shaking the tubes in a 37 °C degree water bath. The pellets are then washed by adding 1 mL of Krebs-Ringer phosphate buffer (KRB, 118 mM NaCl, 5 mM KCl, 4 mM $MgSO_4$, 1 mM $CaCl_2$, 1 mM KH_2PO_4, 16 mM sodium phosphate buffer, and 10 mM glucose), and centrifuging at 8000 × *g* for 4 min at 4 °C. For each future incubation or wash, the synaptosome pellets are resuspended by gentle vortexing for about 5 s, and then collected by centrifugation (5000 × *g* for 4 min at 4 °C). After two washes with KRB, the pellets are fixed with 0.25% paraformaldehyde (PFA). To facilitate resuspension after fixation, the pellets are first resuspended in 200 μL phosphate-buffered saline (PBS; 137 mM NaCl, 2.7 mM KCl, 10 mM NaH_2PO_4, 1.8 mM KH_2PO_4, pH 7.4), to which 250 μL 0.5% PFA solution is then added. After fixation with PFA for 1 h at 4 °C on a rocker, another 1 mL of PBS is added to the tubes, to further decrease the final concentration of PFA and facilitate further resuspension. The pellets are then collected by centrifugation and washed with 1 mL PBS. For labeling of intracellular antigens, the synaptosomal membranes are then permeabilized by incubation with 1 mL of 0.2% Tween 20 in PBS for 15 min at 37 °C. The permeabilization step can be skipped if the antigen of interest is located extracellularly. After permeabilization, the pellets are washed twice with PBS, and then with 500 μL of blocking buffer (2% fetal calf serum in PBS). Afterwards, the pellets are incubated with the appropriate primary antibodies—either specific to the antigen of interest or isotype control antibodies—in blocking buffer for 30 min at room temperature on a rocker. For each experimental sample labeled with a primary antibody selective to the protein of interest, an identical sample, processed simultaneously, is labeled with the respective isotype control antibody, conjugated to the same fluorophore as the primary antibody (when applicable), for use as a negative control during flow cytometric evaluation. After incubation, the pellets are washed with 1 mL 0.2% Tween 20 in PBS. If the primary antibody and the respective isotype control antibody were conjugated to a fluorophore, the pellets are washed again with 1 mL PBL, and resuspended in 1 mL PBS for flow cytometric analysis. When labeling with an unconjugated primary antibody (and isotype control antibody), the pellets are washed once more with 1 mL of 0.2% Tween 20 in PBS, and then incubated with 500 μL of the respective fluorophore-conjugated secondary antibody in blocking buffer for 30 min at room temperature on a rocker. The pellets are then washed once with 1 mL of 0.2% Tween 20 in PBS, and once with 1 mL PBS, and resuspended in 1 mL PBS for flow cytometric analysis.

When more than one primary antibody is used, all of the primary antibodies are combined for the first incubation, and all of the secondary (when applicable) antibodies are combined for the second incubation. When a combination of conjugated and unconjugated primary antibodies is used, the fluorophore-conjugated primary antibodies are added to the fluorophore-conjugated secondary antibodies and used in the second incubation.

For experiments involving several fluorophores, a set of several synaptosome samples, all prepared from the same tissue sample, are used to estimate compensation values for overlapping fluorophores (e.g., FITC and PE). One subset of these samples is individually labeled with only one fluorescent primary antibody (or antibody pair if using unconjugated primary and fluorophore-conjugated secondary antibody) or an isotype control antibody, and another set of samples are labeled with the combination of all the antibodies of interest or isotype control antibodies involved.

2.4 Flow Cytometry Analysis of the Immunolabeled Synaptosome Fraction

The synaptosome suspension is then analyzed using an Apogee A50 (Apogee Flow Systems, Hertfordshire, UK) flow cytometer equipped with 15 mW 488 nm (blue) and 633 nm (red) lasers. For our initial experiments, we used a BD FACSCalibur Flow cytometer (BD Biosciences, CA) equipped with the same combination of lasers. The results obtained with the two machines were not significantly different, but Apogee A50, unlike most other flow cytometers, was developed specifically for the analysis of smaller particles and is thought to provide better discrimination for synaptosomes.

Because our synaptosomal preparation is not gradient-purified, it is likely to contain some large non-synaptosomal elements (such as pieces of myelin, parts of nuclei, extracellular matrix, membranes). To focus our analysis on the population that we believe to be mostly synaptosomal, we use gating on size (Fig. 1a). We use polystyrene (Polysciences Inc., PA) and silica (Spherotech Inc., IL) microspheres of varying diameter (0.5, 0.6, and 1.2 μm) as size standards to restrict all our analysis to particles that are approximately 1 μm. Because the refractory indices of both silica and polystyrene are significantly different from the refractory index of cell-derived particles, synthetic beads of 0.5 μm are expected to scatter approximately as much light as a 1.0 μm synaptosome [64]. Therefore, we use 0.5 and 0.6 μm beads to guide the placement of our size gate. Our size-gated population is then gated again based on the fluorescent signal of interest. This second gate is set on the negative control sample, which is the sample identical to the experimental sample but labeled with the respective isotype control antibody (Fig. 1b, c). This gate is drawn to include the most positive 1 ± 0.1% of the negative (isotype control) sample. The gate is then applied to the experimental sample, and the resulting percentage of positive particles is reported. We found that drawing the positive gate to include 1 ± 0.1% of the total size-gated particle population offers optimal discrimination between the positive and negative sample and provides the best reproducibility in repeated experiments. In our previous studies, we also used a gate of 5 ± 0.1%. Others have previously reported using a similar or more inclusive size gate [65].

2.5 Select Applications

We have previously demonstrated the effectiveness of this technique by looking at the percentage of synaptic particles positive for dopamine transporter (DAT) labeling in the striatum of mice that underwent a toxin-mediated ablation of DAT-containing neurons [62]. Figure 2a shows unpublished data from a recent study in which we used flow cytometry analysis to compare synaptic DAT expression between two models of striatal dopamine loss: DAT-neuron-lesioned mice, in which DAT-expressing neurons are ablated [66], and a tyrosine hydroxylase (*Th*) knockout strain, in which dopamine is lost, but DAT-expressing neurons are spared [67]. Mice with lesioned DAT neurons had a significantly ($p < 0.01$) decreased percentage of DAT-positive particles in striatal synaptosomal preparation (2%), compared to controls (~50%) and to tyrosine hydroxylase knockout mice (40%).

We have also examined synaptosome preparations from postmortem human cortical samples for evidence of synaptosomal accumulation of Aβ [62]. We reported that samples from patients with the highest neuropathological AD burden, as assessed by either Braak stage (Braak stages V–VI), CERAD score (3, or "Frequent"), or Thal phase (Thal phases 4–5), had significantly more Aβ-positive particles than samples with the lowest neuropathologic burden of AD (each $p < 0.05$).

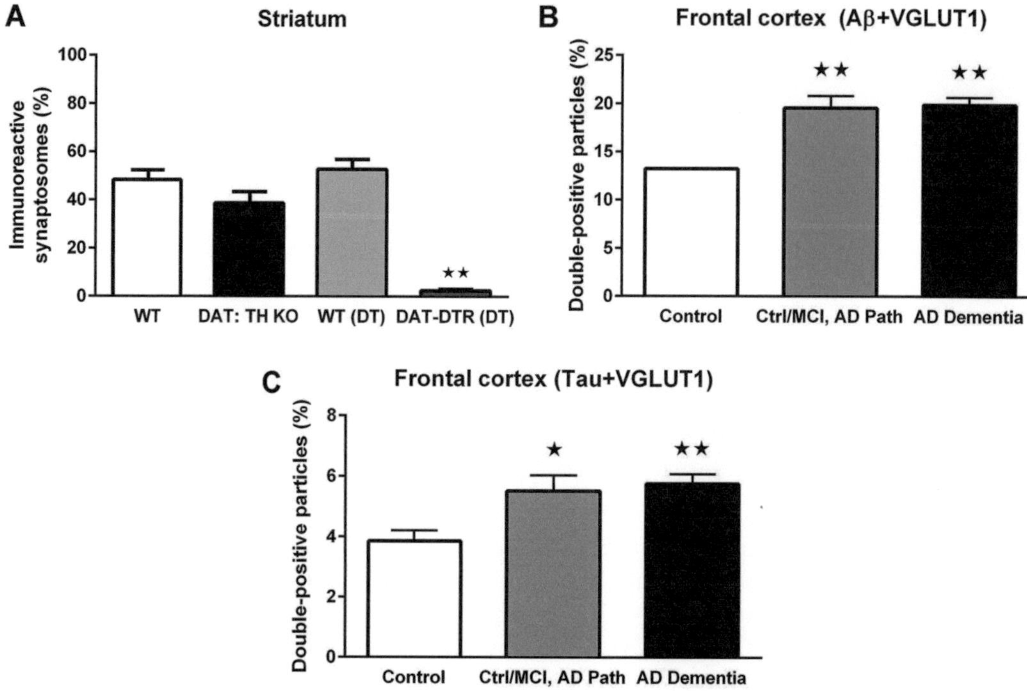

Fig. 2 Flow cytometry analysis of synaptosomal preparations from mouse and human brain tissue. (**a**) Percentage of synaptosomal preparations from mouse striatum immunolabeled with antibody against DAT. Tissue was collected from mice with tyrosine hydroxylase (TH) inactivation in dopamine active transporter (DAT) neurons (DAT: TH KO), mice with diphtheria toxin (DTR)-induced loss of DAT neurons (DAT-DTR), and respective wild-type (WT) control mice. Loss of DAT neurons in DAT-DTR mice resulted in significantly (by ANOVA and post hoc) reduced levels of DAT-positive synaptosomes, while the number of DAT-positive synaptic particles in TH inactivation mice was not significantly different from WT mice. (**b** and **c**) Percentage of synaptosomal preparations from human cortex co-immunolabeled with antibodies against (**b**) Aβ and VGLUT1, and (**c**) PHF-τ and VGLUT1. Cortical tissue was collected from individuals with no dementia and only low levels of AD pathology (control), MCI or normal cognition and moderate-to-high levels of AD pathology (Ctrl/MCI, AD Path), or clinical diagnosis of dementia and a high level of AD pathology (AD dementia). The number of double-positive particles was significantly higher in the AD dementia and the Ctrl/MCI, AD Path group than the control group (each by ANOVA and post hoc). Data are shown as group means ± standard error of mean. $^{*}p < 0.05$ and $^{**}p < 0.01$

Finally, we used a combination of several antibodies to determine if pathological proteins, such as $A\beta_{42}$ and PHF-τ, co-localize with specific types of synapses, and whether the degree of co-labeling is affected by disease state. As shown in Fig. 2b, c, synaptosomes originating from cortical glutamatergic terminals expressing VGLUT1 also show positive signal for $A\beta_{42}$ and PHF-τ; this signal is more pronounced in the AD group.

3 Conclusions and Future Directions

Synaptic pathophysiology is notoriously difficult to study in humans. At the same time, mounting evidence supports a crucial role for synaptic dysfunction in multiple neurodegenerative diseases, among them AD and LBD. Because subtle changes in synaptic function appear to happen early in the disease process, unravelling mechanisms underlying synaptic changes may offer new perspectives for early treatment of these debilitating diseases for which there is currently no cure. Available animal models of AD and LBD, although undoubtedly valuable, fail to fully recapitulate the disease process, making studies in human tissue a necessity.

We and others have previously demonstrated the utility of crude synaptosome preparation obtained from postmortem human brain tissue for studies of synaptic alteration in AD and LBD. Using flow cytometry to analyze the crude fraction makes it possible to omit the gradient purification step, thus significantly decreasing preparation time and, most importantly, conserving valuable tissue resources. Another important benefit of using flow cytometry is its high throughput, which allows for analysis of tens of thousands of particles in a matter of minutes, and the ability to multiplex with several fluorophores.

In addition to human postmortem samples, we have also analyzed synaptosome preparation from nonhuman primates and human surgical specimens, and concluded that the effect of postmortem interval and agonal state, unavoidable in studies involving autopsy tissue, does not significantly influence the results, at least for measurements of SNAP-25, VGLUT1, VGLUT2, SERT, Aβ, and PHF-τ.

One exciting future direction for this method is to use another flow cytometry technique, fluorescent-activated cell sorting (FACS), to isolate synaptosomal particles based on a combination of several characteristics, such as the neuron of origin or association with a particular pathological protein, and use the sorted population for molecular and genetic studies to answer questions about the relative susceptibility of different neuronal populations to pathological processes. Another is to use bead-based immunological isolation techniques to enrich synaptosomes and use the enriched population for molecular and genetic studies.

In summary, applying the techniques of flow cytometry to synaptosomal preparation provides an efficient tool for assessing relative abundance of different types of synapses in multiple regions of human and animal brain, as well as the degree of association of pathological proteins, such as Aβ, PHF-τ, and α-synuclein, with specific nerve terminals.

Acknowledgements

We would like to thank Emily Sherfield, Julie Paladin, Samantha Rice, Leanne Hellstern, Erica Melief, and Allison Beller. Parts of the work presented in this chapter were supported by grants from the NIH: University of Washington Alzheimer's Disease Research Center (ADRC) P50 AG05136, Stanford ADRC P50 AG047366, and the Pacific Udall Center P50 NS062684.

References

1. Goedert M, Spillantini MG (2006) A century of Alzheimer's disease. Science 314(5800):777–781
2. Holtzman DM, Morris JC, Goate AM (2011) Alzheimer's disease: the challenge of the second century. Sci Transl Med 3(77):77sr1
3. Braak H, Braak E (2000) Pathoanatomy of Parkinson's disease. J Neurol 247(Suppl 2):II3–II0
4. Burre J et al (2010) Alpha-synuclein promotes SNARE-complex assembly in vivo and in vitro. Science 329(5999):1663–1667
5. Ingelsson M (2016) Alpha-Synuclein oligomers-neurotoxic molecules in Parkinson's disease and other Lewy body disorders. Front Neurosci 10:408
6. Li JY, Plomann M, Brundin P (2003) Huntington's disease: a synaptopathy? Trends Mol Med 9(10):414–420
7. Koffie RM, Hyman BT, Spires-Jones TL (2011) Alzheimer's disease: synapses gone cold. Mol Neurodegener 6(1):63
8. Selkoe DJ (2002) Alzheimer's disease is a synaptic failure. Science 298(5594):789–791
9. Davies CA et al (1987) A quantitative morphometric analysis of the neuronal and synaptic content of the frontal and temporal cortex in patients with Alzheimer's disease. J Neurol Sci 78(2):151–164
10. DeKosky ST, Scheff SW (1990) Synapse loss in frontal cortex biopsies in Alzheimer's disease: correlation with cognitive severity. Ann Neurol 27(5):457–464
11. Masliah E (2001) Recent advances in the understanding of the role of synaptic proteins in Alzheimer's disease and other neurodegenerative disorders. J Alzheimers Dis 3(1):121–129
12. Scheff SW et al (2007) Synaptic alterations in CA1 in mild Alzheimer disease and mild cognitive impairment. Neurology 68(18):1501–1508
13. Terry RD et al (1991) Physical basis of cognitive alterations in Alzheimer's disease: synapse loss is the major correlate of cognitive impairment. Ann Neurol 30(4):572–580
14. Masliah E et al (1994) Synaptic and neuritic alterations during the progression of Alzheimer's disease. Neurosci Lett 174(1):67–72
15. Koffie RM et al (2009) Oligomeric amyloid beta associates with postsynaptic densities and correlates with excitatory synapse loss near senile plaques. Proc Natl Acad Sci U S A 106(10):4012–4017
16. Meyer-Luehmann M et al (2008) Rapid appearance and local toxicity of amyloid-beta plaques in a mouse model of Alzheimer's disease. Nature 451(7179):720–724
17. Serrano-Pozo A et al (2011) Neuropathological alterations in Alzheimer disease. Cold Spring Harb Perspect Med 1(1):a006189
18. Spires TL et al (2005) Dendritic spine abnormalities in amyloid precursor protein transgenic mice demonstrated by gene transfer and intravital multiphoton microscopy. J Neurosci 25(31):7278–7287
19. Stern EA et al (2004) Cortical synaptic integration in vivo is disrupted by amyloid-beta plaques. J Neurosci 24(19):4535–4540
20. Urbanc B et al (2002) Neurotoxic effects of thioflavin S-positive amyloid deposits in transgenic mice and Alzheimer's disease. Proc Natl Acad Sci U S A 99(22):13990–13995
21. Woodhouse A et al (2005) Does beta-amyloid plaque formation cause structural injury to neuronal processes? Neurotox Res 7(1–2):5–15
22. Roher AE et al (1996) Morphology and toxicity of Abeta-(1-42) dimer derived from neuritic and vascular amyloid deposits of Alzheimer's disease. J Biol Chem 271(34):20631–20635
23. Walsh DM et al (2002) Naturally secreted oligomers of amyloid beta protein potently

inhibit hippocampal long-term potentiation in vivo. Nature 416(6880):535–539

24. Wang HY et al (2000) Amyloid peptide Abeta(1–42) binds selectively and with picomolar affinity to alpha7 nicotinic acetylcholine receptors. J Neurochem 75(3):1155–1161
25. Shankar GM et al (2007) Natural oligomers of the Alzheimer amyloid-beta protein induce reversible synapse loss by modulating an NMDA-type glutamate receptor-dependent signaling pathway. J Neurosci 27(11):2866–2875
26. Lesne S et al (2006) A specific amyloid-beta protein assembly in the brain impairs memory. Nature 440(7082):352–357
27. Shankar GM et al (2008) Amyloid-beta protein dimers isolated directly from Alzheimer's brains impair synaptic plasticity and memory. Nat Med 14(8):837–842
28. Klein WL (2006) Synaptic targeting by a beta oligomers (ADDLS) as a basis for memory loss in early Alzheimer's disease. Alzheimers Dement 2(1):43–55
29. Lambert MP et al (1998) Diffusible, nonfibrillar ligands derived from Abeta1-42 are potent central nervous system neurotoxins. Proc Natl Acad Sci U S A 95(11):6448–6453
30. Viola KL, Klein WL (2015) Amyloid beta oligomers in Alzheimer's disease pathogenesis, treatment, and diagnosis. Acta Neuropathol 129(2):183–206
31. Katsuse O et al (2006) Neurofibrillary tangle-related synaptic alterations of spinal motor neurons of P301L tau transgenic mice. Neurosci Lett 409(2):95–99
32. Ginsberg SD et al (2000) Expression profile of transcripts in Alzheimer's disease tangle-bearing CA1 neurons. Ann Neurol 48(1):77–87
33. Callahan LM, Vaules WA, Coleman PD (1999) Quantitative decrease in synaptophysin message expression and increase in cathepsin D message expression in Alzheimer disease neurons containing neurofibrillary tangles. J Neuropathol Exp Neurol 58(3):275–287
34. Kopeikina KJ, Hyman BT, Spires-Jones TL (2012) Soluble forms of tau are toxic in Alzheimer's disease. Transl Neurosci 3(3):223–233
35. Spires-Jones TL, Hyman BT (2014) The intersection of amyloid beta and tau at synapses in Alzheimer's disease. Neuron 82(4):756–771
36. Roberson ED et al (2011) Amyloid-beta/Fyn-induced synaptic, network, and cognitive impairments depend on tau levels in multiple mouse models of Alzheimer's disease. J Neurosci 31(2):700–711
37. Nisbet RM et al (2015) Tau aggregation and its interplay with amyloid-beta. Acta Neuropathol 129(2):207–220
38. Ittner LM et al (2010) Dendritic function of tau mediates amyloid-beta toxicity in Alzheimer's disease mouse models. Cell 142(3):387–397
39. Ittner LM, Gotz J (2011) Amyloid-beta and tau--a toxic pas de deux in Alzheimer's disease. Nat Rev Neurosci 12(2):65–72
40. Crimins JL et al (2013) The intersection of amyloid beta and tau in glutamatergic synaptic dysfunction and collapse in Alzheimer's disease. Ageing Res Rev 12(3):757–763
41. Zaltieri M et al (2015) Alpha-synuclein and synapsin III cooperatively regulate synaptic function in dopamine neurons. J Cell Sci 128(13):2231–2243
42. Tsigelny IF et al (2012) Role of alpha-synuclein penetration into the membrane in the mechanisms of oligomer pore formation. FEBS J 279(6):1000–1013
43. Uversky VN et al (2001) Stabilization of partially folded conformation during alpha-synuclein oligomerization in both purified and cytosolic preparations. J Biol Chem 276(47):43495–43498
44. Hunn J et al (2015) Patterns and utility of routine surveillance in high grade endometrial cancer. Gynecol Oncol 137(3):485–489
45. Diogenes MJ et al (2012) Extracellular alpha-synuclein oligomers modulate synaptic transmission and impair LTP via NMDA-receptor activation. J Neurosci 32(34):11750–11762
46. Janezic S et al (2013) Deficits in dopaminergic transmission precede neuron loss and dysfunction in a new Parkinson model. Proc Natl Acad Sci U S A 110(42):E4016–E4025
47. Rockenstein E et al (2014) Accumulation of oligomer-prone alpha-synuclein exacerbates synaptic and neuronal degeneration in vivo. Brain 137(Pt 5):1496–1513
48. Nemani VM et al (2010) Increased expression of alpha-synuclein reduces neurotransmitter release by inhibiting synaptic vesicle reclustering after endocytosis. Neuron 65(1):66–79
49. Volles MJ, Lansbury PT Jr (2002) Vesicle permeabilization by protofibrillar alpha-synuclein is sensitive to Parkinson's disease-linked mutations and occurs by a pore-like mechanism. Biochemistry 41(14):4595–4602
50. Chen MK et al (2008) VMAT2 and dopamine neuron loss in a primate model of Parkinson's disease. J Neurochem 105(1):78–90

51. Masoud ST et al (2015) Increased expression of the dopamine transporter leads to loss of dopamine neurons, oxidative stress and l-DOPA reversible motor deficits. Neurobiol Dis 74:66–75
52. Mosharov EV et al (2009) Interplay between cytosolic dopamine, calcium, and alpha-synuclein causes selective death of substantia nigra neurons. Neuron 62(2):218–229
53. Plowey ED, Chu CT (2011) Synaptic dysfunction in genetic models of Parkinson's disease: a role for autophagy? Neurobiol Dis 43(1):60–67
54. Meinertzhagen IA et al (2009) From form to function: the ways to know a neuron. J Neurogenet 23(1–2):68–77
55. Whittaker VP (1993) Thirty years of synaptosome research. J Neurocytol 22(9):735–742
56. Whittaker VP (1959) The isolation and characterization of acetylcholine-containing particles from brain. Biochem J 72:694–706
57. Gylys KH, Fein JA, Cole GM (2000) Quantitative characterization of crude synaptosomal fraction (P-2) components by flow cytometry. J Neurosci Res 61(2):186–192
58. Sokolow S et al (2012) Isolation of synaptic terminals from Alzheimer's disease cortex. Cytometry A 81(3):248–254
59. Sokolow S et al (2012) AD synapses contain abundant Abeta monomer and multiple soluble oligomers, including a 56-kDa assembly. Neurobiol Aging 33(8):1545–1555
60. Sokolow S et al (2015) Pre-synaptic C-terminal truncated tau is released from cortical synapses in Alzheimer's disease. J Neurochem 133(3):368–379
61. Bilousova T et al (2016) Synaptic amyloid-beta oligomers precede p-Tau and differentiate high pathology control cases. Am J Pathol 186(1):185–198
62. Postupna NO et al (2014) Flow cytometry analysis of synaptosomes from post-mortem human brain reveals changes specific to Lewy body and Alzheimer's disease. Lab Investig 94(10):1161–1172
63. Postupna NO et al (2017) Human striatal dopaminergic and regional serotonergic synaptic degeneration with Lewy body disease and inheritance of APOE ε4. Am J Pathol 187:884
64. Chandler WL, Yeung W, Tait JF (2011) A new microparticle size calibration standard for use in measuring smaller microparticles using a new flow cytometer. J Thromb Haemost 9(6):1216–1224
65. Sokolow S et al (2012) Preferential accumulation of amyloid-beta in presynaptic glutamatergic terminals (VGluT1 and VGluT2) in Alzheimer's disease cortex. Neurobiol Dis 45(1):381–387
66. Morgan RG et al (2015) Relative contributions of severe dopaminergic neuron ablation and dopamine depletion to cognitive impairment. Exp Neurol 271:205–214
67. Henschen CW, Palmiter RD, Darvas M (2013) Restoration of dopamine signaling to the dorsal striatum is sufficient for aspects of active maternal behavior in female mice. Endocrinology 154(11):4316–4327

Index

Kathryn M. Murphy (ed.), *Synaptosomes*, Neuromethods, vol. 141,
https://doi.org/10.1007/978-1-4939-8739-9, © Springer Science+Business Media, LLC, part of Springer Nature 2018

MIX
Papier aus verantwortungsvollen Quellen
Paper from responsible sources
FSC® C105338

If you have any concerns about our products,
you can contact us on
ProductSafety@springernature.com

In case Publisher is established outside the EU,
the EU authorized representative is:
Springer Nature Customer Service Center GmbH
Europaplatz 3, 69115 Heidelberg, Germany

Printed by Libri Plureos GmbH
in Hamburg, Germany